New Records of the Great Dying in South China

Series Editors

Weihong He, State Key Laboratory of Biogeology and Environmental Geology, School of Earth Sciences, China University of Geosciences, Wuhan, China

Jianxin Yu, State Key Laboratory of Biogeology and Environmental Geology, School of Earth Sciences, China University of Geosciences, Wuhan, China

Haishui Jiang, China University of Geosciences, Wuhan, China

AF800272

New Record of Great Dying in South China series systematically documents the recently discovered Permian–Triassic biota, including fossil plants, brachiopods, bivalves, radiolarians, conodonts, foraminifers, ostracods, and stratigraphy across the PTB interval. The series focuses the taxonomy, phylogeny, biodiversity, biostratigraphy, evolution of biota, paleoecology and paleogeography. The series summarizes the processes of mass extinction and environmental crisis.

Wei-hong He · G. R. Shi · Jian-xin Yu ·
Dao-Liang Chu · Ke-Xin Zhang ·
Hai-Shui Jiang · Ting-Lu Yang · Yi-Fan Xiao
Editors

Stratigraphy Around the Permian–Triassic Boundary of South China

Springer

Editors
Wei-hong He
State Key Laboratory of Biogeology
and Environmental Geology
School of Earth Sciences
China University of Geosciences
Wuhan, Hubei, China

Jian-xin Yu
State Key Laboratory of Biogeology
and Environmental Geology
School of Earth Sciences
China University of Geosciences
Wuhan, Hubei, China

Ke-Xin Zhang
State Key Laboratory of Biogeology
and Environmental Geology
School of Earth Sciences
China University of Geosciences
Wuhan, Hubei, China

Ting-Lu Yang
Faculty of Geosciences
East China University of Technology
Nanchang, China

G. R. Shi
School of Earth, Atmospheric and Life
Sciences
University of Wollongong
New South Wales, NSW, Australia

Dao-Liang Chu
State Key Laboratory of Biogeology
and Environmental Geology
School of Earth Sciences
China University of Geosciences
Wuhan, Hubei, China

Hai-Shui Jiang
State Key Laboratory of Biogeology
and Environmental Geology
School of Earth Sciences
China University of Geosciences
Wuhan, Hubei, China

Yi-Fan Xiao
State Key Laboratory of Biogeology
and Environmental Geology
China University of Geosciences
Wuhan, Hubei, China

ISSN 2524-4574 ISSN 2524-4582 (electronic)
New Records of the Great Dying in South China
ISBN 978-981-99-9352-9 ISBN 978-981-99-9350-5 (eBook)
https://doi.org/10.1007/978-981-99-9350-5

© The Editor(s) (if applicable) and The Author(s), under exclusive license to Springer Nature Singapore Pte Ltd. 2023

This work is subject to copyright. All rights are solely and exclusively licensed by the Publisher, whether the whole or part of the material is concerned, specifically the rights of translation, reprinting, reuse of illustrations, recitation, broadcasting, reproduction on microfilms or in any other physical way, and transmission or information storage and retrieval, electronic adaptation, computer software, or by similar or dissimilar methodology now known or hereafter developed.
The use of general descriptive names, registered names, trademarks, service marks, etc. in this publication does not imply, even in the absence of a specific statement, that such names are exempt from the relevant protective laws and regulations and therefore free for general use.
The publisher, the authors, and the editors are safe to assume that the advice and information in this book are believed to be true and accurate at the date of publication. Neither the publisher nor the authors or the editors give a warranty, expressed or implied, with respect to the material contained herein or for any errors or omissions that may have been made. The publisher remains neutral with regard to jurisdictional claims in published maps and institutional affiliations.

This Springer imprint is published by the registered company Springer Nature Singapore Pte Ltd.
The registered company address is: 152 Beach Road, #21-01/04 Gateway East, Singapore 189721, Singapore

Paper in this product is recyclable.

Foreword

The end-Permian mass extinction (EPME) was the most severe biodiversity crisis during the Phanerozoic, however, its causes remain highly controversial so far. It has been previously suggested that global regression was the main cause for this largest mass extinction during the 1990s. One of the largest regressions during the Phanerozoic occurred at the end of the Guadalupian and caused sea water to withdraw from most parts of the Pangea supercontinent including the North America, Russian and Siberian Cratons, and many blocks of the Gondwanaland. This regression significantly reduced the habitable area for marine organisms and numerous marine species disappeared in the Lopingian (Late Permian). The EPME was first quantitatively investigated based on a compilation of marine animal families and a 96% extinction rate was documented by Sepkoski (1981), and this rate is still used by many publications and public media although recent investigations indicate that the rate is probably much lower than Sepkoski's statistic data. The large extinction rate of the EPME was mainly due to a coarse time resolution from Sepkoski's data in which he used the Tatarian Series as the last time bin to represent the EPME. The Tatarian actually ranges from late Guadalupian to Early Triassic in the latest Permian timescale, thus all species disappeared from the end-Guadalupian to the lowest Triassic were ascribed to the EPME by Sepkoski (1981).

The scenario of the EPME caused by a global regression has been dramatically changed after the sections in South China were investigated. Differing from most areas on the Pangean super continent, South China was situated in the Paleotethys Ocean which was open to the Panthalassa Ocean in the east. After a short regression during the end-Guadalupian, the South China block received marine deposits again from the Lopingian until Middle Triassic. Thus, numerous well-preserved and continuous marine Permian–Triassic boundary sections were found and South China has become a research hotspot on the EPME and Permian–Triassic boundary which led to three successive GSSPs (the base-Lopingian, base-Changhsingian, and base of Triassic GSSPs) finally defined in South China (Yin et al., 2001; Jin et al., 2006a, b).

Through more than half century's studies on the PTB sections in South China, it has been well recognized that Sepkoski's EPME actually consists of two mass extinctions, one is at the end-Guadalupian and the other at the end-Permian (Jin et al., 1994). It has been now unanimously agreed that the real EPME happened within a very short time (Shen et al., 2019) near the PTB and several lines of evidence suggested that the EPME was associated with intensive volcanism, global warming, and anoxia in the ocean. All these data were benefited from the detailed studies of the PTB sections in South China. Therefore, a standard stratigraphic framework is essential to better understand the patterns and causes of the EPME.

This book timely provides high-quality records of the PTB sections across different depositional settings from terrestrial to marine facies in South China, including detailed descriptions of lithologies, high-resolution correlation of key boundaries and fossil occurrences and ranges at each section. This book also presents the evolution of the fossil communities in both marine and terrestrial ecosystems over space and through time in South China.

July 2023

Shu-Zhong Shen
School of Earth Sciences
and Engineering
Nanjing University
Nanjing, China

References

Jin YG, Shen SZ, Henderson CM, Wang XD, Wang W, Wang Y, Cao CQ, Shang QH (2006a) The global stratotype section and point (GSSP) for the boundary between the Capitanian and Wuchiapingian stage (Permian). Episodes 29:253–262

Jin YG, Wang Y, Henderson CM, Wardlaw BR, Shen SZ, Cao CQ (2006b) The global boundary stratotype section and point (GSSP) for the base of Changhsingian stage (Upper Permian). Episodes 29:175–182

Jin YG, Zhang J, Shang QH (1994) Two phases of the end-Permian mass extinction, Pangea: Global environments and resources. Can Soci Petrol Geol Mem 17:813–822

Sepkoski JJ (1981) A factor analytic description of the Phanerozoic marine fossil record. Paleobiology 7:36–53

Shen SZ, Ramezani J, Chen J, Cao CQ, Erwin DH, Zhang H, Xiang L, Schoepfer SD, Henderson CM, Zheng QF, Bowring SA, Wang Y, Li XH, Wang XD, Yuan DX, Zhang YC, Mu L, Wang J, Wu YS (2019) A sudden end-Permian mass extinction in South China. Geol Soc Am Bull 131:205–223

Yin HF, Zhang KX, Tong JN, Yang ZY, Wu SB (2001) The global stratotype section and point (GSSP) of the Permian–Triassic boundary. Episodes 24:102–114

Acknowledgments

We sincerely thank the senior Prof. Hongfu Yin for his organization and encouragement during the preparation of this book.

We wish to express our sincere thanks to Profs. Jinnan Tong, Qinglai Feng, Xulong Lai, Yongbiao Wang, and Zhongqiang Chen for their guidance to the young scholars (including teachers and postgraduates and most of them as the co-author or co-editor). We also wish to thank Prof. Shunbao Wu for his great contribution to the research on ammonite and bivalve taxonomy in this book and help in supervising students.

We are also deeply grateful to Prof. Shuzhong Shen for generously providing materials (Liangfengya and Penglaitan sections) and other supports (e.g., suggestions on using of some genera names, writing preface for the book).

The authors wish to express their thanks to Zhijun Niu, Yang Zhang, Huiting Wu, Yang Jiang, Yu Du, Xi Sun, Mingliang Yue, Bin Chen, Fei Teng, and Zongyan Zhang, for their help in the fieldwork.

We also thank the Editor Bharath Kumar Dhamodharan for his pushing during drafting the book. And thank Prof. Jiaxin Yan for the discussion with him on the depositional environment.

This project has been supported by NSFC Grant Nos. [42230205, 42172012 (Wei-Hong He); 92055201 (Jian-Xin Yu); 41902005 (Ting-Lu Yang); 41902016 (Yi-Fan Xiao); 41772016, 41372030, 40872008, 40502001 (Wei-Hong He)]; the Foundation of the Geological Survey of China [1212011220529 (Wei-Hong He), DD20221645 (Ke-Xin Zhang)]; and the University of Wollongong for the research of G. R. Shi.

Contents

Contributors

Dao-Liang Chu State Key Laboratory of Biogeology and Environmental Geology, School of Earth Sciences, China University of Geosciences, Wuhan, Hubei, China

Song-Zhu Gu School of Earth Sciences, China University of Geosciences, Wuhan, Hubei, China

Wei-Hong He State Key Laboratory of Biogeology and Environmental Geology, School of Earth Sciences, China University of Geosciences, Wuhan, Hubei, China

Ya-Fei Huang School of Earth Sciences, China University of Geosciences, Wuhan, Hubei, China

Hai-Shui Jiang State Key Laboratory of Biogeology and Environmental Geology, School of Earth Sciences, China University of Geosciences, Wuhan, Hubei, China

G. R. Shi School of Earth, Atmospheric and Life Sciences, University of Wollongong, Wollongong, NSW, Australia

Hai-Jun Song State Key Laboratory of Biogeology and Environmental Geology, School of Earth Sciences, China University of Geosciences, Wuhan, Hubei, China

N. Suzuki Department of Earth Science, Graduate School of Science, Tohoku University, Sendai, Japan

Han Wang School of Earth Sciences, China University of Geosciences, Wuhan, Hubei, China

Yi-Fan Xiao State Key Laboratory of Biogeology and Environmental Geology, School of Earth Sciences, China University of Geosciences, Wuhan, Hubei, China

Ting-Lu Yang Faculty of Geosciences, East China University of Technology, Nanchang, China

Ai-Hua Yuan School of Earth Sciences, China University of Geosciences, Wuhan, Hubei, China

Jian-Xin Yu State Key Laboratory of Biogeology and Environmental Geology, School of Earth Sciences, China University of Geosciences, Wuhan, Hubei, China

Ke-Xin Zhang State Key Laboratory of Biogeology and Environmental Geology, School of Earth Sciences, China University of Geosciences, Wuhan, Hubei, China

Xiong-Hua Zhang School of Earth Sciences, China University of Geosciences, Wuhan, Hubei, China

Introduction

1

Wei-Hong He and G. R. Shi

The research on the Permian–Triassic boundary (PTB) sections of South China mainly started from the 1980s and a series of monographs/books have since been published (Li et al. 1989; Sheng et al. 1984; Yang et al. 1987, 1991; Yao et al. 1980; Zhao et al. 1981). These works described the lithological features of numerous sections bed by bed. However, most of the fossils reported from these sections were not studied systematically in terms of taxonomy, hampering the precise definition, placement and correlation of the PTB. It was not until the early 2000s the Global Stratigraphic Section and Point (GSSP) for the PTB was finally ratified by the International Stratigraphic Commission (ICS), based on the Meishan D section in Zhejiang Province, China (Yin et al. 2001). The ratification of the PTB GSSP in South China marked an important turning point for the studies of the PTB and associated stratigraphy and palaeontology. In the past 20 years, not only have new and more PTB sections, both marine and non-marine, been discovered and documented, Changhsingian–Induan fossils in different taxa have been intensively collected and studied from scores of sections with continuous Permian–Triassic sedimentation successions, resulting in copious publications including monographs (e.g., Chu et al. 2016; Feng et al. 2007; He et al. 2014, 2015, 2019; Jiang et al. 2007, 2011, 2014; Shen et al. 2019; Song et al. 2009, 2011; Sun et al. 2012a; Tian

W.-H. He (✉)
State Key Laboratory of Biogeology and Environmental Geology, School of Earth Sciences, China University of Geosciences, 388 Luma Road, Wuhan, Hubei, China
e-mail: whzhang@cug.edu.cn

G. R. Shi
School of Earth, Atmospheric and Life Sciences, University of Wollongong, Wollongong, NSW 2522, Australia
e-mail: guang@uow.edu.au

© The Author(s), under exclusive license to Springer Nature Singapore Pte Ltd. 2023
W. He et al. (eds.), *Stratigraphy Around the Permian–Triassic Boundary of South China*, New Records of the Great Dying in South China,
https://doi.org/10.1007/978-981-99-9350-5_1

et al. 2014; Wang et al. 2017; Wu et al. 2018; Yang et al. 2013, 2015; Yuan and Shen 2011; Zhang et al. 2007, 2014).

Taking the advantage of abundant new collections and the discovery of new fossils, the taxonomy of many previously described fossils (e.g., conodont, brachiopod, radiolarian and plant fossils) have been reexamined and their classification updated (e.g., Feng et al. 2020a, b; He et al. 2019; Xiao et al. 2018; Yuan et al. 2014). In particular, the biostratigraphic framework built by the detailed studies of the conodont faunas has been constructed and refined more precisely. For example, when the GSSP of the PTB was ratified in 2001 (Yin et al. 2001), only six zones were defined, including *Clarkina subcarinata, Clarkina changxingensis, Clarkina yini, Hindeous latidentatus–Clarkina meishanensis, Hindeous parvus* and *Isarcicella isarcica* zones, in ascending order. In comparison, up to now, nine conodont zones have been recognised spanning the PTB, comprising, in ascending order, the *Clarkina wangi, Clarkina subcarinata, Clarkina changxingensis, Clarkina yini, Clarkina meishanensis, Hindeous changxingensis–Clarkina zhejiangensis, Hindeodus parvus, Isarcicella staechei* and *Isarcicella isarcica* zones (Jiang et al. 2007; Tong et al. 2019; Yuan et al. 2014; Zhang et al. 2007, 2009). Additionally and almost in step with the advancement of the PTB palaeontological and biostratigraphic studies, the carbon isotope-based chemostratigraphic studies have also received considerable attentions and been used as an important tool for constraining and calibrating the PTB (copious publications Cao et al. 2010; Chen et al. 2019; Grasby and Beauchamp 2008; Hermann et al. 2010; Shen et al. 2019; Xie et al. 2007; Yang et al. 2011; Zhang et al. 2016).

Although a huge number of achievements on the end-Permian (or PTB) mass extinction have been made, the pattern of mass extinction is still debated, as are its causes. Most published works considered the mass extinction either as a single pulse or a two-pulse event (Jin et al. 2000; Song et al. 2013; Shen et al. 2019). Even these alternative scenarios may not be thoroughly representative, because an additional and earlier biotic crisis has also been recognised in deep-water facies, based on the study of brachiopod and radiolarian faunas (Fan et al. 2020; Feng et al. 2007; He et al. 2015, 2019). Regarding the causes for the mass extinction, it is even more contentious and remains a global hot topic, though volcanism, anoxia, global warming and acidification are generally considered as the most likely ultimate or proximal culprits (e.g., Burgess et al. 2017; Clarkson et al. 2015; Garbelli et al. 2017; Isozaki 1997; Joachimski et al. 2020; Jurikova et al. 2020; Kump 2018; Sun et al. 2012b; Wignall and Twitchett 1996; Xie et al. 2017; Yang et al. 1991; Zhang et al. 2020).

This book systematically described over 30 stratigraphic sections with continuous Permian–Triassic sedimentation successions. The sections were distributed respectively in terrestrial facies, nearshore clastic shelf, shallow-water carbonate platform and deep-water facies. These sections have been chosen primarily because they have yielded abundant and biostratigraphically important fossils that have been subject to comprehensive taxonomy studies. The aim of this book is to (1) establish a cohesive and highly resolved stratigraphic and biostratigraphic framework of the PTB sections across different depositional facies in South China,

(2) objectively analyze and review the patterns and processes of the community and ecosystem evolution over space and through time in the lead to the end-Permian mass extinction, (3) to refine the spatio–temporal patterns of the mass extinction, and finally (4) to briefly discuss the possible causes of the mass extinction. It is hoped that the information provided in this book will help to advance the understanding of the long-debated Great Dying.

References

Burgess SD, Muirhead JD, Bowring SA (2017) Initial pulse of Siberian Traps sills as the trigger of the end-Permian mass extinction. Nat Commun 8:164. https://doi.org/10.1038/s41467-017-00083-9

Cao CQ, Yang YC, Shen SZ, Wang W, Zheng QF, Summons RE (2010) Pattern of $\delta^{13}C_{carb}$ and implications for geological events during the Permian-Triassic transition in South China. Geol J 45:186–194

Chen Y, Ye Q, Jiang HS, Wignall PB, Yuan JL (2019) Conodonts and carbon isotopes during the Permian–Triassic transition on the Napo platform, South China. J Earth Sci 30:244–257

Chu DL, Yu JX, Tong JN, Benton MJ, Song HJ, Huang YF, Song T, Tian L (2016) Biostratigraphic correlation and mass extinction during the Permian–Triassic transition in terrestrial–marine siliciclastic settings of South China. Glob Planet Change 146:67–88

Clarkson MO, Kasemann SA, Wood RA, Lenton TM, Daines SJ, Richoz S, Ohnemueller F, Meixner A, Poulton SW, Tipper ET (2015) Ocean acidification and the Permo–Triassic mass extinction. Science 348:229–232

Fan JX, Shen SZ, Erwin DH, Sadler PM, MacLeod N, Cheng QM, Hou XD, Yang J, Wang DX, Wang Y, Zhang H, Chen X, Xiang G, Zhang YC, Shi YK, Yuan DX, Chen Q, Zhang LN, Li C, Zhao YY (2020) A high-resolution summary of Cambrian to Early Triassic marine invertebrate biodiversity. Science 367:272–277

Feng QL, He WH, Gu SZ, Meng YY, Jin YX, Zhang F (2007) Radiolarian evolution during the latest Permian in South China. Glob Planet Change 55:177–192

Feng Z, Wei HB, Ye RH, Sui Q, Gou XD, Guo Y, Liu LJ, Yang SL (2020a) Latest Permian peltasperm plant from southwest China and its paleoenvironmental implications. Front Earth Sci 8:559430. https://doi.org/10.3389/feart.2020.559430

Feng Z, Wei HB, Guo Y, He XY, Sui Q, Zhou Y, Lin HY, Gou XD, Lv Y (2020b) From rainforest to herbland: new insights into land plant responses to the end-Permian mass extinction. Earth Sci Rev 204:103153. https://doi.org/10.1016/j.earscirev.2020.103153

Garbelli C, Angiolini L, Shen SZ (2017) Biomineralization and global change: a new perspective for understanding the end-Permian extinction. Geology 45:19–22

Grasby S, Beauchamp B (2008) Intrabasin variability of the carbon-isotope record across the Permian–Triassic transition, Sverdrup Basin, Arctic Canada. Chem Geol 253:141–150

He WH, Shi GR, Zhang Y, Yang TL, Zhang KX, Wu SB, Niu ZJ, Zhang ZY (2014) Changhsingian (latest Permian) deep-water brachiopod fauna from South China. J Syst Paleontol 12:907–960

He WH, Shi GR, Twitchett RJ, Zhang Y, Zhang KX, Song HJ, Yue ML, Wu SB, Wu HT, Yang TL, Xiao YF (2015) Late Permian marine ecosystem collapse began in deeper waters: evidence from brachiopod diversity and body size changes. Geobiology 13:123–138

He WH, Shi GR, Zhang KX, Yang TL, Shen SZ, Zhang Y (2019) Brachiopods around the Permian–Triassic boundary of South China. Springer, Singapore, 261pp

Hermann E, Hochuli PA, Bucher H, Vigran JO, Weissert H, Bernasconi SM (2010) A close-up view of the Permian–Triassic boundary based on expanded organic carbon isotope records from Norway (Trøndelag and Finnmark Platform). Glob Planet Change 74:156–167

Isozaki Y (1997) Permo–Triassic boundary superanoxia and stratified superocean: records from lost deep sea. Science 276:235–238

Jiang HS, Lai XL, Luo GM, Aldridge R, Zhang KX, Wignall P (2007) Restudy of conodont zonation and evolution across the P/T boundary at Meishan section, Changxing, Zhejiang, China. Glob Planet Change 55:39–55

Jiang HS, Lai XL, Yan CB, Aldridge R, Wignall P, Sun YD (2011) Revised conodont zonation and conodont evolution across the Permian–Triassic boundary at the Shangsi section, Guangyuan, Sichuan, South China. Glob Planet Change 77:103–115

Jiang HS, Lai XL, Sun YD, Wignall P, Liu JB, Yan CB (2014) Permian–Triassic conodonts from Dajiang (Guizhou, South China) and their Implication for the age of microbialite deposition in the aftermath of the end-Permian mass extinction. J Earth Sci 25:413–430

Jin YG, Wang Y, Wang W, Shang QH, Cao CQ, Erwin DH (2000) Pattern of marine mass extinction near the Permian—Triassic boundary in South China. Science 289:432–436

Joachimski MM, Alekseev AS, Grigoryan A, Gatovsky YA (2020) Siberian Trap volcanism, global warming and the Permian–Triassic mass extinction: new insights from Armenian Permian–Triassic sections. Geol Soc Am Bull 132:427–443

Jurikova H, Gutjahr M, Wallmann K, Flögel S, Liebetrau V, Posenato R, Angiolini L, Angiolini C, Brand U, Wiedenbeck M, Eisenhauer A (2020) Permian–Triassic mass extinction pulses driven by major marine carbon cycle perturbations. Nat Geosci 13:745–750

Kump L (2018) Climate change and marine mass extinction. Science 362:1113–1114

Li ZS, Zhan LP, Zhu XF, Zhang JH, Huang HQ, Xu DY, Yan Z, Li HM (1989) Study on the Permian–Triassic biostratigraphy and event stratigraphy of northern Sichuan and southern Shaanxi. Geological Publishing House, Beijing, 435 pp [in Chinese]

Sheng JZ, Chen CZ, Wang YG, Rui L, Liao ZT, Bando Y, Ishii K, Nakazawa K, Nakamura K (1984) Permian—Triassic boundary in middle and eastern Tethys. J Fac Sci; Hokkaido University, Series 4: Geology and mineralogy, 21:133–181

Shen SZ, Ramezani J, Chen J, Cao CQ, Erwin DH, Zhang H, Xiang L, Schoepfer SD, Henderson CM, Zheng QF, Bowring SA, Wang Y, Li XH, Wang XD, Yuan DX, Zhang YC, Mu L, Wang J, Wu YS (2019) A sudden end-Permian mass extinction in South China. Geol Soc Am Bull 131:205–223

Song HJ, Tong JN, Chen ZQ (2009) Two episodes of foraminiferal extinction near the Permian—Triassic boundary at the Meishan section, South China. Aust J Earth Sci 56:765–773

Song HJ, Wignall PB, Chen ZQ, Tong JN, Bond DPG, Lai XL, Zhao XM, Jiang HS, Yan CB, Niu ZJ, Chen J, Yang H, Wang YB (2011) Recovery tempo and pattern of marine ecosystems after the end-Permian mass extinction. Geology 39:739–742

Song HJ, Wignall PB, Tong JN, Yin HF (2013) Two pulses of extinction during the Permian–Triassic crisis. Nat Geosci 6:52–56

Sun DY, Tong JN, Xiong YL, Tian L, Yin HF (2012a) Conodont biostratigraphy and evolution across Permian–Triassic boundary at Yangou Section, Leping, Jiangxi Province, South China. J Earth Sci 23:311–325

Sun YD, Joachimski MM, Wignall PB, Yan CB, Chen YL, Jiang HS, Wang LN, Lai XL (2012b) Lethally hot temperatures during the Early Triassic greenhouse. Science 338:366–370

Tian L, Tong JN, Song HJ, Liang L, Yang LR, Song HY, Wang CG, Zhao XM, Sun DY (2014) Foraminiferal evolution and formation of Oolitic limestone near Permian—Triassic boundary at Yangou section, Jiangxi Province. Earth Sci J China Univ Geosci 39:1573–1586

Tong JN, Chu DL, Liang L, Shu WC, Song HJ, Song T, Song HY, Wu YY (2019) Triassic integrative stratigraphy and timescale of China. Sci China Earth Sci 62:189–222

Wang H, Jiang Y, Yang TL, Xiao YF, Wu SB, He WH (2017) End-Permian biotic assemblage of Kejiao, southern Guizhou and its geological implications. Geol Sci Technol Inf 36:168–173 [in Chinese with English abstract]

Wignall PB, Twitchett RJ (1996) Oceanic anoxia and the end Permian mass extinction. Science 272:1155–1158

Wu HT, He WH, Shi GR, Zhang KX, Yang TL, Zhang Y, Xiao YF, Chen B, Wu SB (2018) A new Permian-Triassic boundary brachiopod fauna from the Xinmin section, southwestern Guizhou, south China and its extinction patterns. Alcheringa 42:339–372

Xiao YF, Suzuki N, He WH (2018) Low-latitudinal standard Permian radiolarian biostratigraphy for multiple purposes with Unitary Association, Graphic Correlation, and Bayesian inference methods. Earth Sci Rev 179:168–206
Xie SC, Pancost RD, Huang JH, Wignall PB, Yu JX, Tang XY, Chen L, Huang XY, Lai XL (2007) Changes in the global carbon cycle occurred as two episodes during the Permian–Triassic crisis. Geology 35:1083–1086
Xie SC, Algeo TJ, Zhou WF, Ruan XY, Luo GM, Huang JH, Yan JX (2017) Contrasting microbial community changes during mass extinctions at the Middle/Late Permian and Permian/Triassic boundaries. Earth Planet Sci Lett 460:180–191
Yang ZY, Yin HF, Wu SB, Yang FQ, Ding MH, Xu GR (1987) Permian–Triassic boundary stratigraphy and fauna of South China. Geological Publishing House, Beijing, 378 pp [in Chinese with English abstract]
Yang ZY, Wu SB, Yin HF, Xu GR, Zhang KX (1991) Permo–Triassic events of South China. Geological Publishing House, Beijing, 183 pp [in Chinese with English abstract]
Yang H, Chen ZQ, Wang YB, Tong JN, Song HJ, Chen J (2011) Composition and structure of microbialite ecosystems following the end-Permian mass extinction in South China. Palaeogeogr Palaeoclimatol Palaeoecol 208:111–128
Yang LR, Song HJ, Tong JN, Chu DL, Tian L (2013) The extinciton pattern of fusulinids during the Permian–Triassic crisis at the Kangjiaping section, Cili, Hunan Province. Acta Micropalaeongologica Sinica 30:353–366 [in Chinese with English abstract]
Yang TL, He WH, Zhang KX, Wu SB, Zhang Y, Yue ML, Wu HT, Xiao YF (2015) Palaeoecological insights into the Changhsingian-Induan (latest Permian–earliest Triassic) bivalve fauna at Dongpan, southern Guangxi, South China. Alcheringa 40:98–117
Yao ZQ, Xu JT, Zheng ZG, Zhao XH, Mo ZG (1980) Upper Permian biostratigraphy and the Permian–Triassic boundary in western Guizhou and eastern Yunnan. 1–69. In: Nanjing Institute of Geology and Paleontology (ed), Stratigraphy and Paleontology of the Upper Permian Coral Bearing Formation in Western Guizhou and Eastern Yunnan. Science Press, Beijing [in Chinese]
Yin HF, Zhang KX, Tong JN, Yang ZY, Wu SB (2001) The Global Stratotype Section and Point (GSSP) of the Permian–Triassic boundary. Episodes 24:102–114
Yuan DX, Shen SZ (2011) Conodont succession across the Permian–Triassic boundary of the Liangfengya section, Chongqing, South China. Acta Palaeontologica Sinica 50:420–438 [in Chinese with English abstract]
Yuan DX, Shen SZ, Henderson CM, Chen J, Zhang H, Feng HZ (2014) Revised conodont-based integrated high-resolution timescale for the Changhsingian Stage and end-Permian extinction interval at the Meishan sections, South China. Lithos 204:220–245
Zhang KX, Tong JN, Shi GR, Lai XL, Yu JX, He WH, Peng YQ, Jin YL (2007) Early Triassic conodont–palynological biostratigraphy of the Meishan D Section in Changxing, Zhejiang Province, South China. Palaeogeogr Palaeoclimatol Palaeoecol 252:4–23
Zhang KX, Lai XL, Tong JN, Jiang HS (2009) Progresses on study of conodont sequence for the GSSP section at Meishan, Changxing, Zhejiang Province, South China. Acta Palaeontologica Sinica 48:474–486 [in Chinese with English abstract]
Zhang Y, Shi GR, He WH, Zhang KX, Wu HT (2014) A new Changhsingian (Late Permian) brachiopod fauna from the Zhongzhai section (South China), Part 2: Lingulida, Orthida, Orthotetida and Spiriferida. Alcheringa 38:480–503
Zhang H, Cao CQ, Liu XL, Mu L, Zheng QF, Liu F, Xiang L, Liu LJ, Shen SZ (2016) The terrestrial end-Permian mass extinction in South China. Palaeogeogr Palaeoclimatol Palaeoecol 448:108–124
Zhang FF, Shen SZ, Cui Y, Lenton TM, Dahl TW, Zhang H, Zheng QF, Wang WQ, Krainer K, Anbar AD (2020) Two distinct episodes of marine anoxia during the Permian–Triassic crisis evidenced by uranium isotopes in marine dolostones. Geochim Cosmochim Acta 287:165–179
Zhao JK, Sheng JZ, Yao ZQ, Liang ZL, Chen CZ, Rui L, Liao ZT (1981) The Changhsingian and Permian–Triassic boundary of South China. Nanjing Inst Geol Palaeontol Bull 2:1–112 [in Chinese, with English abstract]

Palaeogeographical Settings of South China in the Changhsingian to Induan and Palaeogeographic Distribution of the Studied Permian–Triassic Boundary Sections

2

Wei-Hong He, Ke-Xin Zhang, G. R. Shi, and Yi-Fan Xiao

2.1 Geographic Distribution of the Studied Permian–Triassic Boundary Sections

The studied sections include Meishan, Daoduishan, Huangzhishan, Majiashan, Yangou, Jinshan, Chibi, Zhaixia, Xiejiaping, Shangsi, Laolongdong, Liangfengya, Rencunping, Kangjiaping, Jinjibang, Guanbachong, Kele, Xiaohebian, Chahe, Zhejue, Jiucaichong, Jinzhong, Mide, Wadu, Tucheng, Zhongzhai, Xinmin, Kejiao, Dajiang, Zuodeng, Penglaitan and Dongpan (totally 32 sections) (abbreviation for section names sees Fig. 2.1 and Appendix A).

W.-H. He (✉) · K.-X. Zhang · Y.-F. Xiao
State Key Laboratory of Biogeology and Environmental Geology, School of Earth Sciences, China University of Geosciences, 388 Luma Road, Wuhan, Hubei, China
e-mail: whzhang@cug.edu.cn

K.-X. Zhang
e-mail: kx_zhang@cug.edu.cn

Y.-F. Xiao
e-mail: yifanxiao@cug.edu.cn

G. R. Shi
School of Earth, Atmospheric and Life Sciences, University of Wollongong, Wollongong, NSW 2522, Australia
e-mail: guang@uow.edu.au

© The Author(s), under exclusive license to Springer Nature Singapore Pte Ltd. 2023

W. He et al. (eds.), *Stratigraphy Around the Permian–Triassic Boundary of South China*, New Records of the Great Dying in South China,
https://doi.org/10.1007/978-981-99-9350-5_2

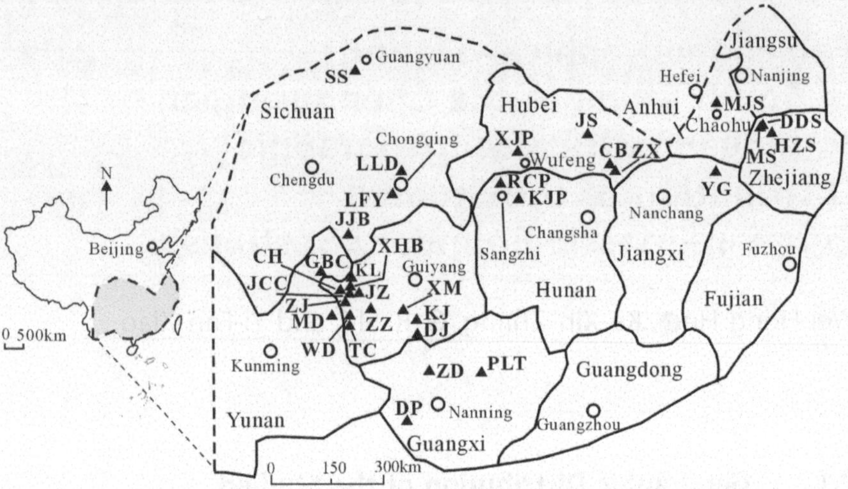

Fig. 2.1 Geographic distribution of the studied Permian–Triassic Boundary sections in South China. Solid triangle indicates the approximate location for each section. MS—Meishan, DDS—Daoduishan; HZS—Huangzhishan; MJS—Majiashan; YG—Yangou; JS—Jinshan; CB—Chibi; ZX—Zhaixia; XJP—Xiejiaping; SS—Shangsi; LLD—Laolongdong; LFY—Liangfengya; RCP—Rencunping; KJP—Kangjiaping; JJB—Jinjibang; GBC—Guanbachong; KL—Kele; XHB—Xiaohebian; CH—Chahe; ZJ—Zhejue; JCC—Jiucaichong; JZ—Jinzhong; MD—Mide; WD—Wadu; TC—Tucheng; ZZ—Zhongzhai; XM—Xinmin; KJ—Kejiao; DJ—Dajiang; ZD—Zuodeng; PLT—Penglaitan; DP—Dongpan

2.2 Palaeogeographical Settings of South China from the Changhsingian to Induan and Palaeogeographic Localities of the Studied Sections

2.2.1 Palaeogeographic Settings of South China from the Changhsingian to Induan

In the Late Permian, South China was a separate block located at the eastern margin of the Palaeotethys Ocean (Fig. 2.2). A branch of the Palaeotethys Ocean, known as the Qinling–Dabie deep sea, separated South China from North China. The deep sea was connected to the Panthalassa Ocean to the east (Fig. 2.2). The other branch of the Palaeotethys Ocean, the Ailaoshan oceanic basin, was located along the southern margin of South China. This basin was also linked to the Panthalassa Ocean eastward (Fig. 2.2).

(1) Palaeogeographic setting of South China in the Changhsingian

In the Changhsingian (Late Permian), the South China block comprised ancient land masses, terrestrial facies, submerged rises, shallow-water marine clastic shelfs, shallow-water marine carbonate platforms, deep-water marine siliceous

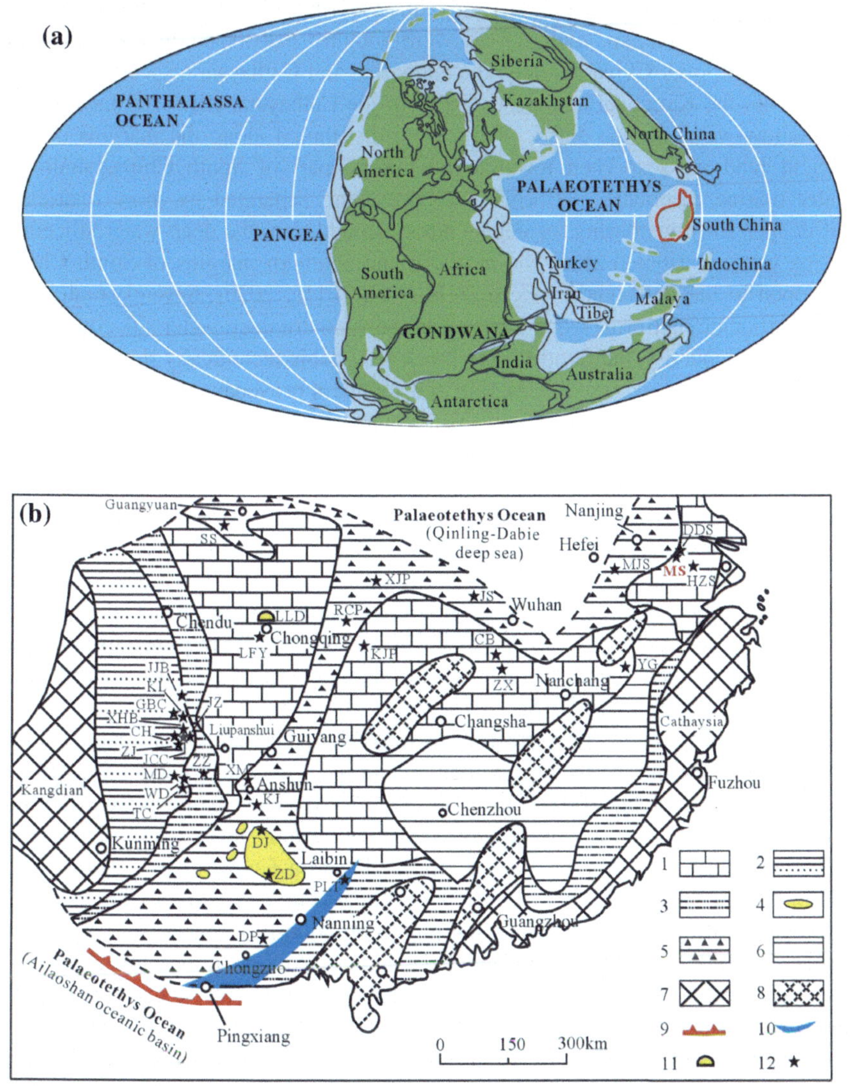

Fig. 2.2 Palaeogeography of South China in the Late Permian. **a**—South China in the Late Permian global palaeogeography (revised after Scotese 2000); **b**—Changhsingian palaeogeography and lithofacies map of South China (revised after He et al. 2015a, 2020). 1—Shallow-water marine carbonate platform; 2—Terrestrial facies; 3—Shallow-water marine clastic shelf; 4—Isolated marine carbonate platform; 5—Deep-water marine siliceous basin; 6—Shallow-water restricted siliceous basin; 7; Ancient land mass; 8—Submerged rise; 9—Subduction of Palaetethys Ocean and possible subduction direction; 10—Rift basin; 11—Reef and microbialite; 12—Studied section (abbreviation for section names sees Appendix A)

basins, and shallow-water restricted siliceous basins (intraplatform basins) (Fig. 2.2). The ancient land masses were roughly distributed in the west and east parts of South China, namely, the Kangdian and Cathaysia respectively. Eastward from the Kangdian and westward from the Cathaysia, terrestrial facies and/or shallow-water marine clastic shelfs were distributed along the margins of the ancient land masses. Then towards the middle part of South China, shallow-water marine carbonate platforms were prevalent, separated by more restricted but deeper siliceous basins (or depocentres). In addition, the deep-water siliceous basins were distributed along the northern and southern margins of South China and open to the Palaeotethys Ocean (He et al. 2015a, 2020). Several small, isolated carbonate platforms were scattered in the southern deep-water siliceous basin (Fig. 2.2). Generally, the South China block was stable tectonically throughout the Late Permian. However, sedimentation and sediment provenances were commonly influenced by syndepositional volcanism, evident in the presence of numerous volcanic ash or volcaniclastic deposits throughout the Changhsingian to Induan interval. It is believed that the volcanic activities were driven by the subduction of the Palaeotethys Ocean. Importantly, this subduction regime initiated the formation of a rift basin along Chongzuo and Laibin due to backarc extension (Fig. 2.2), as evidenced by the presence of Late Permian–earliest Triassic basalt and/or acid pyroclastic rocks in the Chongzuo (unpublished) and Laibin (Shen et al. 2019) areas, as well as evidence from the abundant and widespread presence of volcanic ash beds intercalated in the Changhsingian to Induan sequences across South China (see He et al. 2005, 2010, 2015b, 2019; Wu et al. 2018).

In the Changhsingian, the terrestrial facies is characterized by feldspathic sandstones, siltstones and mudstones, with abundant plant fossils and coal seams (Chu et al. 2016; Yu et al. 2007; Zhang et al. 2016a). The shallow-water marine clastic shelf is dominated by siltstones and calcareous mudstones intercalated with limestones, yielding abundant brachiopods and a few bivalves (Zhang et al. 2013, 2016b). The shallow-water carbonate platform facies (including isolated carbonate platform) is dominated by bioclastic limestones, yielding abundant, diverse brachiopods, foraminifers and a few larger bivalves (or with abundant bioclasts preserved), but lacking ammonoids and radiolarians (or sparse for both of them) (He et al. 2015b, 2016; Shen and He 1991; Song et al. 2011). The deep-water siliceous basin facies is characterized by chert, siliceous mudstones and siliceous limestones, intercalated with calcareous or manganese mudstones, yielding abundant ammonoids, radiolarians and a few small brachiopods, bivalves and foraminifers; laminated mudstones and siliceous limestones often display evidence of Bouma sequences (Bu et al. 2006; Feng et al. 2007; Gu et al. 2007; He et al. 2008, 2011; Jiang et al. 2018; Wang et al. 2018; Xiao et al. 2017, 2022; Yang et al. 2015; Zhang et al. 2009). The shallow-water restricted siliceous basin facies is dominated by chert and siliceous limestone, with a few small brachiopods, bivalves and ammonoids, lacking (or with rare presence of) radiolarians (Liao and Meng 1984; Shen and Zhang 2008; Zhang 1981; Zhou 2007).

(2) Palaeogeographic setting of South China during the Permian–Triassic transition

During the Permian–Triassic transition (equivalent to conodont *Clarkina meishanensis* to *Isarcicella staeschei* Zones), the palaeogeographic framework of South China basically continued from the Changhsingian without any significant changes, except that the western shallow-water clastic shelf was slightly expanded (Fig. 2.3) due to the transgression at that time. For example, the sedimentary facies changed from the terrestrial facies of Changhsingian to the terrestrial–marine transitional facies/shallow-water clastic facies of Permian–Triassic transition at Jinzhong, Mide and Tucheng (Fig. 2.3; see Chu et al. 2016, 2019; Feng et al. 2020; Yu et al. 2007). In addition, microbialite facies also developed in the shallow-water carbonate platform or isolated platform facies during the Permian–Triassic transition (Fig. 2.3; see Huang et al. 2022; Su et al. 2021; Wu et al. 2006).

(3) Palaeogeographic setting of South China in the Induan

Immediately after the end-Permian mass extinction (equivalent to the conodont *Isarcicella isarcica* to *Neospathodus cristagalli*–*Neospathodus dieneri* Zones), although the palaeogeographic framework of South China basically was remained

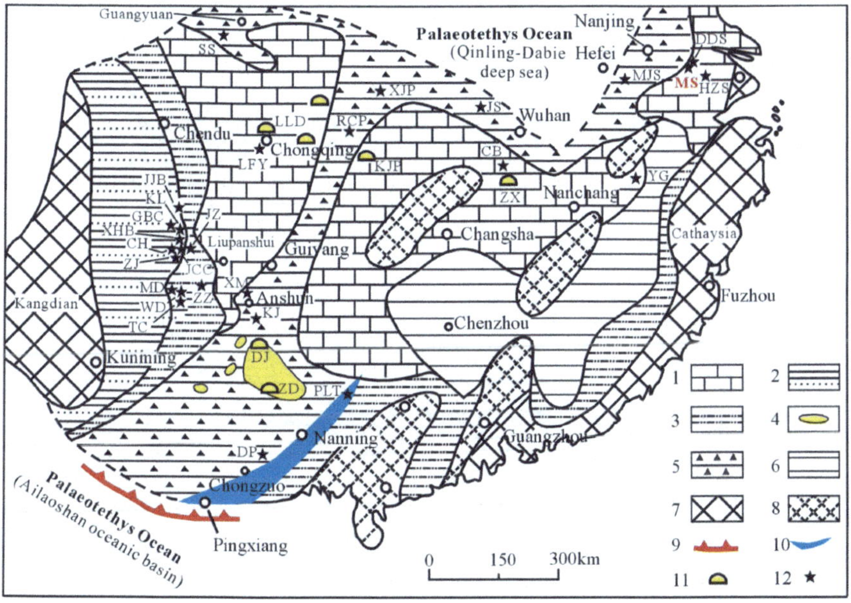

Fig. 2.3 Palaeogeography and lithofacies map of South China in the Permian–Triassic transition (equivalent to *Clarkina meishanensis* to *Isarcicella staeschei* Zones) (revised after He et al. 2015a, 2020). 1—Shallow-water marine carbonate platform; 2—Terrestrial facies; 3—Shallow-water marine clastic shelf; 4—Isolated marine carbonate platform; 5—Deep-water marine siliceous basin; 6—Shallow-water restricted siliceous basin; 7—Ancient land mass; 8—Submerged rise; 9—Subduction of Palaetethys Ocean and possible subduction direction; 10—Rift basin; 11—Microbialite; 12—Studied section (abbreviation for section names sees Appendix A)

with little changes compared to the Changhsingian, the siliceous deposits disappeared. Both deep-water siliceous basins and restricted siliceous basins of the Late Permian were replaced by the calcareous and argillaceous shelf of the Induan (Fig. 2.4). That is to say, the depositional setting of South China became homogenized in the aftermath of the mass extinction, dominated by terrestrial facies, marine silty–argillaeous shelf and marine calcareous–argillaceous shelf facies (Fig. 2.4).

In the lowest Triassic, the terrestrial facies is dominated by purplish red sandstones and siltstones (basal part of the Dongchuan Formation, see Chu et al. 2016; Tong et al. 2019, 2021; Yang et al. 1987; Yu et al. 2007; Zhang et al. 2016a). The marine silty–argillaeous shelf facies is characterized by purplish red to yellowish green siltstones and mudstones, intercalated with thin-bedded limestones (basal part of the Feixianguan Formation; see Tong et al. 2019, 2021; Zhang et al. 2016b, 2016c). The marine calcareous–argillaceous shelf facies is dominated by thin-bedded calcareous mudstones, silty mudstones intercalated with argillaceous limestones (basal part of Taye, Yinkeng or Luolou Formations, see Tong 2015; Tong et al. 2019, 2021; Yang et al. 1987; Zhang et al. 2016c).

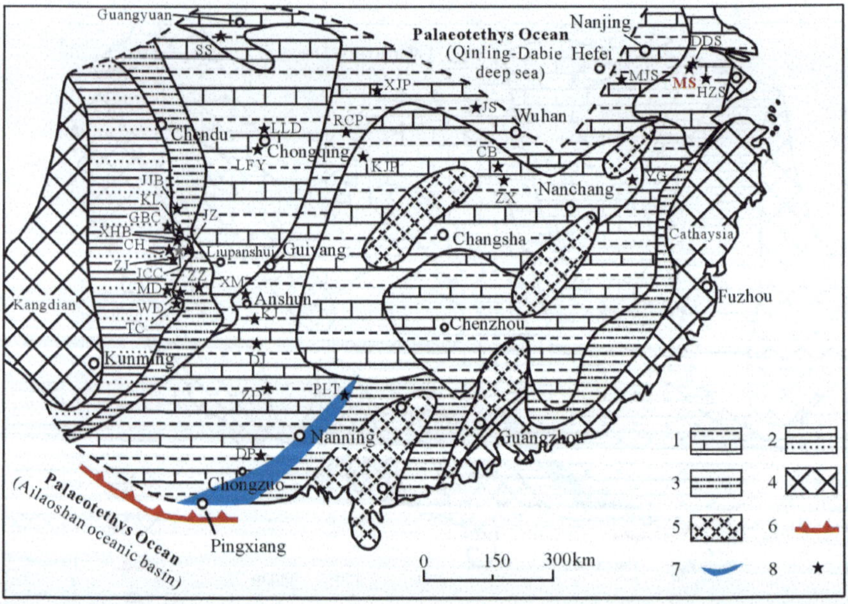

Fig. 2.4 Palaeogeography and lithofacies map of South China in the Induan (equivalent to the conodont *Isarcicella isarcica* to *Neospathodus cristagalli*–*Neospathodus dieneri* Zones). 1—Marine calcareous–argillaceous shelf; 2—Terrestrial facies; 3—Marine silty–argillaeous shelf; 4—Ancient land mass; 5—Submerged rise; 6—Subduction of Palaetethys Ocean and possible subduction direction; 7—Rift basin; 8—Studied section (abbreviation for section names sees Appendix A)

2.2.2 Palaeogeographic Localities of the Studied Sections

(1) Palaeogeographic localities of the studied sections in the Changhsingian

The studied sections including Meishan, Daoduishan, Huangzhishan, Yangou, Chibi, Zhaixia, Kangjiaping, Laolongdong, Liangfengya, Dajiang and Zuodeng, are dominated by bioclastic limestone yielding abundant benthic organisms in the Changhsingian (e.g., foraminifers and brachiopods). They were located on a carbonate platform or along the ramp of a carbonate platform in the Changhsingian (Fig. 2.2). The sections of Majiashan, Jinshan, Xiejiaping, Rencunping, Shangsi, Xinmin, Kejiao and Dongpan, are dominated by chert, siliceous mudstone with abundant cephalopods and radiolarians in the Changhsingian. They were located in the deep-water siliceous basin (Fig. 2.2). The Changhsingian at the sections of Kele, Guanbachong, Xiaohebian, Chahe, Zhejue, Jiucaichong, Jinzhong, Mide and Tucheng, is characterized by feldspathic sandstones, siltstones and mudstones yielding abundant plant fossils and coal seams. These sections were all located in the terrestrial facies (Fig. 2.2). The sections of Jinjibang and Wadu are dominated by siltstones and mudstones intercalated with coal seams, yielding plant fossils, brachiopods and bivalves. Both sections were located in the terrestrial–marine transitional facies. The Changhsingian of Zhongzhai section is mainly composed of siltstone and mudstone intercalated with limestone, with abundant brachiopods. This section was located in the shallow-water clastic shelf (Fig. 2.2). The Changhsingian of Penglaitan section is dominated by limestone and tuffaceous sandstone and mudstone. The section is considered to have been located in a rift basin, due to the intensive volcano eruption in the Late Permian to the Induan (Fig. 2.2).

(2) Palaeogeographic localities of the studied sections in the Permian–Triassic transition

As stated above, no significant palaeogeographic changes occurred from the Changhsingian to the Permian–Triassic transition, but the sections of Jinzhong, Mide and Tucheng spanning the Permian–Triassic transition changed from terrestrial settings to terrestrial–marine transitional facies/shallow-water clastic facies (Fig. 2.3). This change is evidenced by the appearance of siltstone and mudstone with abundant bivalves, brachiopods, ostracods and plant fossils.

(3) Palaeogeographic localities of the studied sections in the Induan

The sections of Guanbachong, Xiaohebian, Chahe, Zhejue, Jiucaichong, Jinzhong, Mide and Tucheng, are dominated by purplish red sandstone and siltstone (basal part of Dongchuan Formation) and were located in the terrestrial facies (Fig. 2.4). The Penglaitan section still remained in a rift basin (Fig. 2.4). The basal part of the Induan at the sections of Jinjibang, Kele, Wadu and Zhongzhai sections, is composed of siltstones and mudstones, yielding abundant bivalves and brachiopods, indicating that they were located in the silty–argillaceous shelf facies (Fig. 2.4). Other studied sections are characterized by argillaceous limestones and calcareous mudstones, with abundant bivalves and ammonoids, suggesting an unequivocal calcareous–argillaceous marine shelf setting (Fig. 2.4).

References

Bu JJ, Wu SB, Zhang HL, Meng YY, Zhang F, Zhang LY (2006) Permian–Triassic Cephalopods from Dongpan Section, Guangxi, and its geological significance. Geol Sci Technol Inf 25:47–51 [in Chinese with English abstract]

Chu DL, Yu JX, Tong JN, Benton MJ, Song HJ, Huang YF, Song T, Tian L (2016) Biostratigraphic correlation and mass extinction during the Permian–Triassic transition in terrestrial–marine siliciclastic settings of South China. Glob Planet Change 146:67–88

Chu DL, Tong JN, Benton MJ, Yu JX, Huang YF (2019) Mixed continental–marine biotas following the Permian–Triassic mass extinction in South and North China. Palaeogeogr Palaeoclimatol Palaeoecol 519:95–107

Feng QL, He WH, Gu SZ, Meng YY, Jin YX, Zhang F (2007) Radiolarian evolution during the latest Permian in South China. Glob Planet Change 55:177–192

Feng Z, Wei HB, Ye RH, Sui Q, Gou XD, Guo Y, Liu LJ, Yang SL (2020) Latest Permian peltasperm plant from southwest China and its paleoenvironmental implications. Front Earth Sci 8:559430. https://doi.org/10.3389/feart.2020.559430

Gu SZ, Feng QL, He WH (2007) The latest Permian deep-water fauna: Latest Changhsingian small foraminifers from southwestern Guangxi, South China. Micropaleontology 53:311–330

He WH, Shen SZ, Feng QL, Gu SZ (2005) A late Changxingian (Late Permian) deep-water brachiopod fauna from the Talung Formation at the Dongpan Section, Southern Guangxi, in South China. J Paleontol 79:927–938

He WH, Zhang Y, Zheng YE, Zhang KX, Gui BW, Feng QL (2008) A late Changhsingian (latest Permian) radiolarian fauna from Chaohu, Anhui and a comparison with its contemporary faunas of South China. Alcheringa 32:199–222

He WH, He WH, Twitchett RJ, Zhang Y, Shi GR, Feng QL, Yu JX, Wu SB, Peng XF (2010) Controls on body size during the Late Permian mass extinction event. Geobiology 8:391–402

He WH, Zhang Y, Zhang Q, Zhang KX, Yuan AH, Feng QL (2011) A latest Permian radiolarian fauna from Hushan, South China and its geological implications. Alcheringa 35:471–496

He WH, Zhang KX, Wu SB, Feng QL, Yang TL, Yue ML, Xiao YF, Wu HT, Zhang Y, Wang GD, Chen B (2015a) End-Permian faunas from Yangtze basin and its marginal region: implications for palaeogeographical and tectonic environments. Earth Sci J China Univ Geosci 40:275–289 [in Chinese with English abstract]

He WH, Shi GR, Twitchett RJ, Zhang Y, Zhang KX, Song HJ, Yue ML, Wu SB, Wu HT, Yang TL, Xiao YF (2015b) Late Permian marine ecosystem collapse began in deeper waters: evidence from brachiopod diversity and body size changes. Geobiology 13:123–138

He WH, Shi GR, Yang TL, Zhang KX, Yue ML, Xiao YF, Wu HT, Chen B, Wu SB (2016) Patterns of brachiopod faunal and body-size changes across the Permian–Triassic boundary: evidence from the Daoduishan section in Meishan area, South China. Palaeogeogr Palaeoclimatol Palaeoecol 448:72–84

He WH, Shi GR, Zhang KX, Yang TL, Shen SZ, Zhang Y (2019) Brachiopods around the Permian–Triassic boundary of South China. Springer, Singapore, 261pp

He WH, Weldon EA, Yang TL, Wang H, Xiao YF, Wu HT, Zhang KX, Wang YB, Wu SB (2020) The palaeoenvironmental and palaeobiogeographical significance of the Late Permian deep-water brachiopod fauna from Dongpan, South China, including descriptions of Micromartinia He & Weldon gen. nov. (Micromartiniidae He & Weldon fam. nov.) and Minutomarginifera nom. nov. J Syst Paleontol 18:885–909

Huang YF, He WH, Liao W, Wang YB, Yi ZX, Yang H (2022) Two pulses of increasing terrestrial input to marine environment during the Permian–Triassic transition. Palaeogeogr Palaeoclimatol Palaeoecol 586:110753. https://doi.org/10.1016/j.palaeo.2021.110753

Jiang Y, Wu HT, Xiao YF, Yang TL, He WH, Wang H, Sun X, Wu SB (2018) Ammonoid zonations of Changhsingian–basal Induan from Kejiao, Huishui County, southern Guizhou Province, South China and regional correlation. Geol Sci Technol Inf 37:151–159 [in Chinese with English abstract]
Liao ZT, Meng FY (1984) Changhsingian of Hutang area in Chenxian of Hunan. J Stratigr 8:120–128 [in Chinese]
Scotese CR (2000) Paleomap project. http://www.scotese.com
Shen SZ, He XL (1991) Brachiopod assemblage of Changxing formation at Zhongliangshan, Chongqing. J Stratigr 15:189–196 [in Chinese]
Shen SZ, Zhang YC (2008) Earliest Wuchiapingian (Lopingian, Late Permian) brachiopods in southern Hunan, South China: implications for the pre-Lopingian crisis and onset of Lopingian recovery/radiation. J Paleontol 82:924–937
Shen SZ, Ramezani J, Chen J, Cao CQ, Erwin DH, Zhang H, Xiang L, Schoepfer SD, Henderson CM, Zheng QF, Bowring SA, Wang Y, Li XH, Wang XD, Yuan DX, Zhang YC, Mu L, Wang J, Wu YS (2019) A sudden end-Permian mass extinction in South China. Geol Soc Am Bull 131:205–223
Song HJ, Tong JN, Chen ZQ (2011) Evolutionary dynamics of the Permian–Triassic foraminifer size: Evidence for Lilliput effect in the end-Permian mass extinction and its aftermath. Palaeogeogr Palaeoclimatol Palaeoecol 308:98–110
Su CM, Chen ZQ, Wang X, Wu SQ, Guo Z (2021) Biotic and palaeoecological variations in the Permian−Triassic boundary microbialite (Xiejiacao, South China): Implication for a two-phase ecological crisis in microbialite ecosystems. Glob Planet Change 207:103679. https://doi.org/10.1016/j.gloplacha.2021.103679
Tong JN (2015) Division and Correlation of Marine Lower-Middle Triassic Strata in East China. Acta Geoscientica Sinica 36:546–558 [in Chinese with English abstract]
Tong JN, Chu DL, Liang L, Shu WC, Song HJ, Song T, Song HY, Wu YY (2019) Triassic integrative stratigraphy and timescale of China. Sci China Earth Sci 62:189–222
Tong JN, Chu DL, Miao X, Shu WC, Guo WW, Wu YY, Su Y, Wu K, Ji KX, Yu YY (2021) Lithostratigraphic subdivision and correlation of the Triassic in China. J Stratigr 45:340–363 [in Chinese with English abstract]
Wang H, Xiao YF, Zhang N, Jiang Y, Sun X, He WH (2018) End-Permian radiolarian fauna from north part of Nanpanjing basin and implications of paleo-water depths. Earth Sci 43:3901–3909 [in Chinese with English abstract]
Wu HT, He WH, Weldon EA (2018) Prelude of benthic community collapse during the end-Permian mass extinction in siliciclastic offshore sub-basin: Brachiopod evidence from South China. Global Planet Change 163:158–170
Wu YS, Jiang HX, Liao TP (2006) Sea-level drops in the Permian−Triassic boundary section at Laolongdong, Chongqing, Sichuan Province. Acta Petrologica Sinica 29:2405–2412 [in Chinese with English abstract]
Xiao YF, Suzuki N, He WH (2017) Applications and limitations of micro-XCT imaging in the studies of Permian radiolarians: a new genus with bi-polar main spines. Acta Palaeontol Pol 62:647–656
Xiao YF, Wang KY, He WH, Suzuki N, Zhang KX, Yang TL, Wu SB (2022) Changhsingian (Lopingian, Permian) radiolarian paleobiogeography on and around the Yangtze Platform. Palaeoworld (in press).https://doi.org/10.1016/j.palwor.2022.07.001
Yang ZY, Yin HF, Wu SB, Yang FQ, Ding MH, Xu GR (1987) Permian–Triassic boundary stratigraphy and fauna of South China. Geological Publishing House, Beijing, 378 pp [in Chinese with English abstract]
Yang TL, He WH, Zhang KX, Wu SB, Zhang Y, Yue ML, Wu HT, Xiao YF (2015) Palaeoecological insights into the Changhsingian-Induan (latest Permian–earliest Triassic) bivalve fauna at Dongpan, southern Guangxi, South China. Alcheringa 40:98–117
Yu JX, Peng YQ, Zhang SX, Yang FQ, Zhao QM, Huang QS (2007) Terrestrial events across the Permian–Triassic boundary along the Yunnan-Guizhou border, SW China. Glob Planet Change 55:193–208

Zhang YX (1981) Late Permian bivalves from Yuanjia of Jiahe, Hunan Province. Acta Palaeontologica Sinica 20:260–265 [in Chinese with English abstract]

Zhang ZY, He WH, Zhang Y, Yang TL, Wu SB (2009) Late Permian—earliest Triassic Ammonoid Sequences from the Rencunping Section, Sangzhi County, Hunan Province, South China and their regional correlation. Geol Sci Technol Inf 28:23–30 [in Chinese with English abstract]

Zhang Y, He WH, Shi GR, Zhang KX (2013) A new Changhsingian (Late Permian) Rugosochonetidae (Brachiopoda) fauna from the Zhongzhai section, southwestern Guizhou Province, South China. Alcheringa 37:223–247

Zhang H, Cao CQ, Liu XL, Mu L, Zheng QF, Liu F, Xiang L, Liu LJ, Shen SZ (2016a) The terrestrial end-Permian mass extinction in South China. Palaeogeogr Palaeoclimatol Palaeoecol 448:108–124

Zhang KX, He WH, Xu YD, Song BW, Luo MS (2016b) Sequences of sedimentary rocks and sedimentary tectonics. Geological Publishing House, Beijing, pp 604 [in Chinese]

Zhang Y, Shi GR, He WH, Wu HT, Lei Y, Zhang KX, Du CC, Yang TL, Yue ML, Xiao YF (2016c) Significant pre-mass extinction animal body-size changes: evidence from the Permian–Triassic boundary brachiopod faunas of South China. Palaeogeogr Palaeoclimatol Palaeoecol 448:85–95

Zhou ZR (2007) Unusually large specimens of the Permian Ammonoid *Guiyangoceras* from Hunan, South China. Paleontol J 41:123–127

Description of the Studied Permian–Triassic Boundary Sections of South China

3

Wei-Hong He, G. R. Shi, Jian-Xin Yu, Dao-Liang Chu, Ke-Xin Zhang, Hai-Shui Jiang, Ting-Lu Yang, Yi-Fan Xiao, Ya-Fei Huang, Ai-Hua Yuan, and Hai-Jun Song

Thirty two sections are described in this chapter, including Jinjibang (JJB), Guanbachong (GBC), Kele (KL), Xiaohebian (XHB), Chahe (CH), Jinzhong (JZ), Zhejue (ZJ), Jiucaichong (JCC), Mide (MD), Wadu (WD) and Tucheng (TC) sections in the terrestrial and terrestrial–marine transitional facies (or shallow-water marine clastic shelf); Zhongzhai (ZZ) in the shallow-water marine clastic rock facies (or clastic shelf); Meishan (MS) and Daoduishan (DDS) along the ramp of the carbonate platform; Huangzhishan (HZS), Yangou (YG), Chibi (CB) and Liangfengya (LFY) sections on the carbonate platform; Zhaixia (ZX), Kangjiaping (KJP), Laolongdong (LLD), Dajiang (DJ), Zuodeng (ZD) in the reef facies on

W.-H. He (✉) · J.-X. Yu · D.-L. Chu · K.-X. Zhang · H.-S. Jiang · Y.-F. Xiao · H.-J. Song
State Key Laboratory of Biogeology and Environmental Geology, School of Earth Sciences, China University of Geosciences, 388 Luma Road, Wuhan, Hubei, China
e-mail: whzhang@cug.edu.cn

J.-X. Yu
e-mail: yujianxin@cug.edn.cn

D.-L. Chu
e-mail: chudl@cug.edu.cn

K.-X. Zhang
e-mail: kx_zhang@cug.edu.cn

H.-S. Jiang
e-mail: jiangliuis@163.com

Y.-F. Xiao
e-mail: yifanxiao@cug.edu.cn

H.-J. Song
e-mail: haijunsong@cug.edu.cn

© Springer Nature Singapore Pte Ltd. 2023

W. He et al. (eds.), *Stratigraphy Around the Permian–Triassic Boundary of South China*, New Records of the Great Dying in South China,
https://doi.org/10.1007/978-981-99-9350-5_3

the carbonate platform /along the margin of carbonate platform; Majiashan (MJS), Jinshan (JS), Xiejiaping (XJP), Shangsi (SS), Rencunping (RCP), Xinmin (XM), Kejiao (KJ) and Dongpan (DP) in the deep-water basinal facies; and Penglaitan (PLT) in the rift basin (see Chap. 2).

3.1 Jinjibang Section in Junlian County

The Jinjibang section is located at the Jinjibang Village, Zhenzhou Town, about 21 km southeast to the Junlian County, Sichuan Province, South China (Fig. 3.1).

The sequence of the Jinjibang section is composed of the Xuanwei Formation and the basal part of the Feixianguan Formation in an ascending order (Fig. 3.2). The Xuanwei Formation is dominated by yellowish green to gray siltstones and mudstones, intercalated with coal seams and tuffaceous sandstones and mudstones. The basal part of the Feixianguan Formation is mainly composed of yellowish green siltstones and mudstones. Abundant plant fossils and a few brachiopods and bivalves were discovered in the section (Figs. 3.2, 3.3).

Unexposed upwards
Feixianguan Formation in the uppermost Permian

33. Yellowish green thick-bedded silty mudstone, with bivalves *Pteria variabilis*, *Claraia wangi*, *Austrotindaria antiqua*, *Unionites canalensis* and the conchostracan *Euestheria gutta*. 200 cm
32. Yellowish green thin-bedded mudstone in the upper part; grayish black thin-bedded mudstone in the lower part. The plant fossil *Tomiostrobus* sp., brachiopods lingulids, and bivalves *Promyalina putiatinensis*, *Pteria variabilis*, *Claraia wangi*, *Eumorphotis* sp., *Austrotindaria antiqua*, *Permophorus bregeri*, *Unionites canalensis*, *Neoschizodus laevigatus*, *Neoschizodus orbicularis* were recorded. 200 cm

G. R. Shi
School of Earth, Atmospheric and Life Sciences, University of Wollongong, New South Wales, Wollongong, NSW 2522, Australia
e-mail: guang@uow.edu.au

T.-L. Yang
Faculty of Geosciences, East China University of Technology, 418 Guanglan Avenue, Nanchang, China
e-mail: yang@geology.hk

Y.-F. Huang · A.-H. Yuan
School of Earth Sciences, China University of Geosciences, 388 Luma Road, Wuhan, Hubei, China
e-mail: 1106821675@qq.com

A.-H. Yuan
e-mail: ahyuan@cug.edu.cn

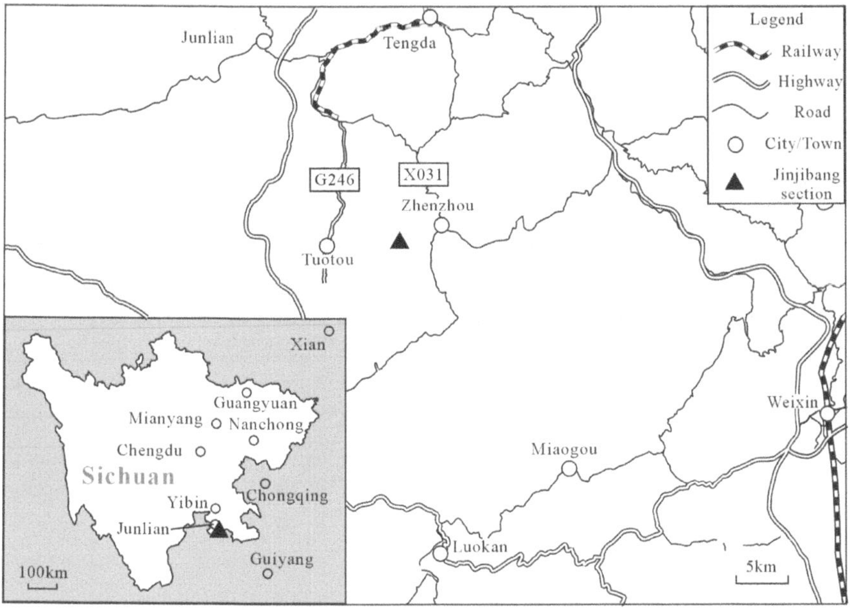

Fig. 3.1 Geographic location of the Jinjibang section. *Note* G246 representing the number (246) of National Highway; X031 representing the number (031) of local major road, the same as below (for other sections)

Conformity
Xuanwei Formation in the Upper Permian

31. Gray medium-bedded argillaceous limestone with the bivalve *Neoschizodus* sp., gastropods and ostracods. 30 cm
30. Gray thin-bedded argillaceous siltstone in the upper part; gray thin-bedded argillaceous siltstone interbedded with mudstone in the lower part; dark gray carbonaceous shale at the basal part. Laminations prevalent throughout. Brachiopods *Orthotetina ruber*, *Neochonetes* sp. and lingulids, bivalves *Bakevellia* sp. and *Bakevellia qinglongensis*, and plant fossils *Lepidophylloides* sp., *Paracalamites* sp., *Pecopteris* sp., *Taeniopteris* sp. and *Sphenopteris* sp. were found. 200 cm
29. Black coal seam at the top part; gray thick-bedded silty mudstone and light gray mudstone in the upper part; yellowish green fine-grained sandstone, tuffaceous siltstone, and silty mudstone in the lower part. Plant fossils *Stigmaria ficoides*, *Paracalamites* sp., *Gigantopteris dictyophylloides*, *Gigantopteris nicotianaefolia*, *Cordaites* sp. and *Cordaites principalis* were examined in the upper part. 431 cm
28. Gray silty mudstone intercalated with coal seams, with ferruginous nodules in the upper part; yellowish green fine-grained tuffaceous sandstone, siltstone and argillaceous siltstone, with ferruginous nodules and laminations in the lower

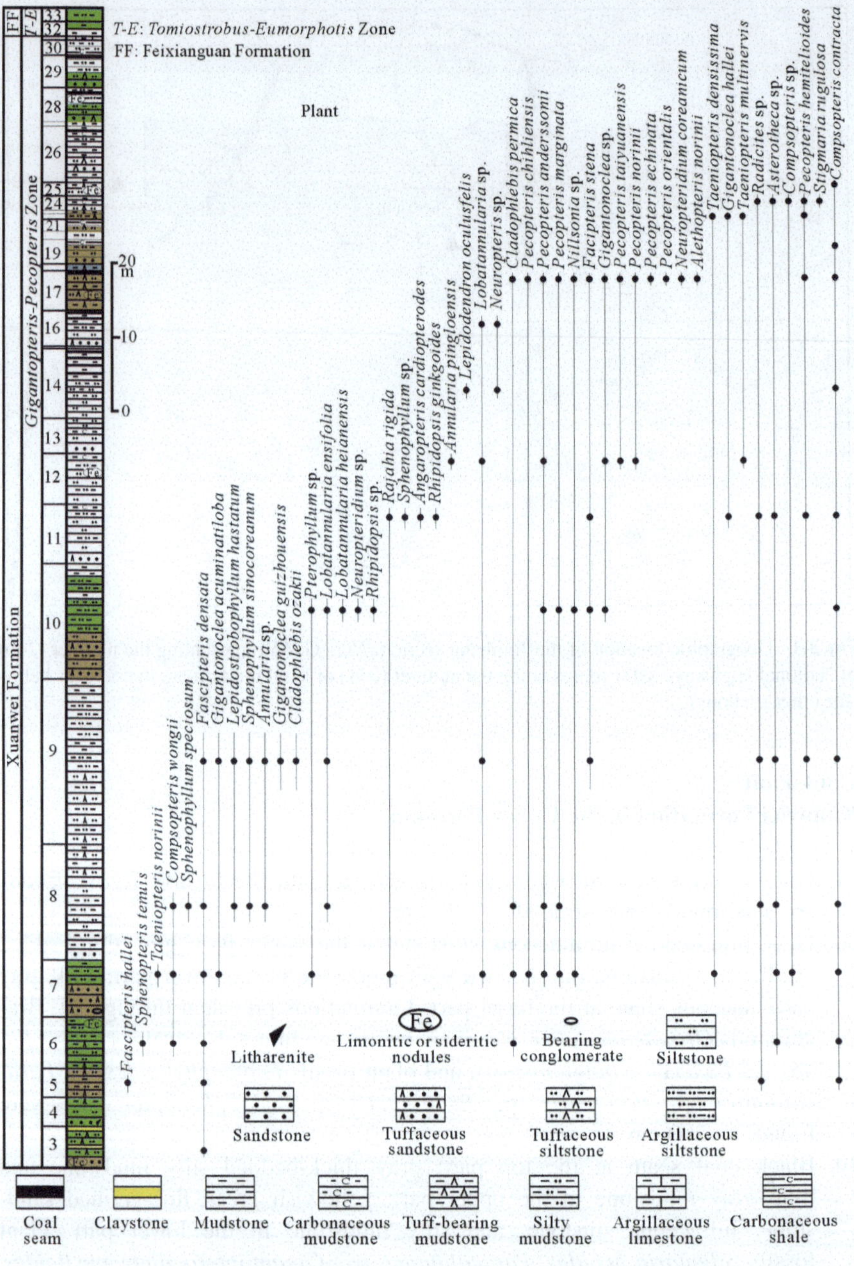

Fig. 3.2 Stratigraphic column and fossil occurrences (plant) at the Jinjibang section (referred to Li et al. 1982)

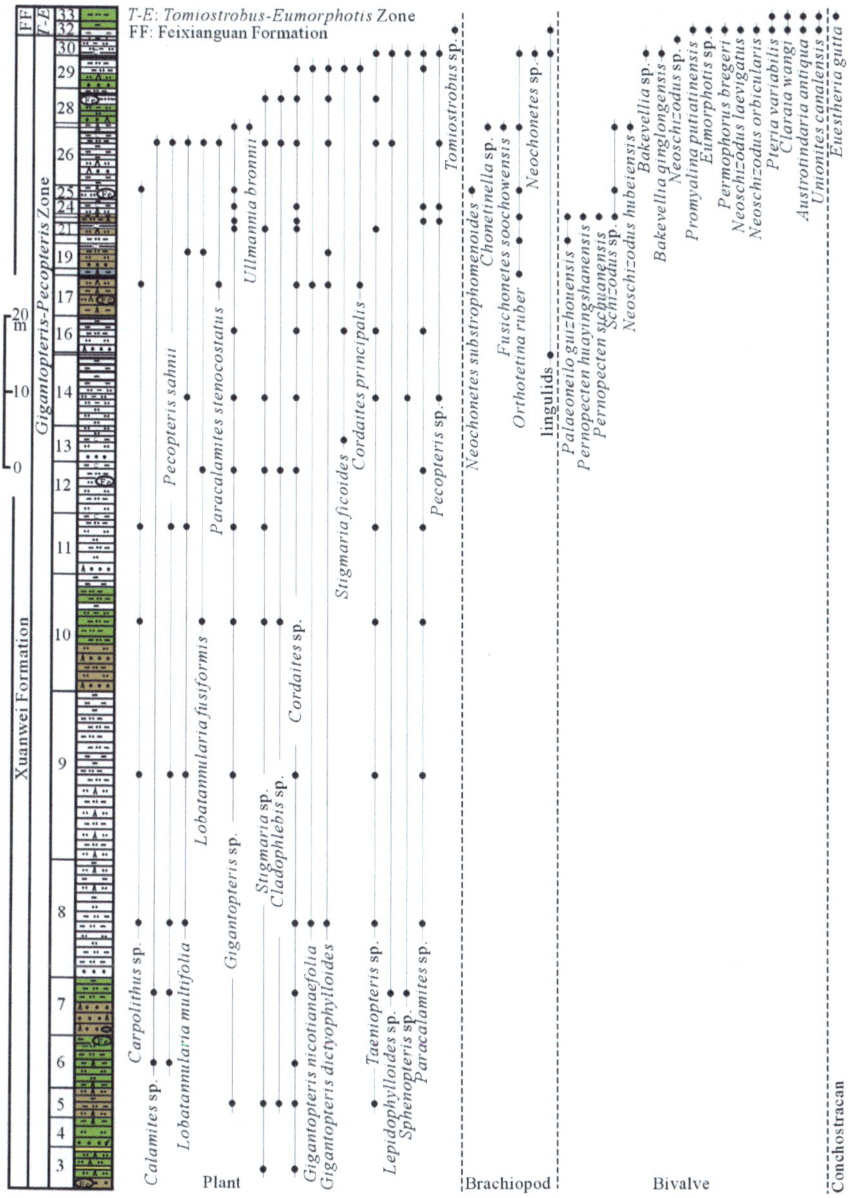

Fig. 3.3 Stratigraphic column and fossil occurrences (plant, brachiopod, bivalve, conchostracan) at the Jinjibang section (fossils from the Xuanwei Formaiton referred to Li et al. 1982). Legends same to Fig. 3.2

part. Plant fossils *Stigmaria* sp., *Gigantopteris dictyophylloides*, *Taeniopteris* sp., *Cladophlebis* sp., and *Cordaites* sp. were identified. 458 cm

27. Gray thin-bedded mudstone and tuffaceous mudstone with brachiopods, bivalves and plant fossils. Brachiopods include *Orthotetina ruber*, *Chonetinella* sp. and *Fusichonetes soochowensis*, bivalves include *Neoschizodus hubeiensis* and *Schizodus* sp., and plant fossils include *Ullmannia bronnii* and *Gigantopteris* sp.. 61 cm
26. Gray mudstone intercalated with silty mudstone, coal seams and carbonaceous shale, with interlayers of fine-grained tuffaceous sandstone and siltstone in the lower part. Plant fossils *Lepidophylloides* sp., *Stigmaria* sp., *Calamites* sp., *Paracalamites stenocostatus*, *Lobatannularia multifolia*, *Lobatannularia fusiformis*, *Gigantopteris dictyophylloides*, *Pecopteris* sp., *Pecopteris sahnii*, *Taeniopteris* sp., *Cladophlebis* sp. and *Cordaites* sp. were found. 757 cm
25. Gray tuffaceous mudstone, with interlayers of siltstone and a few ferruginous nodules in the middle part, and with brachiopods, bivalves and plant fossils in the mudstone. Laminations evident. Brachopods include *Orthotetina ruber* and *Neochonetes substrophomenoides*, the bivalve includes *Schzodus* sp., and plant fossils include *Gigantopteris* sp., *Compsopteris contracta* and *Carpolithus* sp.. 177 cm
24. Gray mudstone and silty mudstone intercalated with carbonaceous mudstone with laminations in the upper part; tuffaceous siltstone in the lower part. Plant fossils *Stigmaria rugulosa*, *Paracalamites* sp., *Gigantopteris* sp., *Pecopteris* sp., *Pecopteris hemitelioides*, *Asterotheca* sp., *Compsopteris* sp., *Cordaites* sp. and *Radicites* sp. were identified. 191 cm
23. Yellowish gray tuffaceous mudstone with ferruginous nodules. The brachiopod *Orthotetina ruber* and bivalves *Pernopecten huayingshanensis*, *Pernopecten sichuanensis*, *Palaeoneilo guizhouensis* and *Schizodus* sp. were examined. 38 cm
22. Yellowish gray silty mudstone in the upper part; yellowish green fine-grained tuffaceous sandstone and siltstone in the middle and lower parts, laminations prevalent in the siltstone. Plant fossils *Paracalamites* sp., *Gigantopteris* sp., *Gigantonoclea hallei*, *Pecopteris* sp., *Pecopteris hemitelioides*, *Taeniopteris multinervis*, *Taeniopteris densissima* and *Cordaites* sp. were identified. 126 cm
21. Yellow fine-grained tuffaceous sandstone and siltstone intercalated with silty mudstone, with gray mudstone intercalated with carbonaceous shale at the top. Plant fossils *Gigantopteris* sp., *Taeniopteris* sp., *Cordaites* sp. and *Stigmaria* sp. were collected. 156 cm
20. Gray thin-bedded silty mudstone. Laminations prevalent throughout. The brachiopod *Orthotetina ruber* and the bivalve *Palaeoneilo guizhouensis* were recorded. 108 cm
19. Yellow fine-grained sandstone, siltstone and silty mudstone, with carbonaceous shale and laminations at the top part, and with normal graded bedding in the upper part and inversal graded bedding in the lower part. The plant fossils

Lobatannularia fusiformis, *Lobatannularia multifolia*, *Compsopteris contracta* and *Gigantopteris dictyophylloides* were found. 329 cm

18. Bluish gray tuffaceous mudstone with wave-shaped cross beddings, with the brachiopod *Orthotetina ruber*. 50 cm
17. Yellow to gray tuffaceous siltstone and mudstone, with interlayers of coal seams and mudstone in the upper part and ferruginous nodules in the lower part. Plant fossils *Gigantopteris dictyophylloides*, *Gigantopteris nicotianaefolia*, *Gigantonoclea* sp., *Pecopteris echinata*, *Pecopteris anderssonii*, *Pecopteris hemitelioides*, *Pecopteris orientalis*, *Pecopteris chihliensis*, *Pecopteris norinii*, *Alethopteris norinii*, *Pecopteris taiyuanensis*, *Pecopteris marginata*, *Cladophlebis permica*, *Neuropteridium coreanicum*, *Facipteris stena*, *Compsopteris contracta*, *Paracalamites stenocostatus*, *Nillsonia* sp., *Cordaites* sp., *Cordaites principalis* and *Carpolithus* sp. were collected in the upper part. 523 cm
16. Gray mudstone interbedded with coal seams in the upper part; gray silty mudstone and fine-grainded tuffaceous sandstone and siltstone, with laminations and pyrite nodules in the lower part. Plant fossils *Lobatannularia* sp., *Gigantopteris* sp., *Neuropteris* sp., *Taeniopteris* sp., *Cordaites* sp., *Stigmaria ficoides* and *Paracalamites* sp. were recorded. 490 cm
15. Black carbonaceous shale with abundant brachiopod lingulids. 26 cm
14. Black coal seams intercalated with light gray mudstone in the upper part; gray silty mudstone and siltstone in the lower part. Plant fossils *Lepidodendron oculusfelis*, *Lobatannularia multifolia*, *Gigantopteris* sp., *Pecopteris* sp., *Compsopteris contracta*, *Neuropteris* sp., *Sphenopteris* sp., *Taeniopteris* sp., *Cordaites* sp. and *Stigmaria* spp. were identified. 936 cm
13. Gray thick- to medium-bedded fine-grained sandstone and siltstone intercalated with light gray mudstone and carbonaceous mudstone, with the plant fossil *Stigmaria ficoides* in the mudstone. 469 cm
12. Gray to light gray silty mudstone and mudstone intercalated with ferruginous nodules in the upper part, and intercalated with thin-bedded carbonaceous mudstone at the top part; gray silty mudstone and siltstone in the lower part. Plant fossils *Gigantopteris* sp., *Gigantonoclea* sp., *Pecopteris taiyuanensis*, *Pecopteris norinii*, *Pecopteris anderssonii*, *Taenioipteris multinervis*, *Cladophlebis* sp., *Lobatannularia* sp., *Lobatannularia fusiformis*, *Annularia pingloensis*, *Paracalamites* sp., *Stigmaria* sp. and *Cordaites* sp. were examined. 682 cm
11. Gray silty mudstone and siltstone intercalated with mudstone and carbonaceous mudstone, with tuffaceous sandstone at the basal part. Plant fossils *Lobatannularia multifolia*, *Sphenophyllum* sp., *Paracalamites* sp., *Angaropteris cardiopterodes*, *Pecopteris hemitelioides*, *Pecopteris sahnii*, *Facipteris stena*, *Asterotheca* sp., *Compsopteris contracta*, *Gigantopteris* sp., *Gigantonoclea hallei*, *Taenioipteris* sp., *Rajahia rigida*, *Rhipidopsis ginkgoides*, *Stigmaria* sp., *Carpolithus* sp. and *Radicites* sp. were collected. 807 cm

10. Yellowish green silty mudstone and light gray mudstone intercalated with argillaceous siltstone in the upper part; yellow silty mudstone, fine-grained tuffaceous sandstone and siltstone in the lower part. Plant fossils *Lobatannularia ensifolia, Lobatannularia heianensis, Lobatannularia fusiformis, Paracalamites* sp., *Stigmaria* sp., *Gigantopteris* sp., *Gigantonoclea* spp., *Compsopteris contracta, Facipteris stena, Neuropteridium* sp., *Pecopteris marginata, Pecopteris anderssonii, Cladophlebis* sp., *Taeniopteris* sp., *Nilssonia* sp., *Pterophyllum* sp., *Rhipidopsis* sp. and *Carpolithus* sp. were recovered in the upper part. 1508 cm
9. Gray mudstone and silty mudstone in the upper part; silty mudstone interbedded with tuffaceous siltstone in the lower part. Plant fossils *Lepidostrobophyllum hastatum, Lobatannularia* sp., *Lobatannularia multifolia, Lobatannularia ensifolia, Annularia* sp., *Paracalamites* sp., *Sphenophyllum sinocoreanum, Gigantonoclea acuminatiloba, Gigantonoclea guizhouensis, Gigantopteris* sp., *Compsopteris contracta, Fascipteris densata, Fascipteris stena, Pecopteris hemitelioides, Pecopteris sahnii, Asterotheca* sp., *Cladophlebis ozakii, Taeniopteris* sp., *Cordaites* sp., *Carpolithus* spp. and *Radicites* sp. were collected. 2224 cm
8. Gray mudstone and tuffaceous siltstone in the upper part; silty mudstone intercalated with siltstone in the lower part, with fine-grained sandstone at the base. Plant fossils *Lepidostrobophyllum hastatum, Sphenophyllum sinocoreanum, Sphenophyllum speciosum, Annularia* sp., *Lobatannularia multifolia, Lobatannularia ensifolia, Paracalamites* sp., *Gigantopteris nicotianaefolia, Gigantopteris dictyophylloides, Fascipteris densata, Compsopteris contracta, Compsopteris wongii, Pecopteris sahnii, Pecopteris anderssonii, Pecopteris marginata, Cladophlebis permica, Asterotheca* sp., *Taeniopteris* sp., *Cordaites* sp., *Carpolithus* spp. and *Radicites* sp. were discovered. 1561 cm
7. Yellowish green mudstone and silty mudstone, with laminations and plant fossils in the upper part; yellow thick-bedded tuffaceous sandstone in the lower part, and with conglomerates at the base. Plant fossils include *Lepidophylloides* sp., *Lobatannularia* sp., *Calamites* sp., *Compsopteris* sp., *Compsopteris contracta, Compsopteris wongii, Fascipteris densata, Gigantonoclea acuminatiloba, Pecopteris chihliensis, Pecopteris sahnii, Pecopteris anderssonii, Pecopteris marginata, Asterotheca* sp., *Cladophlebis permica, Sphenopteris* sp., *Taeniopteris densissima, Taeniopteris norinii, Nilssonia* sp., *Pterophyllum* sp., *Cordaites* sp. and *Rajahia rigida.* 787 cm
6. Yellowish green mudstone and silty mudstone in the upper part, with ferruginous nodules at the top; yellowish green tuffaceous mudstone and siltstone in the lower part. Plant fossils *Calamites* sp., *Compsopteris contracta, Compsopteris wongii, Fascipteris densata, Cladophlebis permica, Pecopteris sahnii, Sphenopteris tenuis, Asterotheca* sp. and *Cordaites* sp. were identified. 704 cm
7. Gray to yellow mudstone and silty mudstone, with interlayers of tuffaceous siltstone in the lower and basal parts. Plant fossils *Fascipteris densata, Fascipteris hallei, Compsopteris contracta, Gigantopteris* sp., *Taeniopteris*

sp., *Cladophlebis* sp., *Cordaites* sp., *Stigmaria* sp. and *Radicites* sp. were collected. 394 cm

8. Yellowish green siltstone, tuffaceous mudstone and litharenite, with fragments of plant fossils. 391 cm
9. Pale claystone, brown basaltic and ferruginous rock and yellowish green tuffaceous siltstone, with plant fossils *Fascipteris densata*, *Cordaites* sp. and *Stigmaria* sp. in the claystone. 546 cm

Note: fossils in beds 3–31 are based on Li et al. (1982), but revised and updated in taxonomy.

3.2 Guanbachong Section in Zhaotong City

The Guanbachong section is located at the southeast corner of the Yudong Reservoir of Leju Village, Zhaotong City, Yunnan Province, South China (Fig. 3.4).

The sequence of the Guanbachong section is composed of the Xuanwei, Kayitou and Dongchuan formations in an ascending order (Fig. 3.5). The Xuanwei Formation is dominated by yellowish to grayish green fine-grained sandstones, siltstones and mudstones, occasionally intercalated with coal seams. The Kayitou Formation is mainly composed of yellowish green to yellow argillaceous siltstones and mudstones. The Dongchuan Formation is characterized by purplish

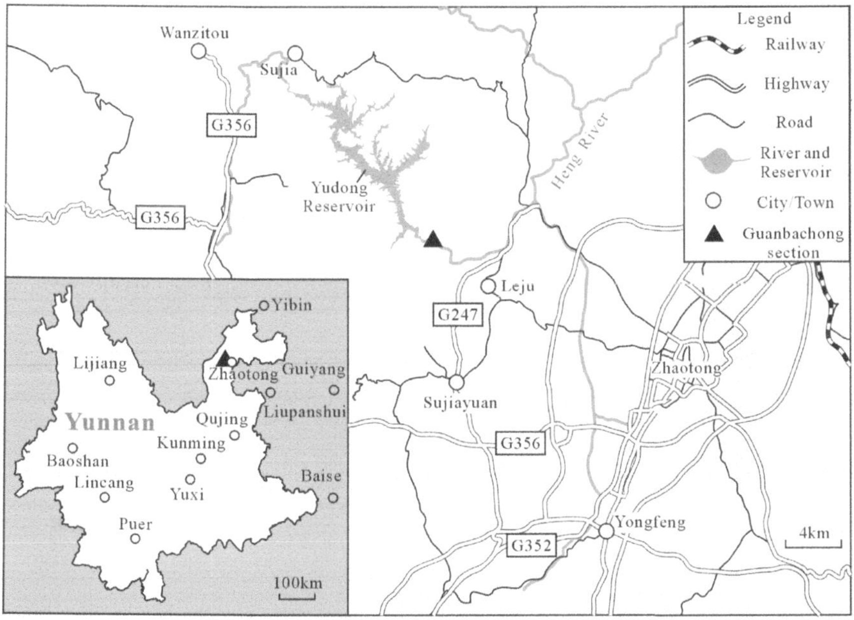

Fig. 3.4 Geographic location of the Guanbachong section

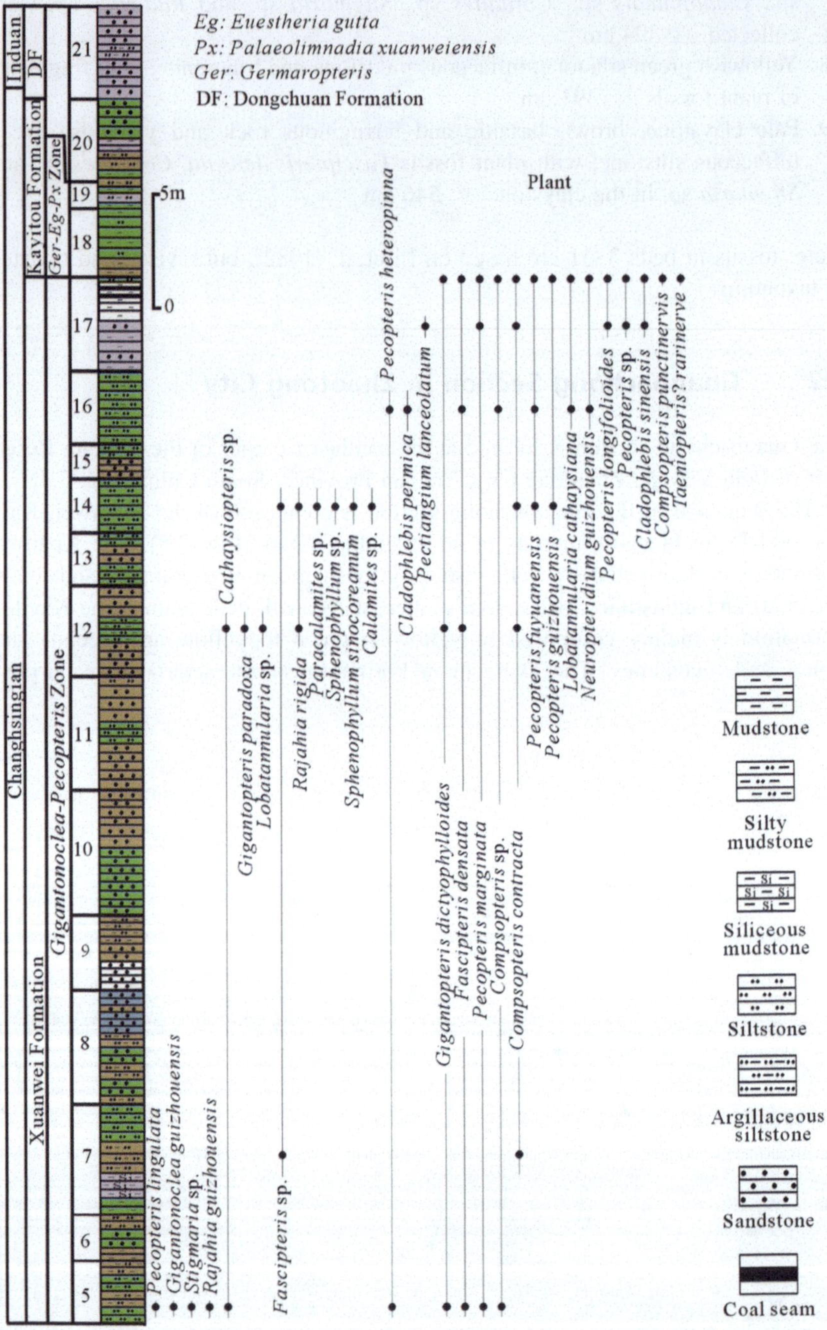

Fig. 3.5 Stratigraphic column and fossil occurrences (plant) at the Guanbachong section (plant fossils referred to Zhang et al. 2016)

red sandstones and siltstones, intercalated with mudstones and calcareous nodules. Abundant plant fossils and a few conchostracans have been found in the section (Figs. 3.5, 3.6).

Dongchuan Formation in the Lower Triassic (Induan)

21. Purplish red sandstone intercalated with argillaceous siltstone. 400 cm

Conformity
Kayitou Formation in the Upper Permian (Changhsingian)

20. Yellowish green sandstone in the upper part; yellow silty mudstone intercalated with yellowish green mudstone in the lower part. 370 cm
19. Grayish yellow argillaceous siltstone interbedded with yellowish green mudstone, with the plant fossil *Germaropteris* sp. and conchostracans *Euestheria gutta* and *Palaeolimnadia xuanweiensis.* 220 cm
18. Yellow to yellowish green mudstone intercalated with siltstone, with the plant fossil *Germaropteris* sp. and conchostracans *Euestheria gutta*, *Euestheria* spp. and *Palaeolimnadia xuanweiensis.* 300 cm

Conformity
Xuanwei Formation in the Upper Permian (Changhsingian)

17. Yellowish green medium- to thin-bedded siltstone (purplish red when weathered) in the uppermost (15 cm thick); black mudstone intercalated with coal seams in the upper part; yellowish green thin-bedded siltstone in the middle part; purplish red thin-bedded mudstone intercalated with brownish yellow medium- to thin-bedded fine-grained sandstone in the lower part. Abundant plant fossils *Compsopteris punctinervis*, *Fascipteris densata*, *Taeniopteris*? *rarinerve*, *Guizhoua gregalis*, *Neuropteridium guizhouensis*, *Pecopteris longifolioides*, *Gigantopteris dictyophylloides*, ?*Distichotheca crossothecoides*, *Pecopteris* sp., *Compsopteris* sp., *Compsopteris contracta*, *Cladophlebis sinensis*, *Lobatannularia heianensis*, *Pecopteris guizhouensis*, *Gigantonoclea longifolia*, *Stigmaria ficoides*, *Paracalamites stenocostatus*, *Pecopteris elegantula*, *Cladophlebis parapermica*, *Pterophyllum eratum*, *Rhipidopsis* cf. *ginkgoides*, *Pecopteris marginata*, *Pecopteris fuyuanensis*, *Marattiopsis*? sp., *Rajahia mirabilis*, *Carpolithus* sp., *Lobatannularia cathaysiana* and *Taeniopteris* cf. *taiyuanensis* were recorded in the siltstone in the uppermost. Plant fossils *Pectiangium lanceolatum*, *Pecopteris longifolioides*, *Pecopteris* sp., *Compsopteris contracta*, *Cladophlebis sinensis* and *Pecopteris marginata* were recorded in the siltstone of the middle part. 405 cm
16. Grayish green medium- to thin-bedded argillaceous siltstone, with abundant plant fossils *Pecopteris fuyuanensis*, *Fascipteris densata*, *Lobatannularia*

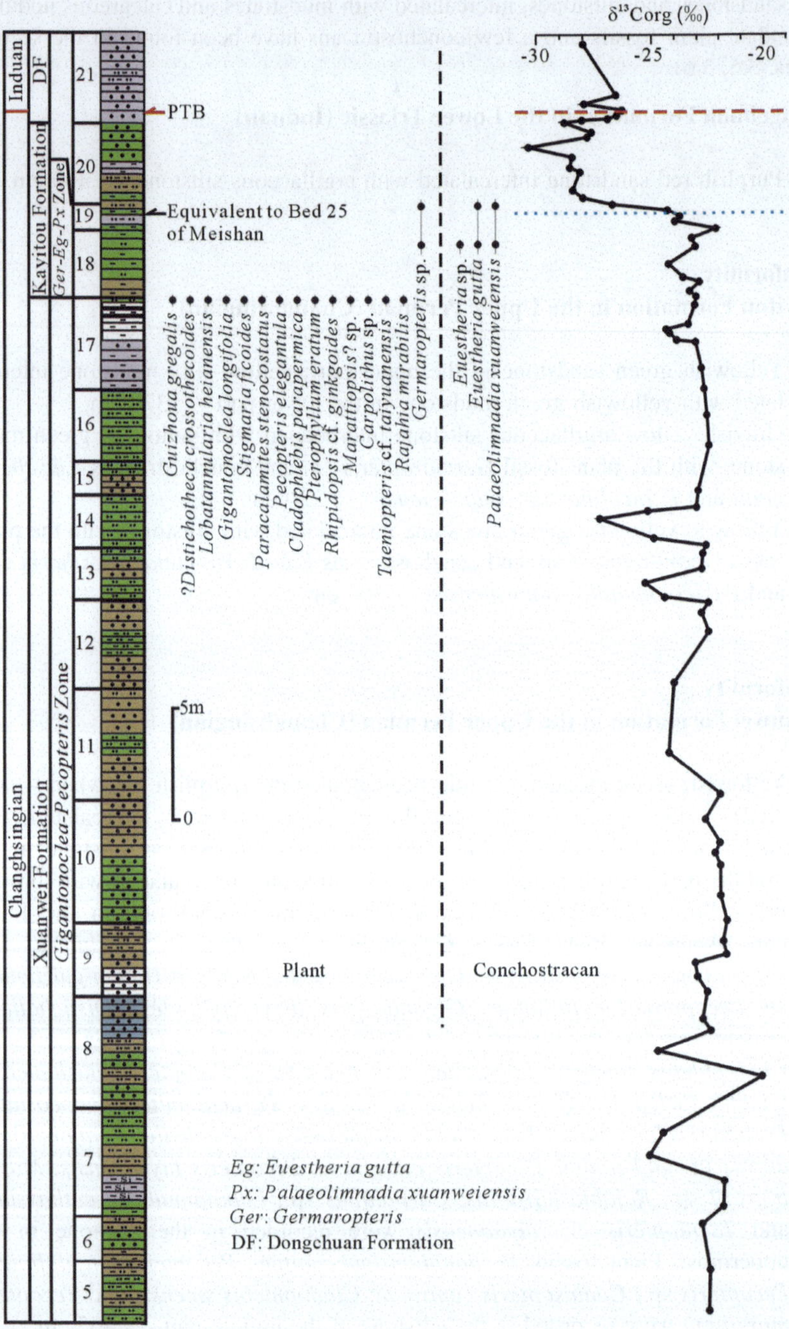

Fig. 3.6 Stratigraphic column and fossil occurrences (plant, conchostracan) at the Guanbachong section. *Note* data of fossils and carbon isotopes referred to Zhang et al. (2016). Legends same to Fig. 3.5

cathaysiana, *Cladophlebis permica*, *Pecopteris heteropinna*, *Compsopteris* sp. and *Neuropteridium guizhouense*. 320 cm

15. Grayish green medium- to thin-bedded siltstone in the upper part; yellow thin-beddded fine-grained sandstone in the lower part. 150 cm
14. Grayish green siltstone with abundant plant fossils *Pecopteris fuyuanensis*, *Paracalamites* sp., *Pecopteris guizhouensis*, *Compsopteris contracta*, *Fascipteris* sp., *Sphenophyllum* sp., *Rajahia rigida*, *Sphenophyllum sinocoreanum* and *Calamites* sp.. 240 cm
13. Yellowish green medium- to thin-bedded siltstone in the upper part; brownish yellow medium- to thin-bedded fine-grained sandstone in the middle part; yellowish green medium- to thin-bedded siltstone in the lower part. Laminations evident in the siltstone. 230 cm
12. Yellow thick- to medium-bedded fine-grained sandstone in the upper part; yellowish green thin-bedded silty mudstone in the middle part; yellow thick- to medium-bedded fine-grained sandstone in the lower part. Abundant plant fossils *Compsopteris contracta*, *Gigantopteris dictyophylloides*, *Gigantopteris paradoxa*, *Cathaysiopteris* sp., *Fascipteris densata*, *Rajahia rigida* and *Lobatannularia* sp. were examined. 400 cm
11. Yellow thin-bedded siltstone at the top part; brownish yellow medium- to thin-bedded fine-grained sandstone in the upper part; grayish green thick- to medium-bedded siltstone in the middle part; yellow thin-bedded fine-grained sandstone in the lower part. 500 cm
10. Brownish yellow (purplish red when weathered) thick-bedded fine-grained sandstone interbedded with thin-bedded fine-grained sandstone, with laminations in the upper part; grayish green thick- to medium-bedded fine-grained sandstone in the lower part. 550 cm
9. Yellow to yellowish green thin-bedded silty mudstone in the upper part; brownish yellow medium- to thin-bedded fine-grained sandstone in the middle part; dark gray to grayish green thin-bedded sandstone in the lower part. 330 cm
8. Bluish gray thin-bedded fine-grained sandstone interbedded with yellow thin-bedded siltstone in the upper part; grayish green medium- to thin-bedded fine-grained sandstone interbedded with brownish yellow thin-bedded argillaceous siltstone in the lower part. 500 cm
7. Grayish green (purplish red when weathered) medium- to thin-bedded siltstone in the upper part; yellow to grayish green thin-bedded silty mudstone in the middle part; purplish red thin-bedded siliceous mudstone overlying yellowish green medium- to thin-bedded mudstone in the lower part. A few plant fossils *Compsopteris contracta* and *Fascipteris* sp. were collected. 480 cm
6. Yellow medium- to thin-bedded siltstone in the upper part; yellowish green medium- to thin-bedded fine-grained sandstone in the middle part; yellow thin-bedded mudstone in the lower part. 200 cm
5. Grayish green medium- to thin-bedded argillaceous siltstone interbedded with yellow medium- to thin-bedded silty mudstone in the upper part; grayish green medium- to thin-bedded siltstone interbedded with yellow thick-

to medium-bedded silty mudstone in the lower part. Abundant plant fossils *Cathaysiopteris* sp., *Gigantopteris dictyophylloides, Fascipteris densata, Pecopteris lingulata, Pecopteris marginata, Pecopteris heteropinna, Gigantonoclea guizhouensis, Stigmaria* sp., *Compsopteris* sp. and *Rajahia guizhouensis* were discovered. 410 cm

4. Dark grayish green to yellowish green thin-bedded mudstone interbedded with yellow thin-bedded siltstone, with yellow thin-bedded mudstone at the basal part. 450 cm
3. Purplish red thick- to medium-bedded siliceous mudstone, with bluish gray silicalite (5 cm thick) at the top. 110 cm
2. Grayish green thin-bedded siltstone overlying yellow medium- to thin-bedded fine-grained sandstone in the upper part; grayish green to purple medium- to thin-bedded siltstone in the middle part; yellowish green thin-bedded fine-grained sandstone overlying dark gray thin-bedded mudstone in the lower part. 360 cm
1. Yellow thin-bedded fine-grained sandstone in the upper part; grayish black thick- to medium-bedded fine-grained sandstone in the middle part; grayish green thick- to medium-bedded conglomerate in the lower part (350 cm thick). 450 cm

Unexposed downwards

Note: Listed fossils are referred to Zhang et al. (2016) (but revised and updated in taxonomy in this book).

3.3 Kele Section in Hezhang County

The Kele section is located nearby the Chunle Village, Kele Town, over 30 km northwest to Hezhang County, Guizhou Province, South China (Fig. 3.7).

The sequence of the Kele section is composed of the Xuanwei and Kayitou formations in an ascending order (Fig. 3.8). The Xuanwei Formation is dominated by yellow to gray fine-grained sandstones, siltstone and mudstones, intercalated with coal seams. The Kayitou Formation is mainly composed of grayish green siltstones and mudstones. A few plant fossils and conchostracans were collected at the Kele section (Fig. 3.8).

Kayitou Formation in the uppermost Permian

43. Grayish green thick-bedded fine-grained sandstone. 340 cm
42. Grayish green thin-bedded mudstone in the upper part; grayish green thick- to medium-bedded mudstone in the middle part; grayish green thin-bedded mudstone in the lower part. 150 cm

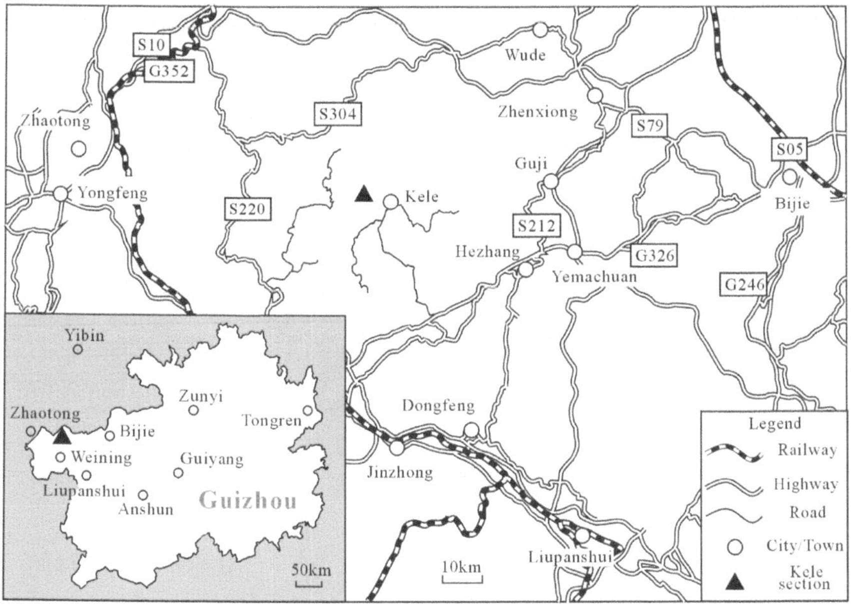

Fig. 3.7 Geographic location of the Kele section. *Note* S10 representing the number (10) of Provincial Highway, the same as below (for other sections)

41. Yellow thick- to medium-bedded medium- to fine-grained sandstone in the upper part; grayish green medium- to thin-bedded silty mudstone with lens-shaped beddings in the lower part; grayish green mudstone at the base (4 cm). 250 cm
40. Yellow thick- to medium-bedded medium- to fine-grained sandstone in the upper part; grayish green medium- to thin-bedded silty mudstone with lens-shaped beddings in the lower part. 200 cm
39. Yellow thin-bedded argillaceous siltstone with ferruginous and manganous nodules (0.5–15 cm in diameter), with brown manganous limestone at the top part (15 cm thick). Abundant brachiopods, bivalves and fragments of plant fossils were found in the siltstone. 450 cm
38. Gray thin-bedded mudstone, with a layer of shelly fossils at the top. 200 cm
37. Pale claystone at the top part (15 cm thick); gray silty mudstone intercalated with two layers of gray lens-shaped siltstone with bivalves in the middle part; pale mudstone (4 cm) with fragments of plant fossils overlying gray thin-bedded silty mudstone in the lower part. Abundant plant fossils (e.g., *Tomiostrobus zelleri*, *Tomiostrobus* spp., *Pecopteris* sp.) were collected at the top part. 130 cm

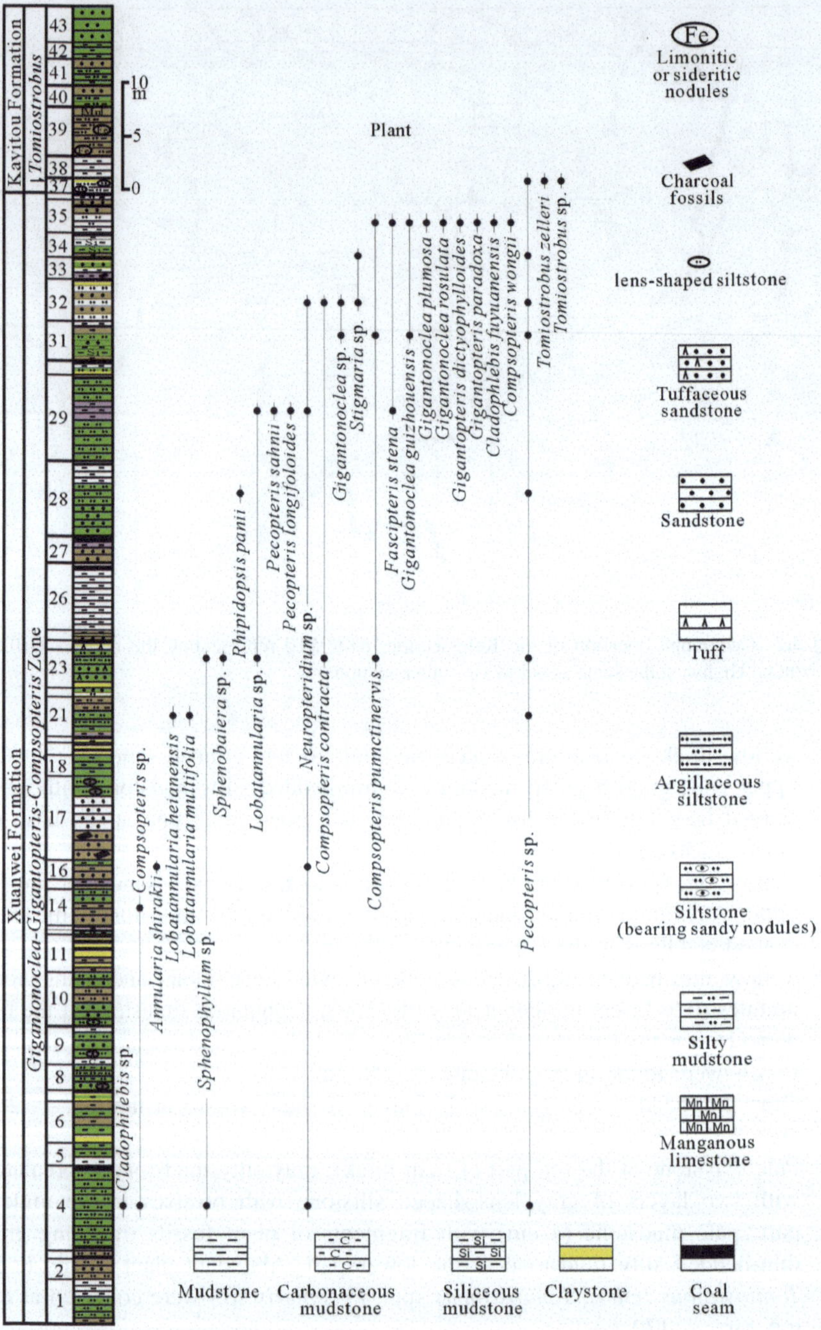

Fig. 3.8 Stratigraphic column and occurrences of plant fossils at the Kele section

Conformity
Xuanwei Formation in the Upper Permian

36. Black coal seam in the upper part (30 cm); gray silty mudstone with abundant root fossils of plants in the lower part. 60 cm
35. Brownish yellow thick- to medium-bedded fine-grained sandstone intercalated with coal seams in the upper part; pale siltstone in the lower part; light gray mudstone with abundant fragments of plant fossils at the base. Plant fossils *Gigantonoclea guizhouensis*, *Gigantonoclea plumosa*, *Gigantonoclea rosulata*, *Gigantopteris dictyophylloides*, *Gigantopteris paradoxa*, *Cladophlebis fuyuanensis*, *Compsopteris wongii*, *Compsopteris punnctinervis* and *Fascipteris stena* were examined in the lower part. 310 cm
34. Black coal seam at the top part (20 cm); pale thin-bedded mudstone in the upper part; dark gray thin-bedded siliceous mudstone in the middle part; dark grayish green siliceous mudstone in the lower part; yellow thin-bedded mudstone at the basal part. Plant fossils *Stigmaria* sp. and *Pecopteris* sp. were found. 225 cm
33. Black coal seam at the top part (20 cm); pale claystone in the upper part; dark grayish green medium-bedded fine-grained sandstone in the middle part; yellow thin-bedded mudstone in the lower part; purplish red mudstone with charcoal fossils at the base part (3 cm). 200 cm
32. Black coal seam (30 cm) overlying pale claystone in the upper part; dark gray thin-bedded siltstone interbedded with brownish yellow fine-grained sandstone in the lower part. Plant fossils *Compsopteris contracta*, *Neuropteridium* sp., *Pecopteris* sp., *Gigantonoclea* sp. and *Stigmaria* sp. were collected. 390 cm
31. Gray thin-bedded mudstone overlying black coal seam at the top part; yellow thin-bedded mudstone in the upper part; grayish green medium-bedded fine-grained sandstone interbedded with grayish green thin-bedded siltstone in the middle part; four cycles were deposited in the lower part and each cycle is characterized by yellowish green mudstone interbedded with dark grayish green medium- to thin-bedded siliceous mudstone; pale to yellow thin-bedded mudstone with abundant fragments of plant fossils at the basal part. Plant fossils *Compsopteris punnctinervis*, *Neuropteridium guizhouensis*, *Gigantonoclea* sp. and *Pecopteris* sp. were discovered. 370 cm
30. Black coal seam (30 cm) overlying a layer of gray thin-bedded mudstone in the upper part; grayish green medium- to thin-bedded siltstone in the lower part; pale to yellow claystone at the base. 130 cm
29. Black coal seam at the top part (15 cm); grayish green medium- to thin-bedded siltstone in the upper part; purplish red thin-bedded mudstone in the middle part; grayish green medium- to thin-bedded siltstone in the lower part. Plant fossils *Pecopteris sahnii*, *Pecopteris longifoloides*, *Fascipteris stena*, *Neuopteridium* sp. and *Lobatannularia* sp. were identified. 800 cm
28. Black coal seam at the top part (10 cm); grayish black thin-bedded mudstone in the upper part; grayish green medium- to thin-bedded siltstone in the middle

part; grayish green thick-bedded fine-grained sandstone in the lower part. Plant fossils *Rhipidopsis panii* and *Pecopteris* sp. were recovered. 700 cm

27. Black coal seam at the top part (50 cm); purplish red thin-bedded mudstone with abundant plant fossils in the upper part; brownish yellow (grayish black when weathered) medium- to thin-bedded, medium- to fine-grained sandstone in the lower part. 250 cm
26. Pale thin-bedded mudstone intercalated with grayish black thin-bedded mudstone, with black coal seam at the top (50 cm). 630 cm
25. Dark gray to yellow mudstone overlying a layer of pale claystone (4 cm) in the upper part; light purple thin-bedded mudstone with fragments of plant fossils in the middle part; gray to yellow thin-bedded mudstone in the lower part; brownish yellow medium- to thin-bedded fine-grained sandstone at the base. 100 cm
24. Black coal seam in the upper part (30 cm); pale claystone in the middle part (35 cm); grayish black tuff in the lower part (20 cm); pale claystone at the base (10 cm). 95 cm
23. Grayish green thin-bedded argillaceous siltstone in the upper part; light green medium-bedded fine-grained tuff-bearing sandstone in the lower part; pale claystone at the basal part (30 cm). Plant fossils *Compsopteris punnctinervis, Compsopteris contracta, Lobatannularia* sp., *Sphenobaira* sp., *Sphenophyllum* sp. and *Pecopteris* sp. were examined. 345 cm
22. Grayish green medium- to thin-bedded fine-grained sandstone with abundant fragments of plant fossils in the upper part (23 cm); black thin-bedded argillaceous siltstone in the middle part (8 cm); dark gray thin-bedded siltstone with abundant fragments of plant fossils in the lower part; pale to purplish red claystone at the base (2 cm). 83 cm
21. Yellow silty mudstone in the upper part; light grayish green thin-bedded argillaceous siltstone in the middle part; dark grayish green thin-bedded mudstone in the lower part; pale to purple mudstone at the base (5 cm). Plant fossils *Lobatannularia heianensis, Lobatannularia multifolia* and *Pecopteris* sp. were collected. 470 cm
20. Black coal bed, with pale to yellow claystone (5 cm) overlying yellowish green mudstone (5 cm) at the base. Abundant fragments of plant fossils were yielded in the claystone. 60 cm
19. Two cycles were deposited by claystone and coal seam in the upper part in descending order: (1) black coal seam (4 cm); (2) purple claystone (3 cm); (3) black coal seam (20 cm); (4) purple claystone (1 cm). Black thin-bedded carbonaceous mudstone in the lower part. 117 cm
18. Yellowish green thin-bedded mudstone. 200 cm
17. Dark green thick-bedded siltstone with brown sandy nodules in the upper part; grayish black thin-bedded fine-grained sandstone in the middle part; yellowish brown thick- to medium-bedded fine-grained sandstone with charcol fossils in the lower part. 790 cm
16. Dark green thin-bedded mudstone in the upper part; brownish yellow mudstone with fragments of plant fossils in the middle part; yellowish brown

medium- to thin-bedded fine-grainded sandstone in the lower part. Plant fossils *Annularia shirakii* and *Neuropteridium* sp. were identified. 180 cm

15. Brownish yellow mudstone overlying purple mudstone in the upper part (15 cm); black coal seam in the middle part (30 cm); pale claystone in the lower part. 85 cm
14. Dark green siltstone overlying fine-grained sandstone in the upper part; yellow thin-bedded mudstone in the middle part; yellowish brown thick- to medium-bedded fine-grained sandstone in the lower part. Plant fossils *Sphenophyllum* sp. and *Compsopteris* sp. were examined. 340 cm
13. Black coal seam in the upper part (15 cm); pale thin-bedded mudstone in the lower part. 40 cm
12. Black coal seam in the upper part (10 cm); light grayish green thick-bedded silty mudstone in the lower part. 45 cm
11. Pale medium- to thin-bedded claystone intercalated with several layers of coal seams, with black coal seam at the top part (10 cm) and yellow thin-bedded mudstone intercalated with pale claystone at the basal part. 350 cm
10. Grayish green medium- to thin-bedded siltstone interbedded with fine-grained sandstone in the upper part; yellow thin-bedded siltstone in the middle part; dark green siltstone interbedded with fine-grained sandstone in the lower part; grayish green medium- to thin-bedded siltstone with silty nodules at the basal part. 500 cm
9. Gray thin-bedded mudstone at the top part; dark green medium- to thin-bedded siltstone interbedded with fine-grained sandstone with abundant fragments of plant fossils in the upper part; grayish green thin-bedded siltstone with purplish red sandy nodules and fragments of plant fossils in the lower part; black carbonaceous mudstone at the basal part (7 cm). 360 cm
8. Dark green thin-bedded argillaceous siltstone, with dark green fine-grained sandstone with brownish yellow sandy nodules at the basal part (30 cm). 240 cm
7. Black coal seam overlying pale claystone at the top part; grayish yellow thin-bedded siltstone and fine-grained sandstone in the upper part; grayish green medium- to thin-bedded mudstone with abundant plant fossils in the middle part; pale to yellow claystone in the lower part (30 cm). 110 cm
6. Dark green (purple red when weathered) medium- to thin-bedded mudstone interbedded with yellowish green mudstone, with pale to yellow medium- to thin-bedded claystone at the basal part (30 cm). 386 cm
5. Green to yellowish green mudstone interbedded with silty mudstone in the upper part; gray mudstone in the lower part. 210 cm
4. Grayish green thin-bedded siltstone intercalated with argillaceous siltstone, with black coal seams at the top and basal parts. The plant fossil *Compsopteris* sp., *Cladophilebis* sp., *Pecopteris* sp. and *Neuropteridium* sp. were examined. 800 cm
3. Light yellow thick-bedded siltstone, with black coal seam at the basal part (5 cm). 75 cm

2. Yellow thin-bedded siltstone interbedded with fine-grained sandstone. 190 cm
1. Gray medium- to thin-bedded mudstone in the upper part; yellowish green medium- to thin-bedded siltstone in the lower part; brownish yellow thin-bedded mudstone at the basal part. 420 cm

Basalt downwards

3.4 Xiaohebian Section in Weining City

The Xiaohebian section is located at the Xiaohebian Village, Halahe Town, Weining City, Guizhou Province, South China (Fig. 3.9).

The sequence of the Xiaohebian section is composed of the Xuanwei, Kayitou and Dongchuan formations in an ascending order (Fig. 3.10). The Xuanwei Formation is dominated by yellowish green to bluish gray fine-grained sandstones, siltstones and mudstones, commonly intercalated with coal seams. The Kayitou Formation is mainly composed of yellowish green to purplish red siltstones. The Dongchuan Formation is characterized by purplish red sandstones, siltstones and mudstones. Abundant plant fossils and a few conchostracan were found in the section (Fig. 3.10).

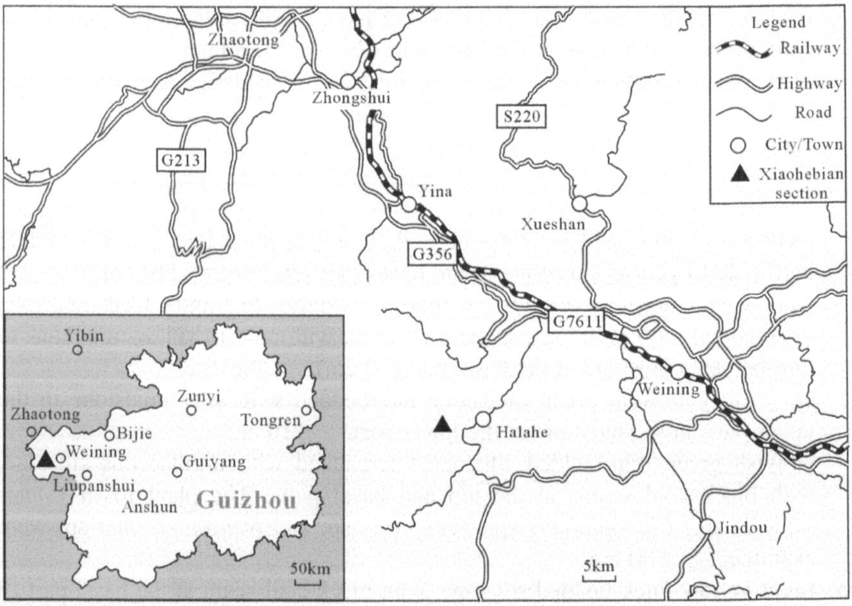

Fig. 3.9 Geographic location of the Xiaohebian section

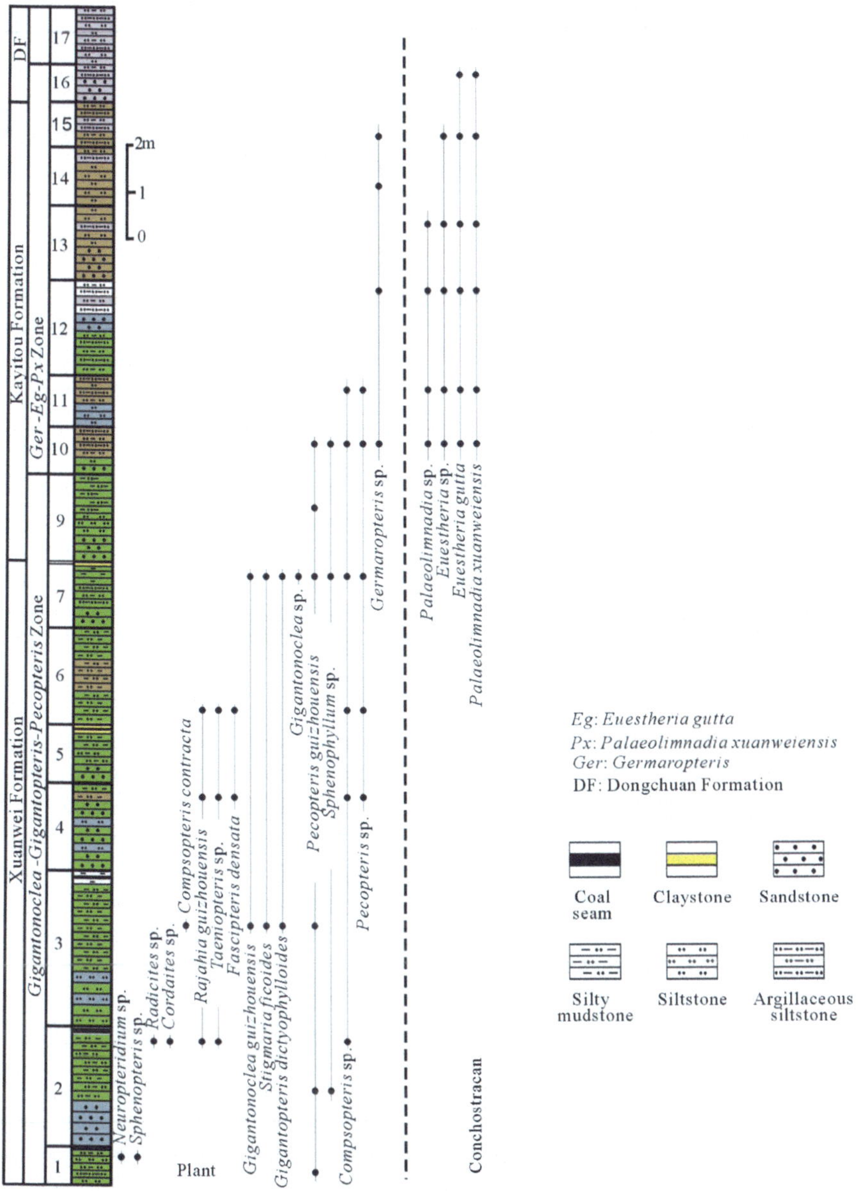

Fig. 3.10 Stratigraphic column and fossil occurrences (plant, conchostracan) at the Xiaohebian section (fossils referred to Chu et al. 2016)

Dongchuan Formation in the Lower Triassic

17. Purplish red siltstone interbedded with argillaceous siltstone. 120 cm
16. Purplish red argillaceous siltstone in the upper part; purplish red thin-bedded fine-grained sandstone in the lower part. Conchostracans *Euestheria gutta* and *Palaeolimnadia xuanweiensis* were identified in the upper part. 80 cm

Conformity
Kayitou Formation in the uppermost Permian

15. Yellow thin-bedded argillaceous siltstone interbedded with purplish red to grayish green argillaceous siltstone. The plant fossil *Germaropteris* spp. and the conchostracans *Euestheria gutta, Euestheria* sp. and *Palaeolimnadia xuanweiensis* were recorded. 95 cm
14. Yellow argillaceous siltstone intercalated with purplish red argillaceous siltstone in the upper part; yellow thin-bedded siltstone in the lower part (80 cm). The plant fossil *Germaropteris* sp. was collected in the lower part. 125 cm
13. Yellow to purplish red thin-bedded siltstone intercalated with argillaceous siltstone in the upper part; yellow thin-bedded fine-grained sandstone in the lower part. The conchostracans *Euestheria gutta, Euestheria* spp., *Palaeolimnadia xuanweiensis* and *Palaeolimnadia* spp. were discovered in the upper part. 160 cm
12. Gray thin-bedded argillaceous siltstone intercalated with purplish red argillaceous siltstone in the upper part; bluish gray thin-bedded fine-grained sandstone in the middle part; yellowish green to yellow thin-bedded argillaceous siltstone in the lower part. The plant fossil *Germaropteris* sp. and the conchostracans *Euestheria gutta, Euestheria* spp., *Palaeolimnadia xuanweiensis* and *Palaeolimnadia* spp. were recorded in the upper part. 200 cm
11. Yellow thin-bedded siltstone and argillaceous siltstone in the upper part; bluish gray thin-bedded siltstone in the lower part. The plant fossils *Pecopteris* sp. and *Compsopteris* sp. and conchostracans *Euestheria gutta, Euestheria* spp., *Palaeolimnadia xuanweiensis* and *Palaeolimnadia* sp. were found in the upper part. 110 cm
10. Yellow thin-bedded argillaceous siltstone in the upper part; grayish green thin-bedded fine-grained sandstone and siltstone in the lower part (40 cm). Plant fossils *Pecopteris guizhouensis, Compsopteris* sp., *Sphenophyllum* sp., *Pecopteris* sp. and *Germaropteris* sp., and the conchostracans *Euestheria gutta, Euestheria* spp., *Palaeolimnadia xuanweiensis* and *Palaeolimnadia* sp. were examined in the upper part. 100 cm
9. Grayish green thin-bedded silty mudstone in the upper part; yellowish green fine-grained sandstone and siltstone in the lower part. The plant fossil *Pecopteris guizhouensis* was recovered in the upper part. 185 cm

Conformity
Xuanwei Formation in the Upper Permian

8. Pale claystone intercalated with black coal seams (coal seams laterally discontinuous). 5 cm
7. Yellowish green thin-bedded mudstone in the upper part; yellowish green thin-bedded argillaceous siltstone in the middle part (50 cm); yellowish green thick-bedded fine-grained sandstone in the lower part. Plant fossils *Gigantonoclea guizhouensis*, *Stigmaria ficoides*, *Gigantonoclea* sp., *Gigantopteris dictyophylloides*, *Pecopteris guizhouensis*, *Sphenophyllum* sp., *Compsopteris* sp. and *Pecopteris* sp. were found in the upper part. 135 cm
6. Yellowish green silty mudstone in the upper part; yellow silty mudstone in the middle part; yellowish green silty mudstone in the lower part. Plant fossils *Fascipteris densata*, *Rajahia guizhouensis*, *Taeniopteris* sp., *Compsopteris* sp. and *Pecopteris* sp. were recorded in the lower part. 205 cm
5. Pale claystone intercalated with black coal seam at the top part; yellowish green thin-bedded argillaceous siltstone and silty mudstone in the upper part (20 cm); grayish green thick- to medium-bedded fine-grained sandstone in the lower part. 125 cm
4. Grayish green silty mudstone with fragments of plant fossils at the top part; yellow thin-bedded siltstone and silty mudstone in the upper part; grayish green thick- to medium-bedded fine-grained sandstone intercalated with bluish gray thin-bedded siltstone in the lower part (150 cm). Plant fossils *Fascipteris densata*, *Rajahia guizhouensis*, *Taeniopteris* sp., *Compsopteris* sp. and *Pecopteris* sp. were examined. 185 cm
3. Grayish black mudstone intercalated with lens-shaped coal at the top part (20 cm); grayish green to gray silty mudstone in the upper part; bluish gray to yellowish green thick-bedded siltstone in the lower part. Plant fossils *Compsopteris contracta*, *Gigantonoclea guizhouensis*, *Stigmaria ficoides*, *Gigantopteris dictyophylloides* and *Pecopteris guizhouensis* were recorded in the upper part. 330 cm
2. Black coal seams at the top part; grayish green thin-bedded argillaceous siltstone interbedded with yellowish green silty mudstone in the upper part; bluish gray (brown when weathered) thick-bedded fine-grained sandstone in the lower part (100 cm). The plant fossils *Compsopteris* sp., *Radicites* sp., *Cordaites* sp., *Rajahia guizhouensis* and *Taeniopteris* sp. were recorded in the upper part, and the plant fossils *Pecopteris guizhouensis* and *Sphenophyllum* sp. were identified in the lower part. 255 cm
1. Black coal seams at the top part; yellowish green siltstone interbedded with argillaceous siltstone in the upper part; grayish green argillaceous siltstone overlying siltstone in the lower part. The plant fossils *Neuropteridium* sp. and *Sphenopteris* sp. were recorded in the upper part and the plant fossil *Pecopteris guizhouensis* was discovered in the lower part.

Unexposed downwards

Note: Fossil lists are referred to Chu et al. (2016) (but updated in taxonomy in this book).

3.5 Chahe Section in Weining County

The Chahe section is located about 8 km northwest to the Chahe Village, Heishitou Town, Weining County, Guizhou Province, South China (Fig. 3.11).

The sequence of the Chahe section is composed of the Xuanwei, Kayitou and Dongchuan formations in an ascending order (Fig. 3.12). The Xuanwei Formation is dominated by yellow, grayish green to bluish gray siltstones, sandstones and coal seams. The Kayitou Formation is mainly composed of brownish yellow siltstones and fine-grained sandstones, intercalated with calcareous mudstone nodules in the upper part. The Dongchuan Formation is characterized by purplish red siltstones and fine-grained sandstones, intercalated with calcareous nodules in the basal part. Extremely abundant plant fossils, spore and pollens and a few conchostracans were identified in the section (Figs. 3.12, 3.13).

Dongchuan Formation in the Lower Triassic

93. Purplish red thick- to medium-bedded siltstone in the upper part; purplish red thick-bedded fine-grained sandstone in the lower part. > 200 cm

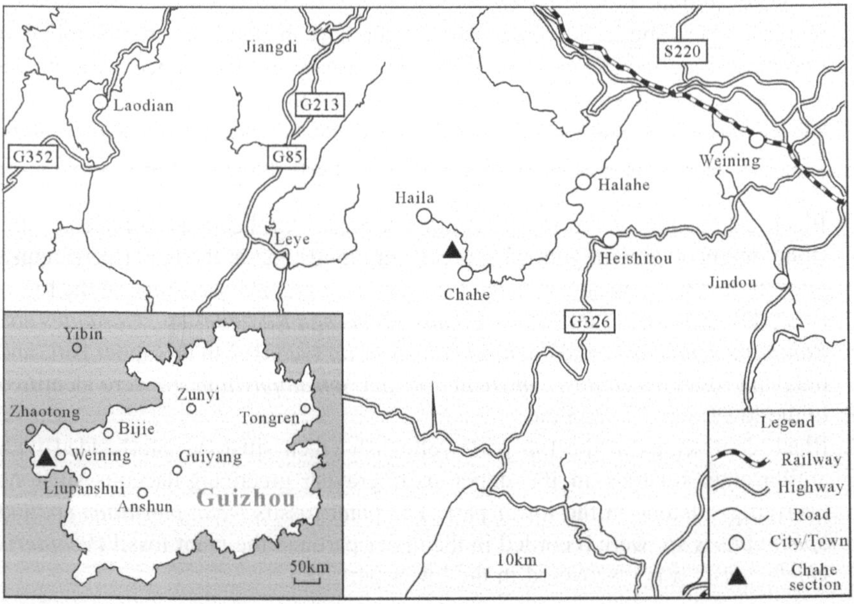

Fig. 3.11 Geographic location of the Chahe section

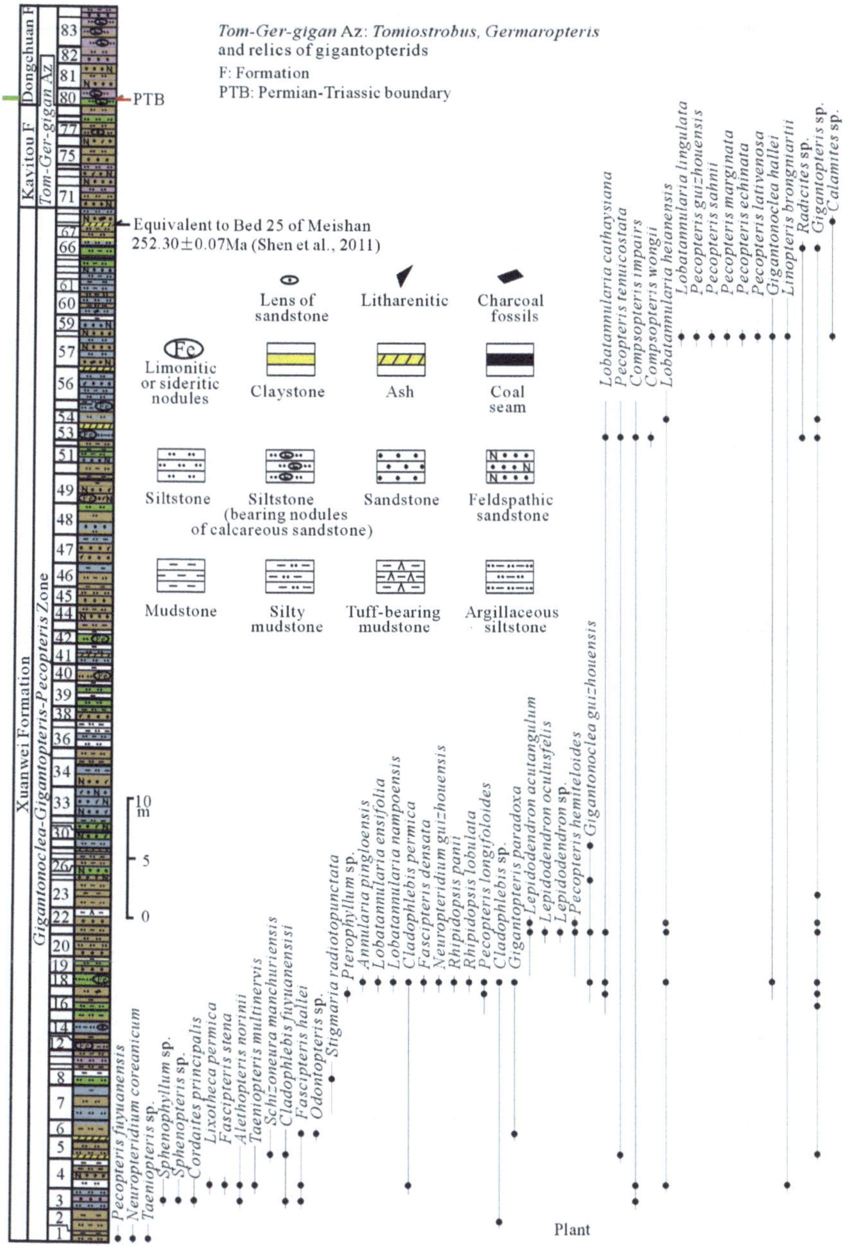

Fig. 3.12 Stratigraphic column and fossil occurrences (plant) at the Chahe section. *Note* plant fossils referred to Chu et al. (2016); the locality of the Permian–Triassic boundary and the horizon equivalent to bed 25 of Meishan (U–Pb date) based on Shen et al. (2011)

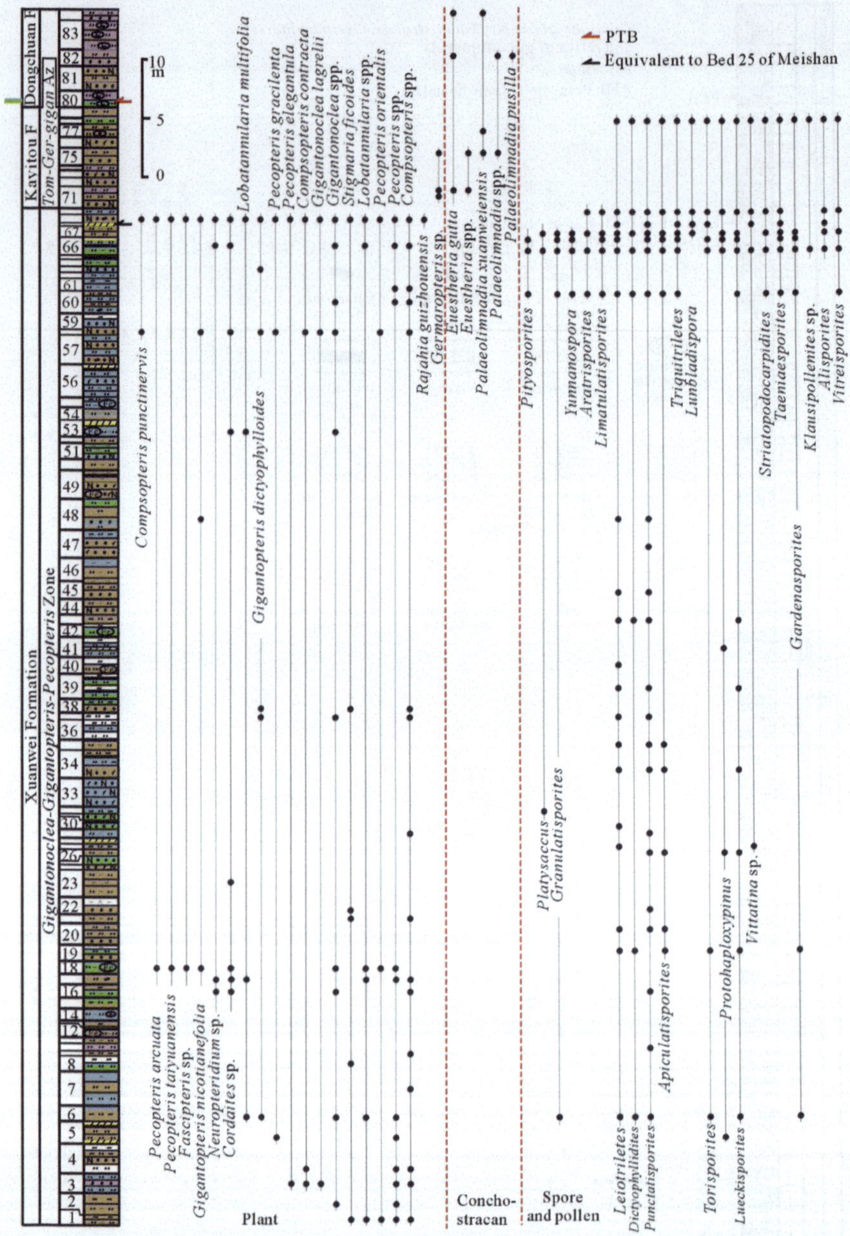

Fig. 3.13 Stratigraphic column and fossil occurrences (plant, conchostracan and spore and pollen) at the Chahe section. *Note* conchostracans referred to Chu et al. (2016); spore and pollen referred to Yu et al. (2022). Legends and abbreviations same to Fig. 3.12

92. Purplish gray thick-bedded, medium- to fine-grained sandstone intercalated with purplish red banding-shaped argillaceous siltstone in the upper part; grayish green thick-bedded, medium- to fine-grained sandstone intercalated with purplish red banding-shaped argillaceous siltstone in the lower part. The cross bedding was observed in the sandstone. 115 cm
91. Purplish red thick-bedded argillaceous siltstone and fine-grained sandstone, with the low-angled cross bedding. 195 cm
90. Grayish green to purplish red thick-bedded feldspathic litharenite intercalated with purplish red thick- to medium-bedded siltstone. The wedge-shaped cross bedding was yielded in the feldspathic litharenite; the small-scaled cross bedding and laminations were yielded in the siltstone. 165 cm
89. Grayish green to yellow thick- to medium-bedded litharenite interbedded with purplish red banding-shaped argillaceous siltstone and purplish gray thick-bedded feldspathic litharenite, with purplish red banding-shaped argillaceous siltstone (bearing abundant calcareous mudstone nodules) in the uppermost. 305 cm
88. Purplish red banding-shaped argillaceous siltstone and fine-grained sandstone, with calcareous nodules in the sandstone. 125 cm
87. Purplish red thick-bedded argillaceous siltstone. 115 cm
86. Purplish gray thin-bedded fine-grained litharenite, with dark purple banding-shaped siltstone in the upper part. 100 cm
85. Purplish red thick-bedded argillaceous siltstone with small-scale wavy cross beddings, laminations and nodules of calcareous sandstone. 100 cm
84. Purplish red siltstone intercalated with nodules of calcareous sandstone, with the cross bedding in the upper part; yellow medium-bedded fine-grained sandstone with purple banding-shaped siltstone in the middle part; purplish red to yellow thin- to medium-bedded banding-shaped siltstone intercalated with nodules of fine-grained sandstone, with the small-scaled wavy cross bedding in the lower part. Conchostracans *Euestheria gutta* and *Palaeolimnadia xuanweiensis* were recovered. 90 cm
83. Purplish red medium- to thin-bedded siltstone intercalated with dark yellow thin-bedded siltstone, with nodules of calcareous sandstone. 250 cm
82. Purplish red medium- to thin-bedded siltstone interbedded with brownish yellow thin-bedded siltstone, with interlayers (1–2 cm in the thickness for each layer) of fine-grained sandstone in the lower part. Conchostracans *Euestheria gutta, Palaeolimnadia pusilla* and *P.* spp. were collected. 140 cm
81. Brownish yellow thick-bedded fine-grained feldspathic sandstone intercalated with banding-shaped siltstone, with purple bandings in the middle and lower parts. Spores and pollens *Bactrosporites* sp. and *Lacrimasporonites* sp. were examined. 210 cm
80. Purple thin-bedded siltstone intercalated with yellow bandings in the upper part; yellowish green siltstone intercalated with purple bandings in the lower part. Nodules of fine-grained calcareous sandstone are intercalated within the siltstone. 140 cm

Conformity
Kayitou Formation in the uppermost Permian

79. Grayish yellow thick-bedded siltstone with rare purplish red bandings (1–3 mm in the thickness for each banding). 80 cm
78. Grayish green thin-bedded siltstone and thick- to medium-bedded fine-grained sandstone, with purple bandings. Spores and pollens *Leiotriletes* spp., *Dictyophyllidites* spp., *Punctatisporites* spp., *Apiculatisporites* spp., *Triquitriletes* spp., *Lunbladispora* spp., *Torisporites* spp., *Protohaploxypinus* spp., *Lueckisporites* spp., *Vittatina* sp., *Striatopodocarpidites* spp., *Taeniaesporites* spp., *Gardenasporites* spp., *Klausipollenites* sp., *Alisporites* spp., and *Vitreisporites* spp. were discovered. 55 cm
77. Yellowish green to brownish yellow thick-bedded siltstone with lens of fine-grained sandstone. The conchostracan *Palaeolimnadia xuanweinsis* was examined. 110 cm
76. Brownish yellow thick-bedded fine- to medium-grained lithic-bearing feldspathic sandstone. 80 cm
75. Yellow thick-bedded fine-grained sandstone in the lower part and siltstone in the upper part. The plant *Germaropteris* sp., and conchostracans *Euestheria* spp., *Palaeolimnadia xuanweiensis* and *P.* spp. were found. 150 cm
74. Purplish red thin-bedded siltstone. 30 cm
73. Yellow thick-bedded lithic-bearing fine-grained feldspathic sandstone. 65 cm
72. Yellow thick-bedded fine-grained feldspathic sandstone intercalated with silty bandings, with purple siltstone at the top (5 cm thick). 75 cm
71. Purple siltstone in the uppermost (5 cm thick); yellow banding-shaped fine-grained sandstone interbedded with banding-shaped siltstone in the middle part; yellow thick-bedded lithic-bearing fine-grained feldspathic sandstone in the lower part. The plant fossil *Germaropteris* sp. and the conchostracans *Euestheria gutta* and *Euestheria* spp. were identified in the uppermost part; the plant fossil *Germaropteris* sp. was recovered in the middle part. 185 cm

Conformity
Xuanwei Formation in the Upper Permian

70. Bluish gray to yellow siltstone, with spores and pollens *Aratrisporites* spp., *Limatulatisporites* spp., *Leiotriletes* spp., *Dictyophyllidites* spp., *Punctatisporites* spp., *Apiculatisporites* spp., *Triquitriletes* spp., *Lunbladispora* spp., *Torisporites* spp., *Protohaploxypinus* spp., *Lueckisporites* spp., *Vittatina* sp., *Striatopodocarpidites* spp., *Taeniaesporites* spp., *Alisporites* spp. and *Vitreisporites* spp.. 50 cm

69. Yellowish gray thick- to medium-bedded fine-grained feldspathic sandstone with abundant charcoal fossils in the upper; bluish gray thick- to medium-bedded argillaceous siltstone with green and yellow bandings in the lower part. Plant fossils *Calamites* sp., *Compsopteris punctinervis, Pecopteris arcuata, P. taiyuanensis, P. gracilenta, P. elegantula, P. orientalis, P.* spp., *Fascipteris* sp., *Gigantopteris nicotianefolia, Gigantopteris dictyophylloides, Gigantonoclea lagrelii, Gigantonoclea* spp., *Neuropteridium* sp., *Cordaites* sp., *Lobatannularia multifolia, L.* spp., *Compsopteris contracta, C.* spp., *Stigmaria ficoides* and *Rajahia guizhouensis* were examined. 70 cm
68. Pale to yellow volcanic ash intercalated with coal seam. Spores and pollens *Leiotriletes* spp., *Punctatisporites* spp., *Triquitriletes* spp., *Lunbladispora* spp., *Torisporites* spp., *Vittatina* sp., *Taeniaesporites* spp. and *Alisporites* spp. were identified. 5 cm
67. Yellow thick-bedded argillaceous siltstone in the upper part and siltstone in the lower part and limonites were observed when weathered. Spores and pollens *Platysaccus* spp., *Granulatisporites* spp., *Yunnanospora* spp., *Aratrisporites* spp., *Limatulatisporites* spp., *Leiotriletes* spp., *Dictyophyllidites* spp., *Punctatisporites* spp., *Apiculatisporites* spp., *Triquitriletes* spp., *Lunbladispora* spp., *Protohaploxypinus* spp., *Lueckisporites* spp., *Striatopodocarpidites* spp., *Taeniaesporites* spp., *Alisporites* spp., *Gardenasporites* spp. and *Vitreisporites* spp. were found. 95 cm
66. Pale claystone and silty mudstone intercalated with brown coal seams in the uppermost; yellow thick-bedded fine-grained feldspathic sandstone in the upper part; bluish gray thick-bedded siltstone in the middle part; grayish green thick-bedded argillaceous siltstone in the lower part. Plant fossils *Radicites* sp., *Gigantopteris* sp., *Neuropteridium* sp., *Cordaites* sp., *Gigantonoclea* spp., *Lobatannularia* spp., *Pecopteris* spp., and *Compsopteris* spp. were examined. Spores and pollens include *Pityosporites* spp., *Platysaccus* spp., *Granulatisporites* spp., *Yunnanospora* spp., *Aratrisporites* spp., *Limatulatisporites* spp., *Leiotriletes* spp., *Dictyophyllidites* spp., *Punctatisporites* spp., *Apiculatisporites* spp., *Triquitriletes* spp., *Torisporites* spp., *Protohaploxypinus* spp., *Lueckisporites* spp., *Vittatina* sp., *Striatopodocarpidites* spp., *Taeniaesporites* spp., *Gardenasporites* spp., *Klausipollenites* sp., *Vitreisporites* spp. and *Alisporites* spp. were recorded. 150 cm
65. Black coal seams interbedded with purple banding-shaped mudstone at the top; grayish green thick- to medium-bedded silty mudstone in the upper part and argillaceous siltstone in the lower part. Spores and pollens *Leiotriletes* sp. and *Polypodiidites fuyuanensis* were recovered. 30 cm
64. Yellowish green to yellow thick-bedded argillaceous siltstone, with dark gray to grayish yellow mudstone at the top. 55 cm
63. Bluish to yellow thick- to medium-bedded fine-grained feldspathic sandstone, with dark gray to grayish green silty mudstone at the top. The plant fossil *Gigantopteris dictyophylloides* was examined. 55 cm
62. Bluish gray to yellow thick- to medium-bedded argillaceous siltstone, with grayish yellow silty mudstone at the top. 50 cm

61. Bluish gray to yellow thick-bedded argillaceous siltstone, with bluish gray to pale mudstone at the top part (5 cm thick). Plant fossils *Pecopteris* spp. and *Compsopteris* spp. were recorded. 105 cm
60. Four cycles were deposited in a descending order: (1) Dark gray silty mudstone in the upper and bluish gray thin-bedded siltstone in the lower (totally 15–20 cm for the cycle). (2) Dark gray silty mudstone in the upper and yellow thick-bedded siltstone in the lower (10 cm for the cycle). (3) Yellow to pale silty mudstone in the upper and grayish yellow thick-bedded fine-grained feldspathic litharenite with small-scaled wedge-shaped cross beddings in the lower. (4) Bluish gray thin-bedded argillaceous siltstone in the upper and grayish yellow to bluish gray thick-bedded siltstone in the lower, with charcoal fossils. Plant fossils include *Gigantonoclea* spp., *Pecopteris* spp. and *Compsopteris* spp.; spores and pollens include *Pityosporites* spp., *Granulatisporites* spp., *Yunnanospora* spp., *Aratrisporites* spp., *Limatulatisporites* spp., *Leiotriletes* spp., *Dictyophyllidites* spp., *Punctatisporites* spp., *Apiculatisporites* spp., *Triquitriletes* spp., *Lueckisporites* spp., *Striatopodocarpidites* spp., *Taeniaesporites* spp., *Gardenasporites* spp. and *Vitreisporites* spp.. 190 cm
59. Two cycles were deposited in a descending order: (1) Grayish yellow thin-bedded mudstone in the upper (10–15 cm thick) and dark green thin-bedded fine-grained litharenite with charcoal fossils in the lower (10 cm). (2) Bluish gray to yellow thick-bedded siltstone in the upper and bluish gray to yellow thick-bedded fine-grained feldspathic sandstone in the lower. 135 cm
58. Grayish yellow thick-bedded fine-grained feldspathic sandstone, with grayish yellow thin-bedded argillaceous siltstone (with laminations) at the top part. Plant fossils *Lobatannularia lingulata, Pecopteris guizhouensis, P. sahnii, P. marginata, P. echinata, P. lativenosa, P. gracilenta, P. orientalis, Gigantonoclea hallei, Gigantonoclea lagrelii, Gigantonoclea* spp., *Linopteris brongniartii, Calamites* sp., *Compsopteris punctinervis, Gigantopteris nicotianefolia, Gigantopteris dictyophylloides, Cordaites* sp., *Compsopteris contracta* and *C.* spp. were identified. 40 cm
57. Two cycles were deposited and each cycle includes: bluish gray thin-bedded siltstone in the upper part, with laminations; grayish yellow thick- to medium-bedded fine- to medium-grained feldspathic sandstone in the lower part, with charcoal fossils. Spores and pollens *Tuberculatosporites impistus* and *Polypodiidites* sp. were examined. 255 cm
56. Black coal seams interbedded with brownish gray volcanic ashes in the uppermost; bluish gray thin-bedded siltstone intercalated with fine- to medium-grained litharenite in the upper part, with laminations in the siltstone; bluish gray to grayish yellow thick-bedded argillaceous siltstone in the middle part; bluish gray to yellow thick- to medium-bedded siltstone in the lower part. 270 cm
55. Bluish gray to grayish yellow thick- to medium-bedded silty mudstone in the upper part; bluish gray to yellow thick-bedded argillaceous siltstone, with limonitic nodules in the lower part. 85 cm

54. Two cycles were deposited in a decending order: (1) bluish gray thin-bedded silty mudstone in the upper; yellow to brown thick- to medium-bedded manganous siltstone in the lower. (2) Yellow thin-bedded argillaceous siltstone in the upper; bluish gray thick- to medium-bedded feldspathic litharenite in the lower. Plant fossils *Lobatannularia heianensis* and *Gigantopteris* sp., and abundant spores and pollens were collected. 105 cm
53. Bluish gray thick- to medium-bedded argillaceous siltstone with abundant limonitic nodules, with pale volcanic ash at the top. Plant fossils *Lobatannularia cathaysiana, L. multifolia, Pecopteris tenuicostata, Compsopteris impairs, Compsopteris wongii, Radicites* sp., *Gigantopteris* sp., *Cordaites* sp. and *Gigantonoclea* spp., and a few spores and pollens were recovered. 140 cm
52. Yellow thick- to medium-bedded argillaceous siltstone, with grayish brown to bluish gray thin-bedded silty mudstone at the top part (5 cm thick). 55 cm
51. Dark gray thin-bedded mudstone at the top part (10–15 cm thick); grayish green thick-bedded argillaceous siltstone in the upper part; bluish gray thick-bedded fine-grained feldspathic sandstone with the small-scaled wedge-shaped cross bedding in the lower part. A few spores and pollen were collected. 130 cm
50. Grayish yellow thick-bedded argillaceous siltstone with wedge-shaped cross beddings. 100 cm
49. Grayish yellow thick-bedded fine-grained feldspathic litharenite (with limonitic nodules), with yellow thin-bedded mudstone intercalated with coal seam at the top part. A few spores and pollen were found. 255 cm
48. Dark grayish green thick- to medium-bedded siltstone at the top part; yellow to bluish gray thin-bedded silty mudstone (10 cm) overlying the yellow thick-bedded siltstone in the upper part; yellow to bluish gray thin-bedded silty mudstone (10 cm) overlying the yellow thick- to medium-bedded argillaceous siltstone (with interlayers of bluish gray thin-bedded fine-grained sandstone in the siltstone) in the lower part. Abundant spores and pollens (e.g., *Leiotriletes exiguus, Punctatisporites elegans*) were discovered. 255 cm
47. Pale silty mudstone at the top part (15–20 cm); bluish gray fine-grained litharenite with limonites in the upper part; yellow thick-bedded medium-grained litharenite with dark purple ferruginous and manganous interlayers in the lower part. Abundant spores and pollens (e.g., *Punctatisporites* sp.) were obtained. 235 cm
46. Two cycles were deposited in a descending order: (1) pale to bluish gray mudstone (35 cm) in the upper and yellow argillaceous siltstone (20 cm) in the lower; (2) bluish gray claystone (10 cm) in the upper and yellow thick-bedded argillaceous siltstone (70 cm) in the lower. Abundant spores and pollens were collected. 200 cm
45. Two cycles were deposited in a descending order: (1) gray to yellow mudstone (< 10 cm) in the upper and bluish green thick-bedded siltstone with pale nodules of silty mudstone in the lower; (2) yellow siltstone intercalated

with pale silty mudstone in the upper and yellowish brown thick-bedded fine-grained sandstone with limonites in the lower. Abundant spores and pollens (e.g., *Leiotrilets* sp., *Punctatisporites pistilus*) were recovered. 145 cm

44. Yellow thick- to medium-bedded argillaceous siltstone in the upper part; yellow to pale silty mudstone (20 cm) in the middle part; yellow to light purple fine-grained feldspathic sandstone in the lower part. 135 cm
43. Pale to yellow medium- to thin-bedded silty mudstone (25 cm) in the upper part; yellow thick- to medium-bedded argillaceous siltstone intercalated with purplish red lens-shaped mudstone in the lower part. Abundant spores and pollens (e.g., *Leiotriletes* sp., *Dictyophyllidites* sp., *Punctatisporites* sp., *Lueckisporites* sp.) were found. 85 cm
42. Pale mudstone at the top part (10 cm); grayish green thin-bedded siltstone with limonitic nodules in the upper part; grayish green thick-bedded siltstone in the lower part. 110 cm
41. Two cycles were deposited in a descending order: (1) yellow to pale mudstone (5–10 cm) in the upper and grayish yellow argillaceous siltstone with brownish red siderites (15 cm) in the lower; (2) bluish gray silty mudstone (40 cm) in the upper and bluish gray siltstone intercalated with yellow or pale volcanic ash in the lower. Abundant spores and pollens (e.g., *Protohaploxypinus* sp.) were obtained. 170 cm
40. Dark gray mudstone in the upper (30 cm); grayish yellow thick- to medium-bedded siltstone with abundant sideritic nodules in the middle; bluish gray thick-bedded fine-grained argillaceous sandstone in the lower. Spores and pollens (e.g., *Leiotriletes* sp.) were collected. 145 cm
39. Four cycles were deposited in a descending order: (1) grayish green to pale mudstone intercalated with coal seams in the upper (25 cm) and grayish green to yellow siltstone in the lower (15 cm); (2) grayish green to pale mudstone intercalated with coal seams in the upper (20 cm) and grayish green to yellow siltstone in the lower (10 cm); (3) grayish green to pale mudstone intercalated with coal seams in the upper (20 cm) and grayish green to yellow siltstone in the lower (50 cm); (4) grayish green to pale mudstone intercalated with coal seams in the upper (30 cm) and grayish green to yellow siltstone in the lower (40 cm). Abundant spores and pollens (e.g., *Leiotriletes* sp., *Punctatisporites* sp., *Lueckisporites* sp.) were recovered. 210 cm
38. Yellow thick-bedded fine- to medium-grained litharenite, with grayish green thin-bedded silty mudstone at the top. Plant fossils *Stigmaria ficoides*, *Compsopteris* sp. and *Gigantopteris dictyophylloides*, and a few spores and pollen were found. 120 cm
37. Grayish brown thick- to medium-bedded siltstone intercalated with pale lens-shaped mudstone, with yellow to pale thin-bedded silty mudstone (< 10 cm) at the top. Plant fossils *Gigantopteris dictyophylloides*, *Gigantonoclea* spp. and *Compsopteris* sp., and abundant spores and pollens (e.g., *Leiotriletes exiguus*, *L.* sp., *Punctatisporites* cf. *latilus*) were discovered. 60 cm
36. Dark gray thin-bedded silty mudstone at the top part (15 cm); bluish gray thick-bedded argillaceous siltstone in the upper part; grayish green silty lens

in the middle part; dark gray thick- to medium-bedded siltstone in the lower part. Abundant spores and pollens were collected. 165 cm

35. Grayish yellow thick-bedded argillaceous siltstone intercalated with pale to grayish green lens-shaped mudstone, with purplish brown to black coal seams intercalated with pale thin-bedded mudstone at the top part (10 cm). Abundant spores and pollens (e.g., *Leiotriletes* sp., *Punctatisporites* sp., *Apiculatisporis* sp.) were recovered. 85 cm
34. Brown silty mudstones intercalated with black coal beddings at the top part (10 cm); grayish yellow argillaceous siltstone in the upper part; bluish gray thick-bedded fine-grained feldspathic litharenites with small-scaled cross beddings in the middle part; grayish yellow thick-bedded medium-grained feldspathic litharenites in the lower part. Abundant spores and pollens (e.g., *Leiotriletes* sp., *Punctatisporites* sp., *Apiculatisporis* sp., *Lueckisporites* sp., *L.* sp.) were collected. 240 cm
33. Bluish gray thick-bedded medium-grained feldspathic litharenites (dark purple when weathered). 235 cm
32. Bluish gray thick- to medium-bedded argillaceous sandstone with large-scaled wedge-shaped cross beddings in the upper part; bluish gray thin-bedded siltstone in the middle part; grayish green thick-bedded argillaceous siltstone in the lower part. The spore and pollen *Platysaccus* sp. was examined. 65 cm
31. Dark grayish green thin-bedded fine-grained feldspathic litharenite intercalated with yellow lens of argillaceous siltstone, with purplish brown coal seams at the top (5–10 cm). 55 cm
30. Dark gray to pale mudstones (10 cm for each layer) intercalated with purplish brown coal seams (10 cm) in the upper part; yellow thick-bedded medium-grained feldspathic litharenite in the middle part; grayish green thin-bedded fine-grained feldspathic litharenite in the lower part. The spore and pollen *Lueckisporites* (?) sp. was found. 75 cm
29. Bluish gray thick-bedded argillaceous siltstone, with yellow silty mudstone at the top. Plant fossils *Compsopteris* spp. and *Gigantonoclea guizhouensis* and abundant spores and pollens (e.g., *Punctatisporites* sp.) were collected. 70 cm
28. Two parts were deposited in a descending order: (1) yellow mudstone in the upper (totally 10–20 cm), black coal seams in the middle (10–10 cm), and pale volcanic ash with cross beddings in the lower (15 cm); (2) yellowish green thick- to medium-bedded argillaceous siltstone. Abundant spores and pollens were recovered. 45 cm
27. Yellow thick-bedded silty mudstone, with bluish gray mudstone at the top (15–20 cm) and dark gray siltstone at the base (18 cm). Spores and pollens *Leiotriletes* sp. and *Vittatina* sp. were examined. 45 cm
26. Yellow thick-bedded argillaceous siltstone, with gray mudstone intercalated with grayish black mudstone at the top. Small-scaled cross beddings were yielded in the mudstone. Abundant spores and pollens (e.g., *Punctatisporites* sp., *Apiculatisporis* sp., *Protohaploxypinus* sp., *Lueckisporites* sp.) were collected. 55 cm

25. Grayish green thick-bedded fine-grained feldspathic sandstone with small-scaled wedge-shaped cross beddings. 70 cm
24. Yellow thick-bedded coarse-grained feldspathic litharenite, with the plant fossil *Gigantopteris guizhouensis*. 40 cm
23. Yellow argillaceous siltstone intercalated with mudstone and siltstone, with plant fossils *Gigantopteris* sp. and *Cordaites* sp.. 275 cm
22. Two cycles were deposited and each cycle is composed of pale, purplish red to yellow tuff-bearing mudstone in the upper and yellow thick-bedded fine-grained argillaceous sandstone in the lower. Plant fossils *Gigantopteris* sp., *Stigmaria ficoides, Lepidodendron acutangulum, Lobatannularia heianensis* and *Pecopteris hemiteloides* and the spore and pollen *Punctatosporites* sp. were discovered. 150 cm
21. Yellow thick-bedded argillaceous siltstone with sideritic nodules, with gray mudstone (5–20 cm) at the top part of the bed. Abundant plant fossils *Stigmaria ficoides, Compsopteris* spp., *Lepidodendron acutangulum, Lepidodendron oculusfelis, Lepidodendron* sp., *Pecopteris hemiteloides, Gigantonoclea guizhouensis, Lobatannularia cathaysiana, Lobatannularia heianensis* and *Gigantopteris* sp., and a few spores and pollens were collected. 75 cm
20. Grayish yellow thick-bedded siltstone in the middle and upper parts; yellowish brown thick- to medium-bedded fine-grained sandstone in the lower part. Abundant spores and pollens (e.g., *Leiotriletes* sp., *Punctatisporites distalis, P.* sp., *Apiculatisporis* sp.) were found. 185 cm
19. Grayish green siltstone intercalated with pale mudstone in the upper part; brownish yellow thick-bedded fine-grained sandstone in the lower part. Abundant spores and pollens (e.g., *Leiotriletes* sp., *Dictyophyllidites* sp., *Apiculatisporis variocorneus, A.* sp., *Torispora* sp., *Gardenasporites* sp., *Lueckisporites* sp.) were recovered. 140 cm
18. Grayish green thick-bedded argillaceous siltstone intercalated with yellowish green lens of fine-grained sandstone (with small-scaled wedge-shaped cross beddings and sideritic nodules), with pale lens-shaped mudstone at the top part. Plant fossils *Pecopteris arcuata, P. taiyuanensis, P. orientalis, P. longifoloides, P.* spp., *Fascipteris densata, Fascipteris* sp., *Gigantopteris nicotianefolia, Gigantopteris paradoxa, Gigantopteris* sp., *Gigantonoclea guizhouensis, Gigantonoclea hallei, Gigantonoclea* spp., *Cordaites* sp., *Lobatannularia ensifolia, L. nampoensis, L. cathaysiana, L. heianensis, L.* spp., *Annularia pingloensis, Cladophlebis permica, Cladophlebis* sp., *Neuropteridium guizhouensis* and *Rhipidopsis panii, Rhipidopsis lobulata* were examined. 110 cm
17. Grayish brown mudstone (with cross beddings) overlying pale mudstone in the upper part of the bed; brownish yellow thick-bedded siltstone and fine-grained sandstone with black charcoal fossils in the lower part. Plant fossils *Pterophyllum* sp., *Pecopteris longifoloides, Pecopteris* spp., *Gigantonoclea guizhouensis, Cordaites* sp., *Lobatannularia multifolia, Lobatannularia* spp., *Compsopteris* spp., *Neuropteridium* sp. and *Gigantopteris* sp. were discovered. 85 cm
16. Grayish green thick-bedded argillaceous siltstone intercalated with a layer (3 cm thick) of yellow mudstone. Plant fossils *Gigantopteris* sp., *Neuropteridium*

sp., *Cordaites* sp., *Gigantonoclea* spp. and *Compsopteris* spp. and the spore and pollen *Punctatisporites pistilus* were obtained. 110 cm

15. Dark gray silty mudstone at the top (5–10 cm); grayish green thick- to medium-bedded siltstone in the upper; yellow argillaceous siltstone in the lower. 90 cm
14. Bluish gray argillaceous siltstone intercalated with yellowish green lens of fine-grained sandstone, with grayish yellow thin-bedded silty mudstone at the top part (0–5 cm, laterally discontinuous). 105 cm
13. Grayish black mudstone with laminations at the top part (2–3 cm thick); pale thin-bedded mudstone in the upper part (10–20 cm); dark gray siderites in the middle part (10 cm); yellow medium- to thin-bedded argillaceous siltstone and silty mudstone in the lower part. 50 cm
12. Gray mudstone intercalated with purplish brown coal seams in the upper part (15–20 cm); yellow and purplish red thick-bedded argillaceous siltstone in the lower part; dark gray siderites at the base (10–20 cm thick). Abundant spores and pollens were discovered. 80 cm
11. Yellow and pale mudstone in the upper part and yellow argillaceous siltstone in the lower part, with interlayers of purplish siltstone in the bed. 55 cm
10. Purplish red thick-bedded siltstone, with dark gray mudstone at the top (15 cm) and siltstone at the base (10 cm). Abundant spores and pollens (e.g., *Punctatisporites* sp.) were found. 65 cm
9. Yellow silty mudstone intercalated with a layer of argillaceous siltstone. Laminations evident in the mudstone. The plant fossils *Compsopteris* spp. were examined. 40 cm
8. Dark gray mudstone at the top part (15 cm thick); yellow mudstone in the upper part (3 cm thick); grayish green thick-bedded siltstone in the lower part. The plant fossils *Stigmaria ficoides* and *Stigmaria radiatopunctata* were examined and abundant spores and pollens were recovered. 130 cm
7. Bluish gray mudstone at the top part (20 cm); yellow silty mudstone in the upper part (40 cm); bluish gray thick-bedded siltstone in the lower part. The plant fossil *Compsopteris* spp. and abundant spores and pollens were found. 285 cm
6. Yellow thick-bedded silty mudstone, with bluish gray silty mudstone (3 cm) at the top. Plant fossils *Fascipteris hallei*, *Odontopteris* sp., *Gigantopteris paradoxa*, *Lobatannularia multifolia*, *Gigantopteris dictyophylloides*, *Gigantonoclea* spp., *Lobatannularia* spp. and *Pecopteris* spp. were examined and abundant spores and pollen (e.g., *Leiotriletes exiguus*, *L. concavus*, *L.* sp., *Dictyophyllidites mortoni*, *D. discretus*, *Punctatisporites pistilus*, *Punctatosporites scabellus*, *P.* sp., *Granulatisporites* sp., *Torispora* cf. *laevigata*, *T. securis*, *T.* sp., *Gardenasporites* sp., *Lueckisporites* sp.) were examined. 135 cm
5. Black thick-bedded coal seams intercalated with yellow siltstone, with yellow volcanic ash and silty mudstone at the top part and brown volcanic ash at the base, and with laminations in the mudstone. Plant fossils *Schizoneura*

manchuriensis, *Cladophlebis fuyuanensis*, *Pecopteris tenuicostata*, *Gigantopteris* sp., *Pecopteris gracilenta* and *Pecopteris* spp., and abundant spores and pollens (e.g., *Protohaploxypinus* sp.) were found. 185 cm

4. Gray mudstone at the top part (20 cm thick); grayish yellow silty mudstone in the upper part; brownish yellow thick-bedded medium-grained feldspathic sandstone in the middle part; dark gray thick-bedded siltstone in the lower part. Plant fossils *Lixotheca permica*, *Fascipteris stena*, *Alethopteris norinii*, *Taeniopteris multinervis*, *Fascipteris hallei*, *Cladophlebis permica*, *Compsopteris impairs*, *Linopteris brongniartii*, *Lobatannularia heianensis*, *Compsopteris contracta*, *Pecopteris* spp. and *Compsopteris* spp. and a few spores and pollens were discovered. 250 cm
3. Bluish gray thick- to medium-bedded siltstone intercalated with dark purplish red medium-bedded fine-grained sandstone. Plant fossils *Sphenophyllum* sp., *Sphenopteris* sp., *Cordaites principalis*, *Alethopteris norinii*, *Cladophlebis fuyuanensisi*, *Fascipteris hallei*, *Compsopteris impairs*, *Pecopteris elegantula*, *Compsopteris contracta*, *Gigantonoclea lagrelii*, *Pecopteris* spp. and *Compsopteris* sp. were identified. 165 cm
2. Two cycle were deposited in a decending order: (1) yellow thin-bedded silty mudstone in the upper and thick-bedded argillaceous siltstone in the lower; (2) yellow thin-bedded silty mudstone in the upper and thick- to medium-bedded argillaceous siltstone in the lower. Plant fossils *Gigantonoclea* sp., *Pecopteris* spp. and *Cladophlebis* sp. were recovered. 180 cm
1. Four cycles were deposited and each cycle is composed of pale to yellow mudstone in the upper (totally from 3 to 20 cm for the thickness of mudstone in each cycle) and yellow argillaceous silststone in the lower (totally 30–50 cm for the thickness of siltstone in each cycle). Siderites was intercalated within the siltstone of the uppermost cycle. Downwards to the basal part of the bed, it consists yellow thick-bedded silty mudstone intercalated with thin-bedded fine-grained feldspathic sandstone. Plant fossils *Neuropteridium coreanicum*, *Taeniopteris* sp., *Pecopteris fuyuanensis*, *P.* spp., *Compsopteris* spp., *Stigmaria ficoides*, *Lobatannularia* sp. and *Pecopteris orientalis* were examined. > 200 cm

Note: Listed plant fossils are referred to Chu et al. (2016) (but updated in taxonomy in this book).

3.6 Jinzhong Section in Weining County

The Jinzhong section is located about 5 km south to the Jinzhong Town, Weining County, Guizhou Province, South China (Fig. 3.14).

The sequence of the Jinzhong section is composed of the Xuanwei Formation, Kayitou Formation and the basal part of the Dongchuan Formation in an ascending order (Fig. 3.15). The Xuanwei Formation is dominated by grayish green thick-bedded silitstones intercalated with coal seams. The Kayitou Formation is mainly

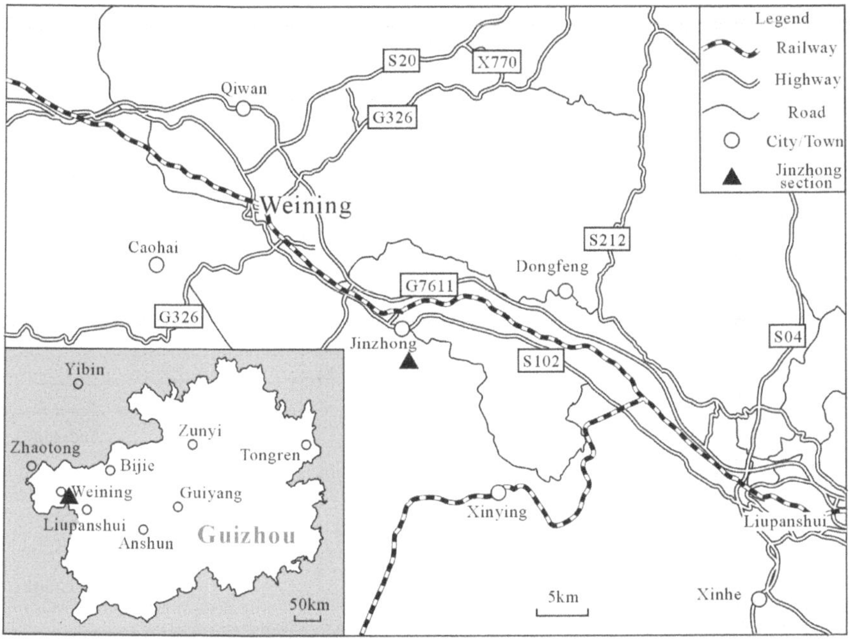

Fig. 3.14 Geographic location of the Jinzhong section

composed of yellowish green thin-bedded siltstones and silty mudstones. The basal part of the Dongchuan Formation is characterized by purple thick-bedded sandstones intercalated with silty mudstones. A few plant fossils, conchostracans, bivalves and brachiopods were yielded in the section (Fig. 3.15).

Dongchuan Formation in the Lower Triassic

17. Purplish red thick-bedded fine-grained sandstone. 930 cm
16. Purplish red thick- to medium-bedded fine-grained sandstone. 820 cm

Conformity
Kayitou Formation in the uppermost Permian to the Lower Triassic

15. Yellowish green to yellow thin-bedded siltstone intercalated with purplish red silty mudstone, with conchostracans *Euestheria gutta* and *Euestheria* spp.. 980 cm
14. Yellow thin-bedded siltstone and argillaceous siltstone, with the plant fossil *Germaropteris* sp., and conchostracans *Euestheria gutta* and *Palaeolimnadia xuanweiensis.* 500 cm
13. Yellow thin-bedded siltstone and argillaceous siltstone, with conchostracans *Euestheria gutta* and *Palaeolimnadia xuanweiensis.* 620 cm

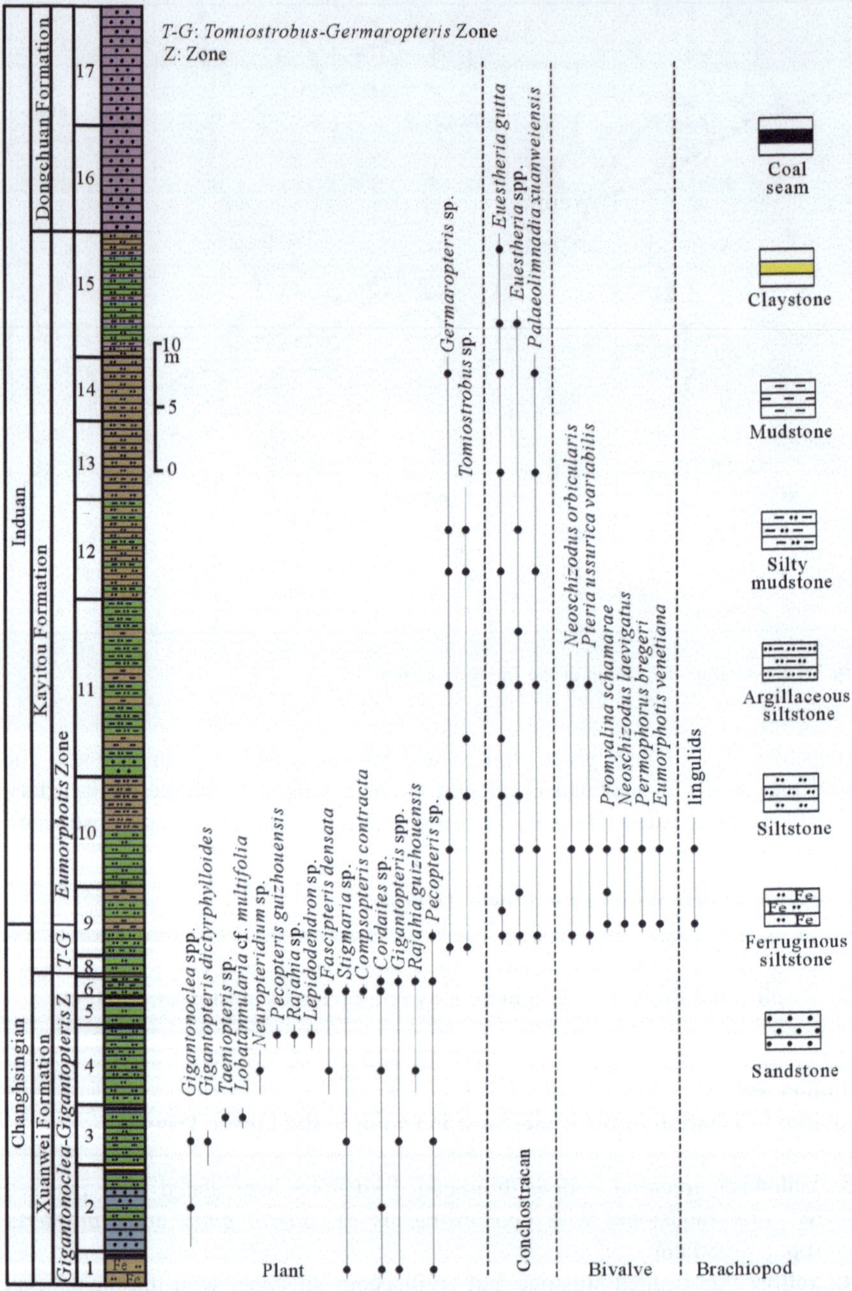

Fig. 3.15 Stratigraphic column and fossil occurrences (plant, conchostracan, bivalve and brachiopod) at the Jinzhong section (fossils referred to Chu et al. 2016)

12. Yellow thin-bedded siltstone and yellowish green medium- to thin-bedded silty mudstone, with plant fossils *Germaropteris* sp. and *Tomiostrobus* sp., and conchostracans *Euestheria gutta*, *Euestheria* sp. and *Palaeolimnadia xuanweiensis*. 780 cm
11. Yellowish green to yellow thin-bedded silty mudstone intercalated with mudstone, with yellowish green thin-bedded sandstone at the basal part. The plant fossils *Germaropteris* sp. and *Tomiostrobus* sp., conchostracans *Euestheria gutta*, *Euestheria* sp. and *Palaeolimnadia xuanweiensis*, bivalves *Neoschizodus orbicularis*, *Pteria ussurica variabilis*, *Promyalina schamarae*, *Neoschizodus laevigatus*, *Permophorus bregeri* and *Eumorphotis venetiana*, and the brachiopod lingulids were discovered. 1395 cm
10. Yellowish green to yellow silty mudstone in the upper part and yellowish green thin-bedded siltstone in the lower part. The plant fossils *Germaropteris* sp. and *Tomiostrobus* sp., conchostracans *Euestheria gutta*, *Euestheria* sp. and *Palaeolimnadia xuanweiensis*, and bivalves *Neoschizodus orbicularis* and *Pteria ussurica variabilis* were identified. 875 cm
9. Yellowish green to yellow thin-bedded argillaceous siltstone and mudstone, with plant fossils *Pecopteris* sp., *Germaropteris* sp. and *Tomiostrobus* spp., conchostracans *Euestheria gutta*, *Euestheria* sp. and *Palaeolimnadia xuanweiensis*, bivalves *Neoschizodus orbicularis*, *Pteria ussurica variabilis*, *Promyalina schamarae*, *Neoschizodus laevigatus*, *Permophorus bregeri* and *Eumorphotis venetiana*, and the brachiopod lingulids. 540 cm
8. Yellowish green thick- to medium-bedded siltstone. 130 cm

Conformity
Xuanwei Formation in the Upper Permian

7. Black coal seams in the upper part and pale claystone in the lower part. 5 cm
6. Yellowish green thin-bedded silty mudstone, with plant fossils *Cordaites* sp., *Gigantopteris* spp., *Rajahia guizhouensis* and *Pecopteris* sp. in the upper part and *Fascipteris densata*, *Stigmaria* sp. and *Compsopteris contracta* in the lower part. 146 cm
5. Grayish green thick- to medium-bedded siltstone, with black coal seams intercalated with pale claystone at the top part. 225 cm
4. Grayish green thick-bedded siltstone intercalated with argillaceous siltstone, with black coal seams at the top part. The plant fossils *Neuropteridium* sp., *Pecopteris guizhouensis*, *Rajahia* sp., *Lepidodendron* sp., *Fascipteris densata*, *Cordaites* sp. and *Rajahia guizhouensis* were found. 625 cm
3. Grayish green thick- to medium-bedded siltstone intercalated with silty mudstone, with black coal seams at the top part. The plant fossils *Gigantonoclea* spp., *Gigantopteris dictyrphylloides*, *Taeniopteris* sp., *Lobatannularia* cf. *multifolia*, *Stigmaria* sp., *Gigantopteris* spp. and *Pecopteris* sp. were examined. 480 cm

2. Grayish green thick-bedded siltstone and bluish gray thick-bedded sandstone, with pale claystone and black coal seams at the top part, with plant fossils *Gigantonoclea* spp. and *Cordaites* sp.. 685 cm
1. Grayish green to yellow thick-bedded ferruginous siltstone, with black coal beds at the top part. The plant fossils *Gigantonoclea* sp., *Gigantopteris* sp., *Cordaites* sp., *Pecopteris* sp. and *Stigmaria* sp. were discovered.

Unexposed at the base

Note: Listed plant fossils are based on Chu et al. (2016), but updated in taxonomy.

3.7 Zhejue Section in Weining County

The Zhejue section outcrops along the 326 National Highway, 7.2 km northeast to the Zhejue Town, Weining County, Guizhou Province, South China (Fig. 3.16).

The sequence of the Zhejue section is composed of the Xuanwei Formation and the basal part of the Kayitou Formation in an ascending order (Fig. 3.17). The Xuanwei Formation is dominated by yellow, grayish green to bluish gray sandstones and argillaceous siltstones. The basal part of the Kayitou Formation is composed of grayish green silty mudstones and siltstones. Abundant spores and pollens and plant fossils were found in the section (Figs. 3.17, 3.18). Additionally, the Dongchuan Formation dominated by purplish red sandstones is outcropped in the area and overlies the Kayitou Formation.

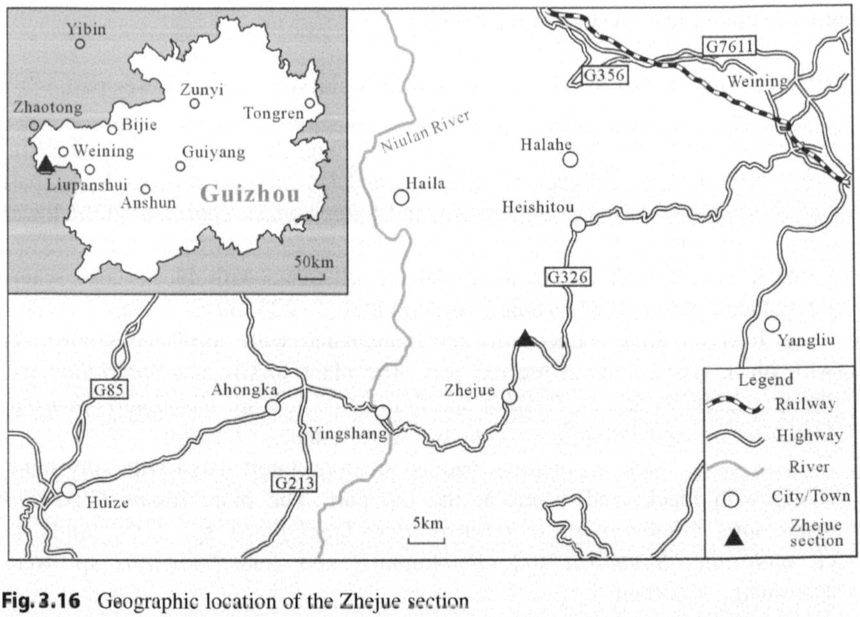

Fig. 3.16 Geographic location of the Zhejue section

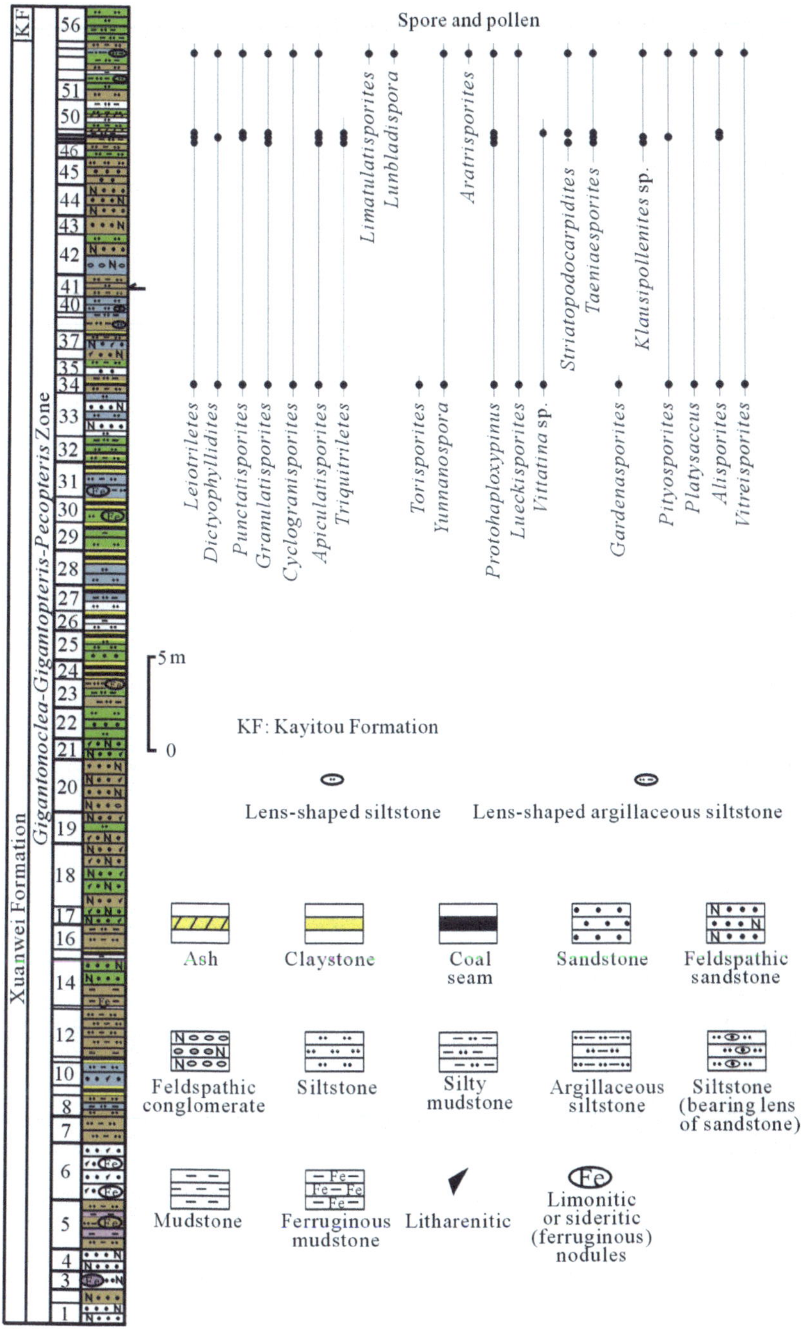

Fig. 3.17 Stratigraphic column and fossil occurrences (spore and pollen) at the Zhejue section (referred to Yu et al. 2022)

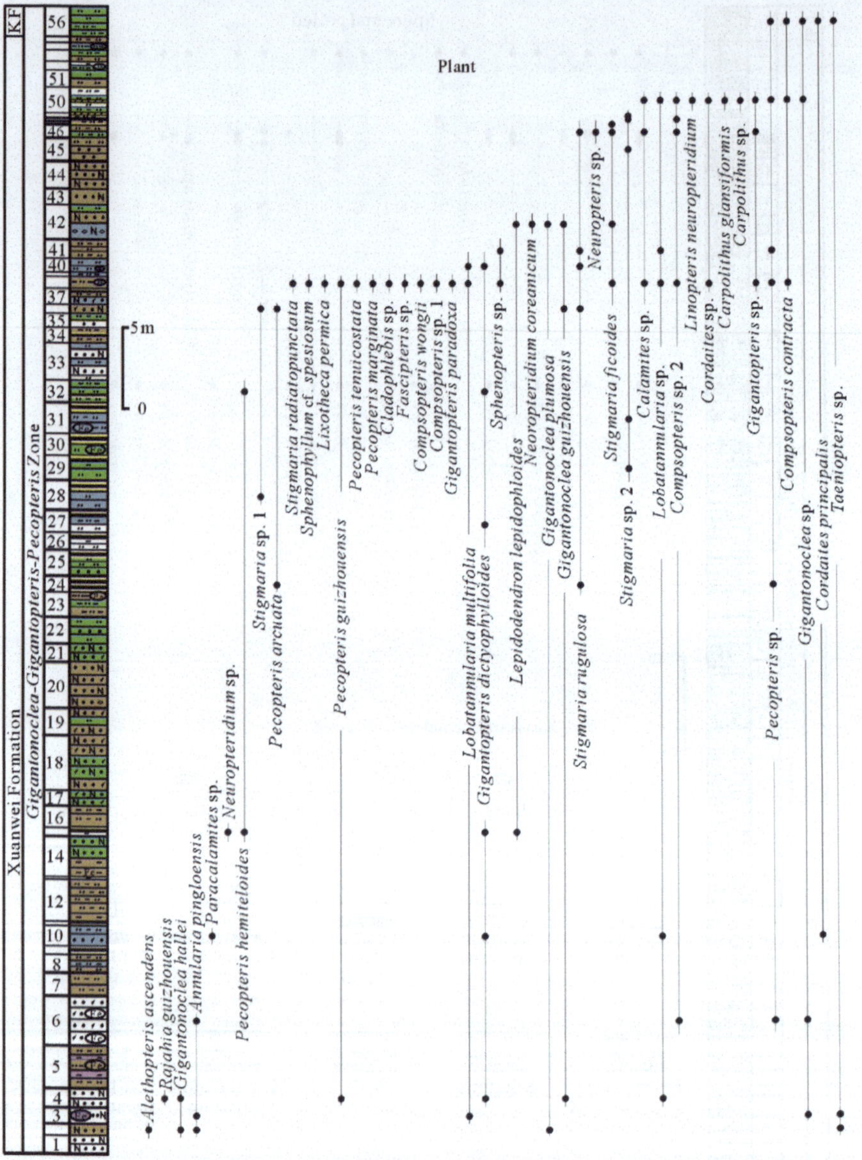

Fig. 3.18 Stratigraphic column and fossil occurrences (plant) at the Zhejue section. Legends same to Fig. 3.17

Kayitou Formation in the Upper Permian

56. Grayish green thin-bedded siltstone in the upper part; grayish green thin-bedded silty mudstone in the lower part. Plant fossils *Compsopteris contracta*, *Gigantonoclea* sp., *Pecopteris* sp., *Cordaites principalis* and *Taeniopteris* sp. were examined. 160 cm

Conformity
Xuanwei Formation in the Upper Permian

55. Grayish green thin-bedded silty mudstone in the upper part; yellow medium-bedded argillaceous siltstone in the lower part. 60 cm
54. Bluish gray thin-bedded silty mudstone in the upper part; dark green thin-bedded to lens-shaped argillaceous siltstone in the lower part. Spores and pollens *Leiotriletes*, *Dictyophyllidites*, *Punctatisporites*, *Granulatisporites*, *Cyclogranisporites*, *Apiculatisporites*, *Limatulatisporites*, *Lunbladispora*, *Yunnanospora*, *Aratrisporites*, *Protohaploxypinus*, *Lueckisporites*, *Striatopodocarpidites*, *Taeniaesporites*, *Klausipollenites* sp., *Pityosporites*, *Platysaccus*, *Alisporites* and *Vitreisporites* were identified. 40 cm
53. Grayish green to yellow medium-bedded siltstone, intercalated with silty mudstone in the upper part. 70 cm
52. Grayish green to dark green thin-bedded mudstone and silty mudstone, intercalated with lens-shaped siltstone. 55 cm
51. Grayish green to yellow thick-bedded siltstone. 95 cm
50. Pale thin-bedded silty mudstone at the top part (<10 cm thick); light grayish green thick- to medium-bedded argillaceous siltstone in the upper part; pale volcanic ash in the middle part (5 cm); grayish green thin-bedded siltstone intercalated with argillaceous siltstone in the lower part. Abundant laminations and sideritic nodules were yielded in the bed. Plant fossils *Calamites* sp., *Lobatannularia* sp., *Compsopteris* sp. 2, *Linopteris neuropteridium*, *Cordaites* sp., *Carpolithus glansiformis*, *Carpolithus* sp., *Gigantopteris* sp., *Pecopteris* sp., *Compsopteris contracta* and *Gigantonoclea* sp. were collected. 160 cm
49. Pale thin-bedded silty claystone intercalated with a layer (10 cm thick) of brown to purple volcanic ash in the upper part, with nodule-shaped silty claystone and coal seams in the lower part. The plant fossil *Stigmaria* sp. 2, and spores and pollens *Leiotriletes*, *Punctatisporites*, *Granulatisporites*, *Apiculatisporites*, *Triquitriletes*, *Protohaploxypinus*, *Vittatina* sp., *Striatopodocarpidites*, *Taeniaesporites* and *Alisporites* were identified. 35 cm
48. Brown to yellow thin-bedded silty claystone intercalated with several layers of coal seams, with plant fossils *Stigmaria* sp. 2 and *Compsopteris* sp. 2, and spores and pollens *Leiotriletes*, *Dictyophyllidites*, *Punctatisporites*,

Granulatisporites, Apiculatisporites, Triquitriletes, Protohaploxypinus, Taeniaesporites, Klausipollenites sp., *Pityosporites* and *Alisporites.* 20 cm

47. Pale claystone, with the plant fossil *Stigmaria ficoides*, and spores and pollens *Leiotriletes, Granulatisporites, Apiculatisporites, Triquitriletes, Protohaploxypinus, Striatopodocarpidites, Taeniaesporites* and *Klausipollenites* sp.. 10 cm
46. Grayish green to yellow thin-bedded argillaceous siltstone, with plant fossils *Stigmaria rugulosa, Neuropteris* sp., *Stigmaria ficoides, Calamites* sp., *Compsopteris* sp. 2 and *Gigantopteris* sp.. 70 cm
45. Grayish yellow thin-bedded siltstone in the upper part; grayish yellow thin-bedded fine-grained sandstone with the plant fossil *Stigmaria* sp. 2 in the lower part. 145 cm
44. Grayish yellow medium- to thin-bedded medium-grained feldspathic sandstone. 160 cm
43. Yellow thick-bedded fine-grained feldspathic sandstone with the trough cross bedding and wedge-shaped cross bedding, intercalated with conglomerates. 100 cm
42. Bluish gray thick-bedded fine-grained feldspathic conglomerate (with abundant black lithoclasts) intercalated with yellow thick-bedded medium-grained feldspathic sandstone, and with grayish green medium-bedded siltstone (20 cm thick). The plant fossils *Lepidodendron lepidophloides, Neuropteridium coreanicum, Stigmaria ficoides, Gigantonoclea guizhouensis* and *G. plumosa* were identified. 220 cm
41. Yellow medium- to thin-bedded argillaceous siltstone intercalated with yellow thin-bedded siltstone. The plant fossils *Sphenopteris* sp., *Stigmaria rugulosa, Lobatannularia* sp. and *Pecopteris* sp. were examined. 120 cm
40. Bluish gray to dark gray medium- to thin-bedded siltstone intercalated with lens of feldspathic sandstone, with plant fossils *Lobatannularia multifolia, Gigantopteris dictyophylloides* and *Stigmaria rugulosa.* 80 cm
39. Brown and bluish gray silty mudstone, with abundant fragments of plant fossils. 25 cm
38. Yellow medium-bedded silty mudstone with bluish gray lens-shaped siltstone. The plant fossils *Stigmaria radiatopunctata, Sphenophyllum* cf. *spesiosum, Lixotheca permica, Pecopteris guizhouensis, Pecopteris tenuicostata, Pecopteris marginata, Cladophlebis* sp., *Fascipteris* sp., *Compsopteris wongii, Compsopteris* sp. 1, *Gigantopteris paradoxa, Lobatannularia multifolia, Sphenopteris* sp., *Stigmaria ficoides, Calamites* sp., *Lobatannularia* sp., *Compsopteris* sp. 2, *Cordaites* sp., *Gigantopteris* sp., *Pecopteris* sp. and *Compsopteris contracta* were collected. 70 cm
37. Grayish brown claystone intercalated with a layer of black coal bed (20 cm) in the upper part; yellow medium-bedded argillaceous siltstone in the middle part (20 cm); bluish gray thick- to medium-bedded medium- to fine-grained feldspathic litharenite in the lower part (40 cm). 99 cm

36. Yellow thick- to medium-bedded fine-grained ferruginous and manganous feldspathic litharenite with large-scale wedge-shaped cross beddings. Plant fossils *Stigmaria* sp. 1, *Pecopteris arcuata*, *Gigantonoclea guizhouensis* and *Stigmaria rugulosa* were discovererd. 65 cm
35. Light yellowish green thin-bedded argillaceous siltstone in the upper part; gray thin-bedded fine-grained sandstone in the middle part; bluish gray thin-bedded argillaceous siltstone in the lower part. The wedge-shaped cross bedding was observed in the uppermost. 80 cm
34. Bluish gray to yellow thin-bedded silty mudstone at the top part; a layer of brown claystone (< 10 cm) overlying coal seams (3–5 cm) in the upper part; grayish yellow thin-bedded fine-grained sandstone and gray thin-bedded argillaceous siltstone in the lower part. Spores and pollens *Leiotriletes*, *Dictyophyllidites*, *Punctatisporites*, *Granulatisporites*, *Cyclogranisporites*, *Apiculatisporites*, *Triquitriletes*, *Torisporites*, *Yunnanospora*, *Protohaploxypinus*, *Lueckisporites*, *Vittatina* sp., *Gardenasporites*, *Pityosporites*, *Platysaccus*, *Alisporites* and *Vitreisporites* were identified. 80 cm
33. Two cycles were deposited and each cycle includes: bluish gray thick- to medium-bedded siltstone in the upper; medium- to fine-grained feldspathic sandstone in the middle; bluish gray thick- to medium-bedded siltstone in the lower. The wedge-shaped and trough cross beddings were yielded in the siltstone. 235 cm
32. Yellowish green silty mudstone in the upper part (20 cm); a layer of coal bed (5–10 cm) overlying the pale claystone in the middle part; grayish green medium- to thin-bedded argillaceous silstone and siltstone in the lower part, with sideritic nodules. In the uppermost, the plant fossils *Gigantopteris dictyophylloides*, and *Pecopteris hemitelioides* were found. 130 cm
31. Yellow claystone at the top part (10 cm); a layer of coal bed in the upper part (20 cm); grayish brown claystone in the middle part; dark to bluish gray medium- to thin-bedded argillaceous siltstone, with sideritic nodules in the lower part. The plant fossil *Stigmaria* sp. 2. was examined. 180 cm
30. Brown claystone at the top part (10 cm); black coal beds in the upper part (25 cm); light brown claystone in the middle part; dark grayish green siltstone with sideritic nodules in the lower part. 130 cm
29. Yellow claystone at the top part (8 cm); black coal beds in the upper part (30 cm); dark grayish green mudstone with fragments of plant fossil in the middle part; dark grayish green siltstone in the lower part. The plant fossil includes *Stigmaria* sp. 2. 150 cm
28. Bluish gray silty claystone in the upper part (60 cm); black coal seams (35 cm) overlying gray claystone in the middle part; bluish gray siltstone in the lower part (50 cm). The plant fossil *Stigmaria* sp. 1 was collected. 180 cm
27. Pale claystone at the top (10 cm); black coal seams intercalated with yellow thin-bedded argillaceous siltstone in the upper part (40–70 cm thick); bluish gray medium-bedded silty mudstone in the middle part (25 cm); grayish black siltstone in the lower part (10 cm). The plant fossil *Gigantopteris dictyophylloides* was recovered. 130 cm

26. Pale claystone in the uppermost (30 cm); black coal seams intercalated with pale silty claystone in the upper part (10–50 cm thick); gray mudstone in the middle part (30 cm); dark gray thick- to medium-bedded siltstone in the lower part (30–60 cm thick). 105 cm
25. Gray to grayish yellow claystone interbedded with silty claystone in the upper part (60 cm); black coal seam (10 cm) overlying pale claystone (20 cm) in the middle part; grayish green thin-bedded fine-grained sandstone interbedded with siltstone in the lower part (50 cm), with silty nodules in the sandstone. 165 cm
24. Pale claystone (20 cm) overlying yellow claystone (10 cm) in the upper part; black coal beds in the middle part (20 cm); yellow claystone (5 cm) overlying pale claystone (35 cm) in the lower part. Plant fossils *Pecopteris arcuata*, *Stigmaria rugulosa* and *Pecopteris* sp. were found. 90 cm
23. Yellow thick- to medium-bedded silty mudstone intercalated with grayish green silty mudstone, with limonitic nodules in the upper part. 150 cm
22. Grayish green thick- to medium-bedded siltstone in the upper part; grayish green medium- to thin-bedded fine-grained sandstone in the lower part, with grayish green siltstone at the base. 165 cm
21. Grayish green thick-bedded feldspathic litharenite. 110 cm
20. Yellow thick- to medium-bedded coarse-grained feldspathic sandstone, abundantly with coarse-grained lithoclasts and silty pebbles. 280 cm
19. Grayish green to yellow thick-bedded coarse- to medium-grained feldspathic litharenite intercalated with grayish green medium-bedded siltstone. 170 cm
18. Grayish green to yellow thick-bedded medium-grained feldspathic litharenite. 340 cm
17. Grayish green thick-bedded fine-grained feldspathic litharenite. 90 cm
16. Yellow medium- to thin-bedded silty mudstone in the upper part; yellow medium- to thin-bedded argillaceous siltstone with wedge-shaped cross beddings in the lower part. 135 cm
15. Brownish yellow claystone at the top part (5 cm); black coal seams intercalated with mudstone, with abundant fragments of plant fossils in the upper part (15 cm); bluish gray claystone in the lower part (25 cm). Plant fossils *Neuropteridium* sp., *Pecopteris hemitelioides*, *Gigantopteris dictyophylloides* and *Lepidodendron lepidophloides* were discovered. 45 cm
14. Grayish green thin-bedded fine-grained feldspathic sandstone in the upper part; yellow medium-bedded ferruginous mudstone in the lower part. 250 cm
13. Dark grayish green mudstone with fragments of plant fossils. 10 cm
12. Yellow thick- to medium-bedded argillaceous siltstone, with grayish yellow thick- to medium-bedded mudstone (20 cm) at the base. 260 cm
11. Black coal seams with grayish brown claystone (3 cm) at the base. 20 cm
10. Bluish gray thick- to medium-bedded argillaceous siltstone in the upper part; bluish gray thick- to medium-bedded fine-grained litharenite with laminations in the lower part. The plant fossils *Paracalamites* sp., *Gigantopteris*

dictyophylloides, *Lobatannularia* sp. and *Cordaites principalis* were examined. 120 cm

9. Light brown claystone at the top part (8 cm); black coal seam in the upper part (12 cm); pale to yellow claystone in the lower part (35 cm). 55 cm
8. Grayish yellow thick- to medium-bedded argillaceous siltstone with nodules of silty mudstone in the upper part (30 cm); bluish gray thick- to medium-bedded silty mudstone in the middle part (30 cm); grayish yellow thick- to medium-bedded siltstone in the lower part (50 cm). 110 cm
7. Yellow thick-bedded argillaceous siltstone (brown when weathered). 140 cm
6. Dark gray thick-bedded finc-grained litharenite with brownish red ferruginous and manganous lens and cross beddings. Plant fossils *Annularia pingloensis*, *Compsopteris* sp. 2, *Pecopteris* sp. and *Gigantonoclea* sp. were recovered. 310 cm
5. Yellow thick-bedded argillaceous siltstone intercalated with yellow, purple to bluish gray thin-bedded mudstone (3–5 cm thick for each layer of mudstone). Gray fine-grained sandy lens and ferruginous nodules were yielded in the siltstone. 260 cm
4. Grayish yellow thick- to medium-bedded fine-grained feldspathic sandstone with cross beddings. Plant fossils *Rajahia guizhouensis*, *Gigantonoclea hallei*, *Pecopteris guizhouensis*, *Gigantopteris dictyophylloides*, *Gigantonoclea guizhouensis* and *Lobatannularia* sp. were found. 120 cm
3. Gray thick-bedded fine-grained feldspathic sandstone with purplish red ferruginous nodules. Plant fossils *Annularia pingloensis*, *Lobatannularia multifolia*, *Gigantonoclea* sp. and *Taeniopteris* sp. were collected. 90 cm
2. Grayish yellow thick- to medium-bedded fine-grained feldspathic sandstone, with coal lens. Plant fossils *Alethopteris ascendens*, *Gigantonoclea hallei*, *Annularia pingloensis*, *Gigantonoclea plumosa* and *Taeniopteris* sp. were examined. 70 cm
1. Gray thick- to medium-bedded fine-grained feldspathic sandstone, with laminations. 110 cm

3.8 Jiucaichong Section in Weining City

The Jiucaichong section is located at the Jiucaichong Village, about 10 km south to the Heishitou Town, Weining City, Guizhou Province, South China (Fig. 3.19).

The sequence of the Jiucaichong section is composed of the Xuanwei, Kayitou formations and the basal part of the Dongchuan Formation in an ascending order (Fig. 3.20). The Xuanwei Formation is dominated by yellowish green to gray siltstones and mudstones, occasionally intercalated with coal seams. The Kayitou Formation is mainly composed of yellow to gray siltstones, occasionally intercalated with fine-grained sandstones. The basal part of the Dongchuan Formation is characterized by grayish green fine-grained sandstones intercalated with purple

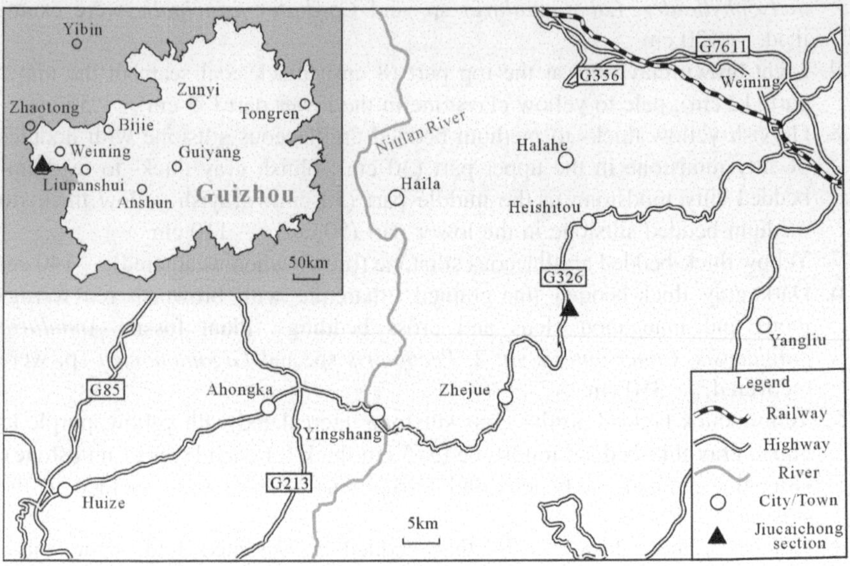

Fig. 3.19 Geographic location of the Jiucaichong section

fine-grained sandstones. A few plant fossils and conchostracans were discovered in the section (Fig. 3.20).

Dongchuan Formation in the Lower Triassic
Not described upwards

32. Grayish green fine-grained sandstone intercalated with purple band-shaped fine-grained sandstone. 200 cm

Conformity
Kayitou Formation in the uppermost Permian to the Lower Triassic

31. Yellow medium- to thin-bedded argillaceous siltstone intercalated with yellowish brown band-shaped mudstone. 100 cm
30. Yellow medium- to thin-bedded siltstone and fine-grained sandstone. 130 cm
29. Grayish green medium-bedded siltstone interbedded with yellow thin-bedded argillaceous siltstone. 120 cm
28. Grayish green medium-bedded siltstone interbedded with yellow thin-bedded argillaceous siltstone. Conchostracans *Palaeolimnadia xuanweiensis*, *Palaeolimnadia pusilla*, *Palaeolimnadia globosa* and *Euestheria gutta* were identified. 60 cm

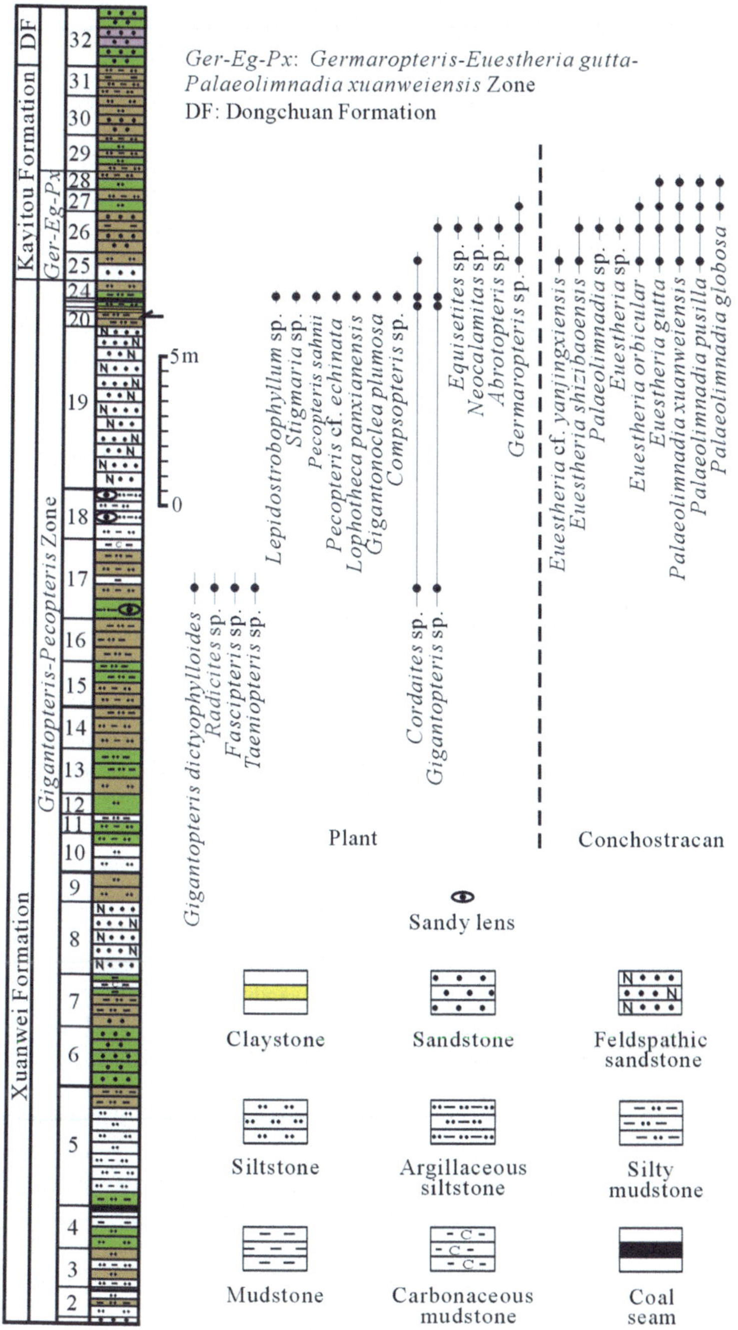

Fig. 3.20 Stratigraphic column and fossil occurrences at the Jiucaichong section (fossils referred to Chu et al. 2016)

27. Grayish green medium-bedded siltstone interbedded with yellow thin-bedded argillaceous siltstone. The plant fossil *Germaropteris* sp. and conchostracans *Palaeolimnadia xuanweiensis, Palaeolimnadia pusilla, Palaeolimnadia globosa, Euestheria gutta* and *Euestheria orbicula* were examined. 75 cm
26. Yellow thin-bedded fine-grained sandstone intercalated with yellow band-shaped mudstone. Plant fossils *Equisetites* sp., *Neocalamites* sp., *Abrotopteris* sp., *Germaropteris* sp., *Gigantopteris* sp. and conchostracans *Palaeolimnadia xuanweiensis, Palaeolimnadia pusilla, Palaeolimnadia* spp., *Euestheria gutta, Euestheria orbicular, Euestheria shizibaoensis* and *Euestheria* spp. were discovered. 140 cm
25. Yellow thick- to medium-bedded silty mudstone in the upper part; dark gray thick-bedded fine-grained sandstone interbedded with siltstone in the lower part. The plant fossils *Germaropteris* sp. and *Cordaites* sp., and conchostracans *Euestheria gutta, Euestheria orbicular, Euestheria* cf. *yanjingxiensis, Euestheria shizibaoensis, Palaeolimnadia xuanweiensis* and *Palaeolimnadia pusilla* were found. 90 cm

Conformity
Xuanwei Formation in the Upper Permian

24. Dark yellow siltstone in the upper part; grayish green silty mudstone in the lower part. Plant fossils *Lepidostrobophyllum* sp., *Stigmaria* sp., *Pecopteris sahnii, Pecopteris* cf. *echinata, Lophotheca panxianensis, Gigantopteris* sp., *Gigantonoclea plumosa, Cordaites* sp. and *Compsopteris* sp. were collected. 60 cm
23. Black coal seams. 7 cm
22. Yellowish green (red when weathered) thick- to medium-bedded argillaceous siltstone, with plant fossils *Cordaites* sp. and *Gigantopteris* sp.. 20 cm
21. Pale claystone intercalated with coal seams. 8 cm
20. Dark yellow (brown when weathered) thick-bedded silty mudstone. 50 cm
19. Gray thick-bedded fine-grained feldspathic sandstone. 540 cm
18. Gray thick-bedded argillaceous siltstone with sandy lens. 170 cm
17. Black carbonaceous mudstone at the top part (5 cm); yellow (brown when weathered) thick- to medium-bedded silty mudstone in the upper part (50 cm); yellow thick- to medium-bedded argillaceous siltstone intercalated with a layer of pale (redish brown when weathered) mudstone in the middle part (150 cm); yellowish green silty mudstone with sandy lens in the lower part (60 cm). Plant fossils *Gigantopteris dictyophylloides, Radicites* sp., *Fascipteris* sp., *Taeniopteris* sp., *Cordaites* sp. and *Gigantopteris* sp. were identified. 265 cm
16. Dark yellow thick-bedded silty mudstone. 140 cm
15. Yellowish green silty mudstone in the upper part; yellow argillaceous siltstone in the lower part. 150 cm

14. Yellow thin-bedded silty mudstone in the upper part; yellow thick-bedded argillaceous siltstone in the lower part. 140 cm
13. Yellowish green thick-bedded silty mudstone in the upper part; yellow thick-bedded siltstone in the lower part. 150 cm
12. Grayish green (brown when weathered) thick-bedded siltstone. 70 cm
11. Gray thin-bedded silty mudstone in the upper part; yellowish green thick-bedded argillaceous siltstone in the lower part. 65 cm
10. Yellowish green thick-bedded argillaceous siltstone in the upper part; grayish brown thick-bedded siltstone in the lower part. 130 cm
9. Yellow thick-bcdded siltstone. 100 cm
8. Gray (brown when weathered) thick-bedded fine-grained feldspathic sandstone. 240 cm
7. Yellow thick-bedded silty mudstone, with grayish green thin-bedded mudstone intercalated with grayish black thin-bedded carbonaceous mudstone at the top, and yellow thick-bedded fine-grained sandstone at the base. 170 cm
6. Yellowish green (brown when weathered) thick-bedded fine-grained sandstone. 200 cm
5. Yellow thin-bedded silty mudstone in the upper part; gray thin-bedded siltstone in the middle part; dark gray thick-bedded argillaceous siltstone in the lower part; yellowish green thick-bedded silty mudstone at the basal part. 400 cm
4. Black coal seams at the top part; pale thick- to medium-bedded mudstone in the upper part; yellowish green thick-bedded siltstone in the lower part. 140 cm
3. Two cycles were deposited in descending order: yellow thick-bedded siltstone (35 cm); pale to brownish yellow thin-bedded argillaceous siltstone (5 cm); yellow thick-bedded siltstone (80 cm); pale to brownish yellow thin-bedded argillaceous siltstone (5 cm). 125 cm
2. Dark grayish green thick-bedded siltstone intercalated with yellow thick- to medium-bedded silty mudstone, with black coal seams at the top (4 cm). 100 cm
1. Gray to pale fine-grained sandstone with dark gray to yellow mudstone at the top. 5 cm

Unexposed downwards

Note: Listed fossils are referred to Chu et al. (2016) (but updated in taxonomy in this book).

3.9 Mide Section in Xuanwei City

The Mide section is located about 3 km east to the Mide Village, Haidai Town, Xuanwei City, Yunnan Province, South China (Fig. 3.21).

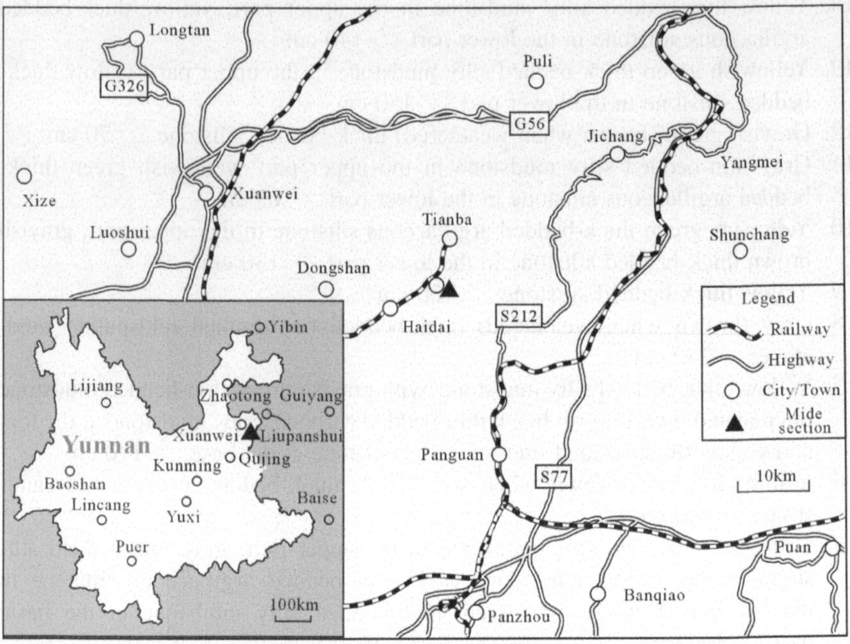

Fig. 3.21 Geographic location of the Mide section

The sequence of the Mide section is composed of the Xuanwei Formation and Kayitou Formation in an ascending order (Fig. 3.22). The Xuanwei Formation is dominated by yellowish green to yellow argillaceous siltstones and silty mudstones intercalated with claystones. The Kayitou Formation is mainly composed of yellowish green to grayish brown argillaceous siltstones and mudstones. Additionally, the Dongchuan Formation dominated by purplish red sandstones is outcropped in the area and overlies the Kayitou Formation. A few bivalves, ostracods, brachiopods and ammonoids were discovered in the Kayitou Formation and abundant plant fossils and spores and pollens were yielded in the Xuanwei Formation and the lowermost part of the Kayitou Formation (Figs. 3.22, 3.23).

Kayitou Formation in the Uppermost Changhsingian to to the Lower Triassic (Induan)
Unexposed at the top

29. Five cycles were deposited and each cycle includes: grayish yellow silty mudstone in the upper part; yellowish green medium-bedded fine-grained sandstone in the lower part.
28. Grayish green medium-bedded argillaceous siltstone interbedded with silty mudstone. Bivalves *Towapteria* sp., *T. scythica*, *Bakevellia* sp., *Pteria* sp., *P. ussurica variabilis*, *Unionites canalensis*, *U. fassaensis* and *U.* sp., the

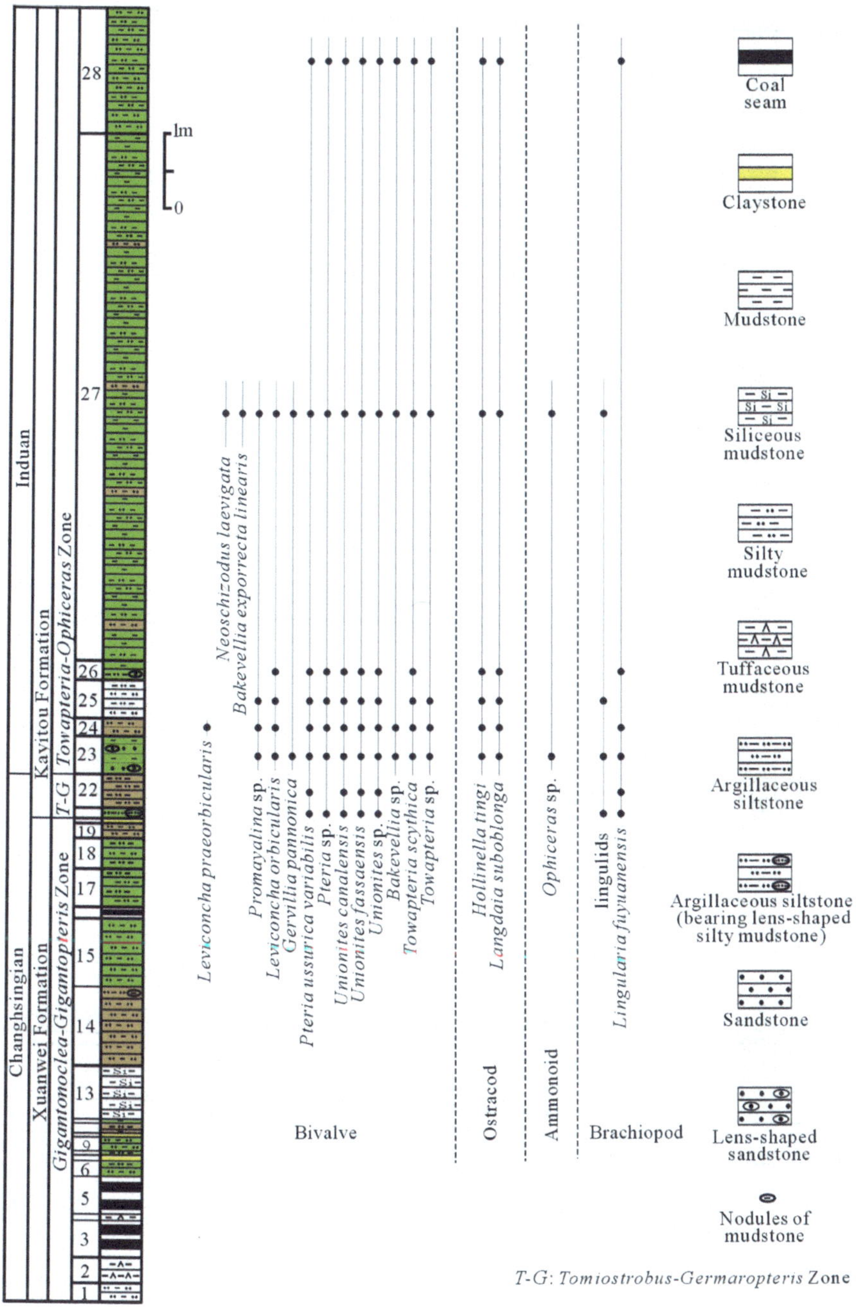

Fig. 3.22 Stratigraphic column and fossil occurrences (bivalve, ostracod, ammonoid and brachiopod) at the Mide section

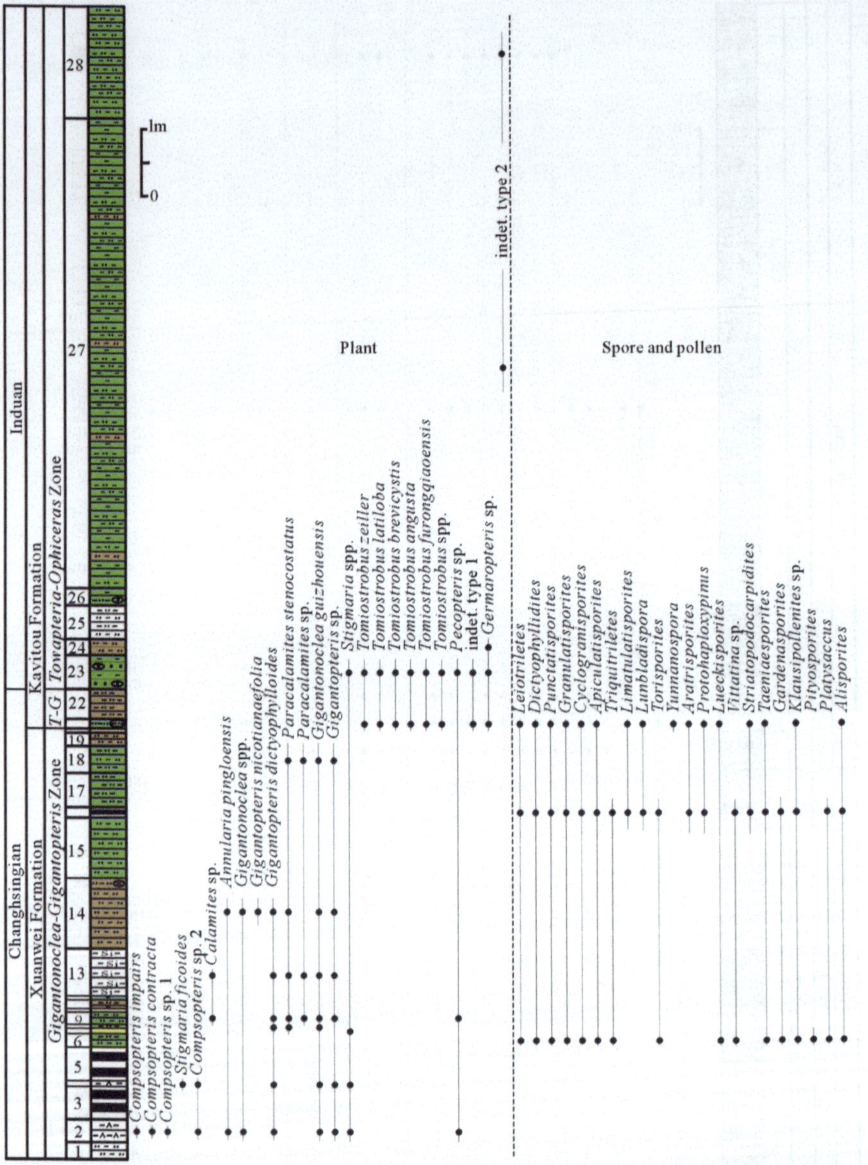

Fig. 3.23 Stratigraphic column and fossil occurrences (plant, spore and pollen) at the Mide section. *Note* spore and pollen referred to Yu et al. (2022); legends same to Fig. 3.22

brachiopod *Lingularia fuyuanensis*, Ostracods *Hollinella tingi* and *Langdaia suboblonga*, and the plant fossil indet. type 2. were found. 200 cm

27. Grayish green mudstone interbedded with silty mudstone, intercalated with four layers of grayish black argillaceous siltstone, with bivalves *Neoschizodus laevigata*, *Bakevellia* sp., *B. exporrecta linearis*, *Promayalina* sp., *Towapteria* sp., *Towapteria scythica*, *Pteria ussurica variabilis*, *Pteria* spp., *Unionites canalensis*, *Unionites fassaensis*, *Unionites* sp., *Leviconcha orbicularis* and *Gervillia pannonica*, the brachiopod lingulids, the ammonoid *Ophiceras* sp., ostracods *Hollinella tingi* and *Langdaia suboblonga*, and the plant fossil indet. type 2. 700 cm
26. Grayish green mudstone interbedded with silty mudstone, intercalated with four layers of grayish black lens-shaped sandstone. Bivalves *Towapteria scythica*, *Pteria ussurica variabilis*, *Pteria* sp., *Unionites canalensis*, *Unionites fassaensis*, *Unionites* sp. and *Leviconcha orbicularis*, the brachiopod *Lingularia fuyuanensis*, and ostracods *Hollinella tingi* and *Langdaia suboblonga* were identified. 24 cm
25. Grayish brown to black argillaceous siltstone interbedded with silty mudstone. Bivalves *Promayalina* sp., *Towapteria scythica*, *Towapteria* sp., *Pteria ussurica variabilis*, *Pteria* spp., *Unionites canalensis*, *Unionites fassaensis*, *Unionites* sp. and *Leviconcha orbicularis*, the brachiopod lingulids, and ostracods *Hollinella tingi* and *Langdaia suboblonga* were collected. 50 cm
24. Grayish brown to yellow medium-bedded argillaceous siltstone, with bivalves *Bakevellia* sp., *Promayalina* sp., *Towapteria scythica*, *Towapteria* sp., *Pteria ussurica variabilis*, *Pteria* spp., *Unionites canalensis*, *Unionites fassaensis*, *Unionites* sp., *Leviconcha orbicularis* and *Leviconcha praeorbicularis*, the brachiopod *Lingularia fuyuanensis*, ostracods *Hollinella tingi* and *Langdaia suboblonga*, and the plant fossil *Germaropteris* sp.. 23 cm
23. Yellow to grayish green mudstone interbedded with pale to yellow lens-shaped sandstone. Bivalves *Bakevellia* sp., *Promayalina* sp., *Towapteria scythica*, *Towapteria* sp., *Pteria ussurica variabilis*, *Pteria* spp., *Unionites canalensis*, *Unionites fassaensis*, *Unionites* sp., *Leviconcha orbicularis* and *Gervillia pannonica*, brachiopods *Lingularia fuyuanensis* and lingulids, the ammonoid *Ophiceras* sp., ostracods *Hollinella tingi* and *Langdaia suboblonga*, and plant fossils *Stigmaria* spp., *Tomiostrobus zeiller*, *Tomiostrobus latiloba*, *Tomiostrobus brevicystis*, *Tomiostrobus angusta*, *Tomiostrobus furongqiaoensis*, *Tomiostrobus* spp., *Pecopteris* sp., indet. type 1, *Germaropteris* sp. were found. 50 cm
22. Yellow thick-bedded silty mudstone, with abundant plant fragments. The brachiopod *Lingularia fuyuanensis*, and bivalves *Pteria ussurica variabilis*, *Unionites canalensis*, *Unionites fassaensis* and *Unionites* sp. were recovered. 45 cm
21. Yellow to grayish green argillaceous siltstone interbedded with lens-shaped silty mudstone. Bivalves *Pteria ussurica variabilis*, *Pteria* sp., *Unionites canalensis*, *Unionites fassaensis* and *Unionites* sp., brachiopods *Lingularia fuyuanensis* and lingulids, plant fossils *Tomiostrobus zeiller*, *Tomiostrobus*

latiloba, *Tomiostrobus brevicystis*, *Tomiostrobus angusta*, *Tomiostrobus furongqiaoensis*, *Tomiostrobus* spp., *Germaropteris* sp., indet. type 1, and spores and pollens *Leiotriletes*, *Dictyophyllidites*, *Punctatisporites*, *Granulatisporites*, *Cyclogranisporites*, *Apiculatisporites*, *Limatulatisporites*, *Lunbladispora*, *Yunnanospora*, *Aratrisporites*, *Protohaploxypinus*, *Lueckisporites*, *Striatopodocarpidites*, *Taeniaesporites*, *Klausipollenites* sp. and *Alisporites* were examined. 16 cm

Conformity
Xuanwei Formation in the Upper Permian (Changhsingian)

20. Pale to grayish yellow claystone. 4 cm
19. Yellow argillaceous siltstone. 20 cm
18. Grayish green to yellow medium-bedded silty mudstone interbedded with argillaceous siltstone, with plant fossils *Paracalamites stenocostatus*, *Paracalamites* sp., *Gigantonoclea guizhouensis* and *Gigantopteris* sp.. 40 cm
17. Grayish green medium-bedded silty mudstone, with abundant plant fragments. 50 cm
16. Black coal seams, with abundant spores and pollens *Leiotriletes*, *Dictyophyllidites*, *Punctatisporites*, *Granulatisporites*, *Cyclogranisporites*, *Apiculatisporites*, *Triquitriletes*, *Limatulatisporites*, *Lunbladispora*, *Torisporites*, *Aratrisporites*, *Protohaploxypinus*, *Vittatina* sp., *Striatopodocarpidites*, *Taeniaesporites*, *Gardenasporites*, *Klausipollenites* sp., *Platysaccus* and *Alisporites.* 15 cm
15. Grayish green argillaceous siltstone, with plant fossils. 90 cm
14. Yellow argillaceous siltstone intercalated with nodules of mudstone at the top part; pale argillaceous siltstone in the upper part (60 cm); light yellow argillaceous siltstone in the middle part (20 cm); argillaceous siltstone in the lower part (25 cm). Plant fossils *Annularia pingloensis*, *Gigantonoclea* spp., *Gigantopteris nicotianaefolia*, *Gigantopteris dictyophylloides*, *Paracalamites stenocostatus*, *Gigantonoclea guizhouensis* and *Gigantopteris* sp. were collected. 105 cm
13. Grayish black thick- to medium-bedded siliceous mudstone with abundant plant fossils *Calamites* sp., *Gigantopteris dictyophylloides*, *Paracalamites stenocostatus*, *Paracalamites* sp., *Gigantonoclea guizhouensis* and *Gigantopteris* sp.. 70 cm
12. Grayish green medium-bedded mudstone. 5 cm
11. Yellow silty mudstone intercalated with light yellow mudstone. 15 cm
10. Grayish yellow to pale claystone. 3 cm
9. Grayish green argillaceous siltstone with abundant plant fossils *Calamites* sp., *Gigantopteris dictyophylloides*, *Paracalamites stenocostatus*, *Paracalamites* sp., *Gigantonoclea guizhouensis*, *Gigantopteris* sp. and *Pecopteris* sp.. 18 cm

8. Grayish green to pale silty mudstone with abundant plant fossils *Gigantopteris dictyophylloides*, *Paracalamites stenocostatus* and *Gigantonoclea guizhouensis*. 5 cm
7. Grayish yellow to pale claystone, with the plant fossil *Stigmaria* sp.. 8 cm
6. Grayish green silty mudstone interbedded with argillaceous siltstone, with abundant plant fossils and spores and pollens *Leiotriletes*, *Dictyophyllidites*, *Punctatisporites*, *Granulatisporites*, *Cyclogranisporites*, *Apiculatisporites*, *Triquitriletes*, *Torisporites*, *Lueckisporites*, *Vittatina* sp., *Taeniaesporites*, *Gardenasporites*, *Klausipollenites* sp., *Pityosporites*, *Platysaccus* and *Alisporites*. 22 cm
5. Black coal seams. 50 cm
4. Pale tuffaceous mudstone with plant fossils *Stigmaria ficoides*, *Compsopteris* sp. 2, *Gigantopteris dictyophylloides*, *Gigantonoclea guizhouensis*, *Gigantopteris* sp. and *Stigmaria* spp. 8 cm
3. Black coal seams. 50 cm
2. Pale tuffaceous mudstone with plant fossils *Compsopteris impairs*, *Compsopteris contracta*, *Compsopteris* sp. 1, *Compsopteris* sp. 2, *Annularia pingloensis*, *Gigantonoclea* spp., *Gigantopteris dictyophylloides*, *Gigantonoclea guizhouensis*, *Gigantopteris* sp., *Pecopteris* sp., *Stigmaria* sp. 35 cm
1. Gray to grayish black argillaceous siltstone, with abundant plant fossils.

Unexposed at the Bottom

Note: Listed spore and pollen is after Yu et al. (2022).

3.10 Wadu Section in Weining County

The Wadu section is located nearby the Wadu Village, about 60 km southwest to the Liupanshui City, Guizhou Province, South China (Fig. 3.24).

The sequence of the Wadu section is composed of the Xuanwei Formation and the lower part of Feixianguan Formation in an ascending order (Fig. 3.25). The Xuanwei Formation is dominated by yellowish green silitstones intercalated with yellow mudstones and black coal seams. The lower part of Feixianguan Formation is mainly composed of bluish gray to purplish red siltstones and argillaceous siltstones. Abundant plant fossils were found in the Xuanwei Formation and a large amount of bivalves were discovered in the basal part of Feixianguan Formation (Fig. 3.25).

Feixianguan Formation in the Lower Triassic (Induan)
Unexposed upwards

25. Purplish red siltstone and argillaceous siltstone, intercalated with bluish gray argillaceous siltstone in the upper part; purplish red siltstone (yellow when

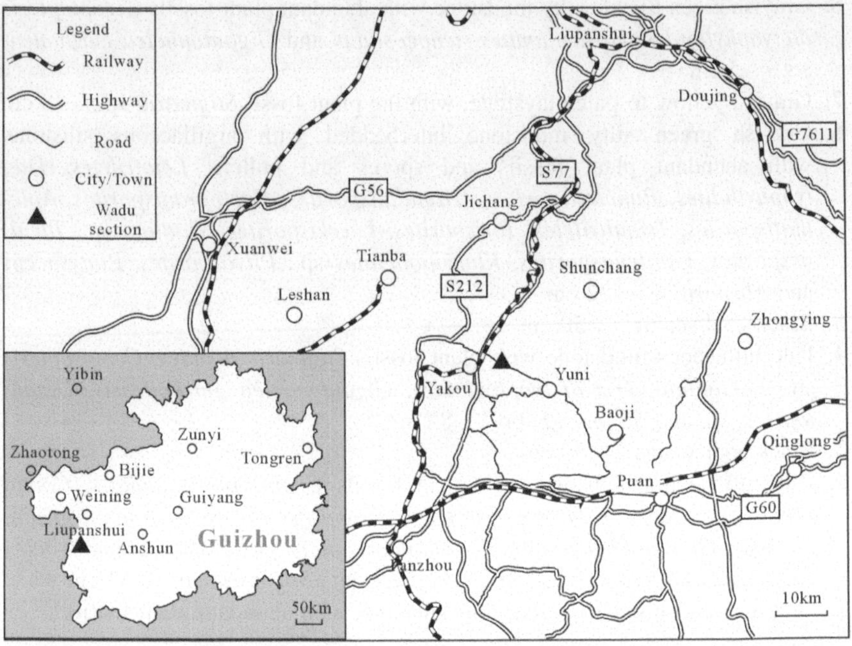

Fig. 3.24 Geographic location of the Wadu section

weathered) in the middle part; bluish gray fine-grained sandstone and siltstone (purplish red when weathered) in the lower part.

24. Purplish red argillaceous siltstone intercalated with bluish gray lens of sandstone. 65 cm
23. Bluish gray to gray fine-grained sandstone, siltstone and argillaceous siltstone. 860 cm
22. Bluish gray siltstone intercalated with purplish red argillaceous siltstone, with bluish gray lens of sandstone. The bivalve *Pteria ussurica variabilis* was identified. 910 cm
21. Bluish gray siltstone interbedded with fine-grained sandstone, intercalated with purplish red argillaceous siltstone. Bivalves *Pteria ussurica variabilis*, *Promyalina schamarae*, *Promyalina putiatinensis* and *Neoschizodus orbicularis* were discovered. 850 cm
20. Bluish gray medium-bedded siltstone interbedded with argillaceous siltstone (purplish red when weathered). Bivalves *Pteria ussurica variabilis*, *Astartella* sp. and *Unionites fassaensis* were collected. 960 cm
19. Grayish green medium-bedded siltstone interbedded with argillaceous siltstone, with grayish green siltstone (50 cm) at the base. Bivalves *Pteria ussurica variabilis*, *Promyalina schamarae*, *Promyalina putiatinensis*, *Neoschizodus orbicularis*, *Astartella* sp., *Unionites fassaensis*, *Claraia*

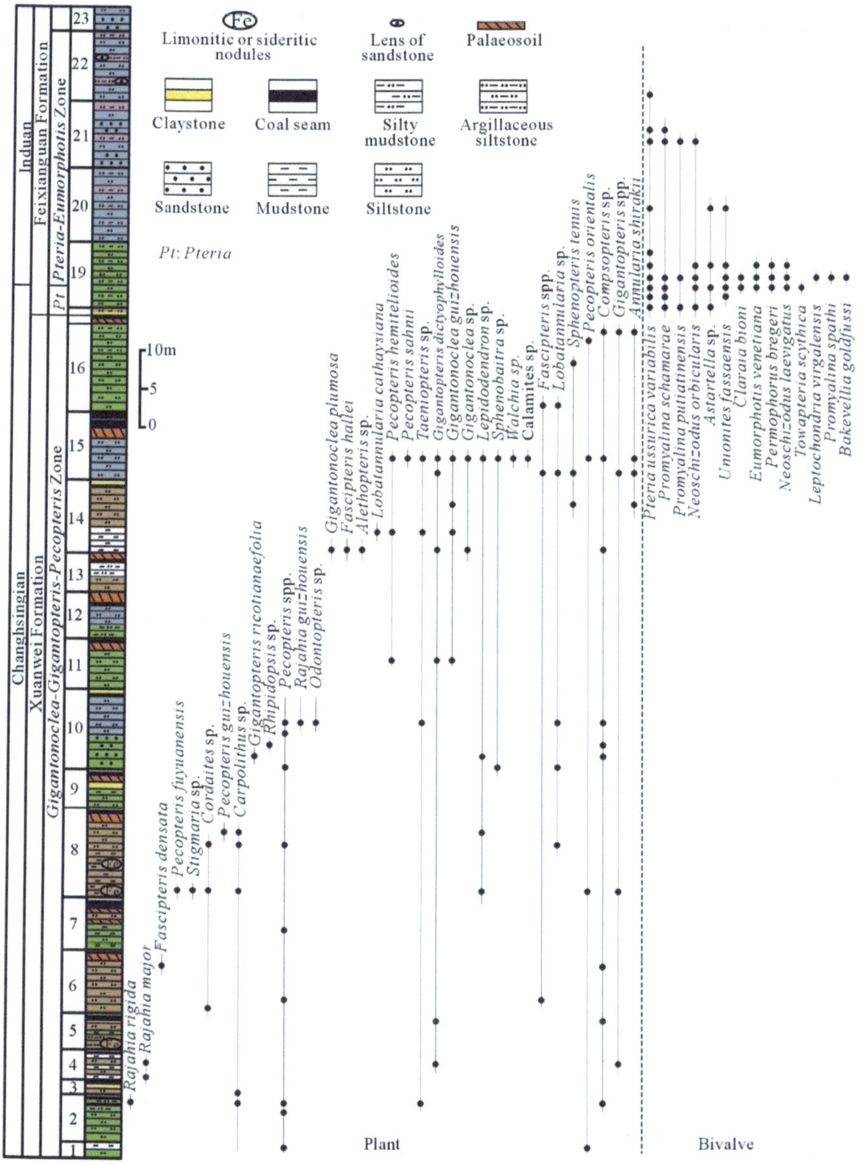

Fig. 3.25 Stratigraphic column and fossil occurrences at the Wadu section (fossils referred to Chu et al. 2016)

bioni, *Eumorphotis venetiana*, *Permophorus bregeri*, *Neoschizodus laevigatus*, *Towapteria scythica*, *Leptochondria virgalensis*, *Promyalina spathi* and *Bakevellia goldfussi* were found. 850 cm

18. Yellow argillaceous siltstone, with claystone at the base. 105 cm

Conformity
Xuanwei Formation in the Upper Permian (Changhsingian)

17. Black thick-bedded coal, with a thin layer of palaeosoil at the base. 110 cm
16. Yellow to grayish green thick- to medium-bedded siltstone and argillaceous siltstone in the upper part; yellowish green thick- to medium-bedded siltstone in the lower part. The plant fossils *Fascipteris* spp., *Lobatannularia* sp., *Sphenopteris tenuis*, *Pecopteris orientalis*, *Compsopteris* sp., *Gigantopteris* spp. and *Annularia shirakii* were identified. 1140 cm
15. Black thick-bedded coal (220 cm thick) overlying a layer (30 cm thick) of palaeosoil in the upper part; bluish gray thick-bedded siltstone in the lower part. Plant fossils *Pecopteris hemitelioides*, *Pecopteris sahnii*, *Taeniopteris* sp., *Gigantopteris dictyophylloides*, *Gigantonoclea guizhouensis*, *Gigantonoclea* sp., *Lepidodendron* sp., *Sphenobaitra* sp., *Walchia* sp., *Calamites* sp., *Fascipteris* spp., *Lobatannularia* sp., *Sphenopteris tenuis*, *Pecopteris orientalis*, *Compsopteris* sp., *Gigantopteris* spp. and *Annularia shirakii* were examined. 885 cm
14. Pale claystone overlying black coal bed at the top part; yellow thick- to medium-bedded siltstone in the upper part; gray thick-bedded mudstone in the lower part. Plant fossils *Gigantonoclea plumosa*, *Fascipteris hallei*, *Alethopteris* sp., *Lobatannularia cathaysiana*, *Pecopteris hemitelioides*, *Taeniopteris* sp., *Gigantopteris dictyophylloides*, *Gigantonoclea guizhouensis*, *Gigantonoclea* sp., *Sphenopteris tenuis*, *Compsopteris* sp. and *Annularia shirakii* were found. 940 cm
13. A layer of palaeosoil (30 cm thick) overlying a coal bed (20 cm thick) at the top part; thin-bedded silty mudstone in the upper part; yellow siltstone in the lower part. 500 cm
12. A layer of palaeosoil (50 cm thick) overlying a coal bed (30 cm thick) at the top part; bluish gray thick-bedded siltstone in the upper part; grayish green thin-bedded silty mudstone in the lower part. Abundant bivalves and brachiopods were yielded at the top part. 600 cm
11. Yellowish green thick-bedded siltstone, with a layer of coal (40 cm) overlying the palaeosoil (20 cm) in the uppermost. Plant fossils *Pecopteris hemitelioides*, *Gigantopteris dictyophylloides* and *Gigantonoclea guizhouensis* were collected. 660 cm
10. Pale claystone and black coal in the uppermost; bluish gray thick- to medium-bedded siltstone in the upper part; grayish green thick-bedded sandstone in the lower part. Plant fossils *Gigantopteris nicotianaefolia*, *Rhipidopsis* sp.,

Pecopteris spp., *Rajahia guizhouensis*, *Odontopteris* sp., *Taeniopteris* sp., *Lepidodendron* sp., *Sphenobaitra* sp., *Lobatannularia* sp. and *Compsopteris* sp. were discovered. 1040 cm

9. Black coal bed in the upper part (100 cm); a layer (15 cm) of palaeosoil and pale claystone in the middle part; grayish green to yellow mudstone and siltstone in the lower part, with nodules of mudstone. 495 cm
8. Black coal bed (30 cm) overlying a layer of palaeosoil (10 cm) at the top part; yellow thick- to medium-bedded siltstone in the upper part; yellow to grayish brown mudstone with sideritic nodules in the lower part. Plant fossils *Pecopteris fuyuanensis*, *Stigmaria* sp., *Cordaites* sp., *Pecopteris guizhouensis*, *Carpolithus* sp., *Pecopteris* spp., *Lepidodendron* sp., *Lobatannularia* sp., *Pecopteris orientalis* and *Gigantopteris* spp. were examined. 1160 cm
7. Black thick coal bed in the upper part (160 cm); palaeosoil intercalated with yellow siltstone in the middle part (40 cm); grayish green thick-bedded mudstone intercalated with thin-bedded siltstone in the lower part. The plant fossil *Pecopteris* spp. and abundant brachiopods and bivalves were found in the lower part. 675 cm
6. Yellow thick-bedded siltstone, with black coal seams (20 cm) overlying a layer of grayish brown palaeosoil (25 cm) at the top part. Plant fossils *Fascipteris densata*, *Cordaites* sp., *Pecopteris* spp., *Fascipteris* spp. and *Compsopteris* sp. were examined. 825 cm
5. Black coal bed in the upper part (120 cm); grayish brown claystone (7 cm), yellow thick-bedded siltstone (125 cm), grayish green mudstone (10 cm) and yellow thick-bedded argillaceous siltstone (80 cm) in a descending order in the middle part; grayish brown silty mudstone intercalated with yellow silty mudstone with sideritic nodules, in the lower part (200 cm). Plant fossils *Gigantopteris dictyophylloides* and *Compsopteris* sp. were identified. 475 cm
4. Black coal bed (50 cm) overlying grayish brown mudstone (20 cm) in the upper part; yellow siltstone intercalated with grayish brown mudstone in the middle part (215 cm); grayish brown mudstone in the lower part. Plant fossils *Rajahia major*, *Gigantopteris dictyophylloides* and *Gigantopteris* spp. were found. 390 cm
3. Black coal bed (78 cm) overly grayish brown claystone (5 cm) in the upper part; yellow thick-bedded siltstone in the lower part, with the plant fossil *Carpolithus* sp.. 190 cm
2. Grayish green silty mudstone and black carbonaceous mudstone or coal seams in the upper part; grayish green thick-bedded siltstone in the lower part. Plant fossils *Rajahia rigida*, *Carpolithus* sp., *Pecopteris* spp., *Taeniopteris* sp., *Pecopteris orientalis* and *Compsopteris* sp. were collected. 610 cm
1. Grayish purple mudstone in the upper part; yellowish green thick- to medium-bedded siltstone intercalated with grayish green siltstone in the lower part. Plant fossils *Pecopteris* spp. and *Pecopteris orientalis* were examined.

Unexposed at the base

Note: Listed fossils are based on Chu et al. (2016), but updated in taxonomy.

3.11 Tucheng Section in Panzhou City

The Tucheng section is located at the Qingshui Village, about 40 km northeast to the Panzhou City, Guizhou Province, South China (Fig. 3.26).

The sequence of the Tucheng section is composed of the Xuanwei Formation and Kayitou Formation in an ascending order (Fig. 3.27). The Xuanwei Formation is dominated by yellow siltstones and mudstones intercalated with black coal seams. The Kayitou Formation is mainly composed of yellow to grayish green siltstones. Abundant plant fossils were collected in the section (Fig. 3.27). Additionally, the Dongchuan Formation dominated by purplish red sandstones is outcropped in the area and overlies the Kayitou Formation.

Kayitou Formation in the uppermost Permian to the Lower Triassic
Unexposed upwards

23. Yellowish green silty mudstone, yielding abundant bivalves, brachiopods and ostracods in the upper part; grayish green to brownish red thin-bedded siltstone with plant fossils in the lower part. Plant fossils include *Sphenopteris* sp., *Pecopteris* sp., *Germaropteris martinsii*, *Germaropteris lobutalum*, *Germaropteris* sp., *Gigantonoclea guizhouensis*, *Gigantopteris dictyophylloides*, *Gigantopteris* sp., *Gigantonoclea* sp. and *Sphenopteris tenuis*.

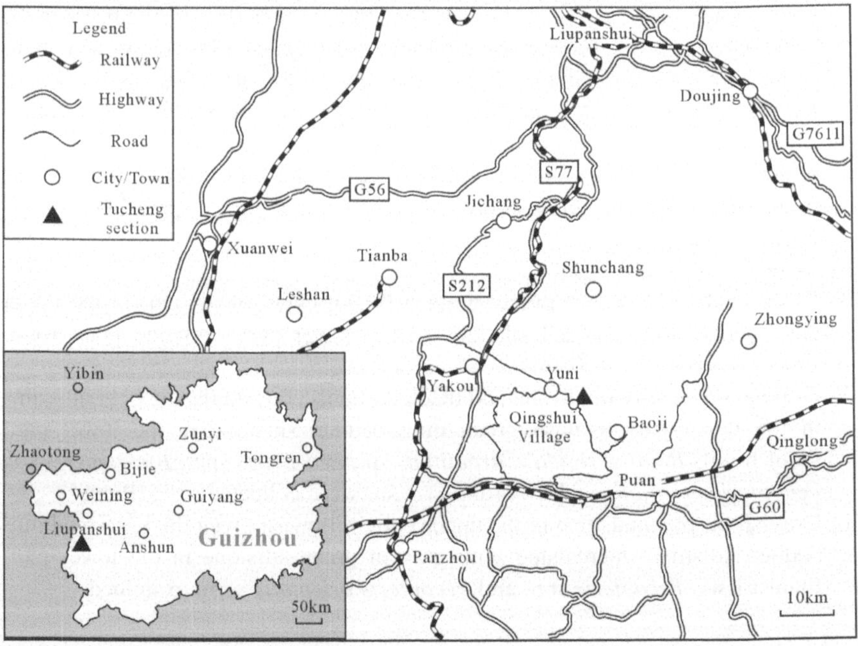

Fig. 3.26 Geographic location of the Tucheng section

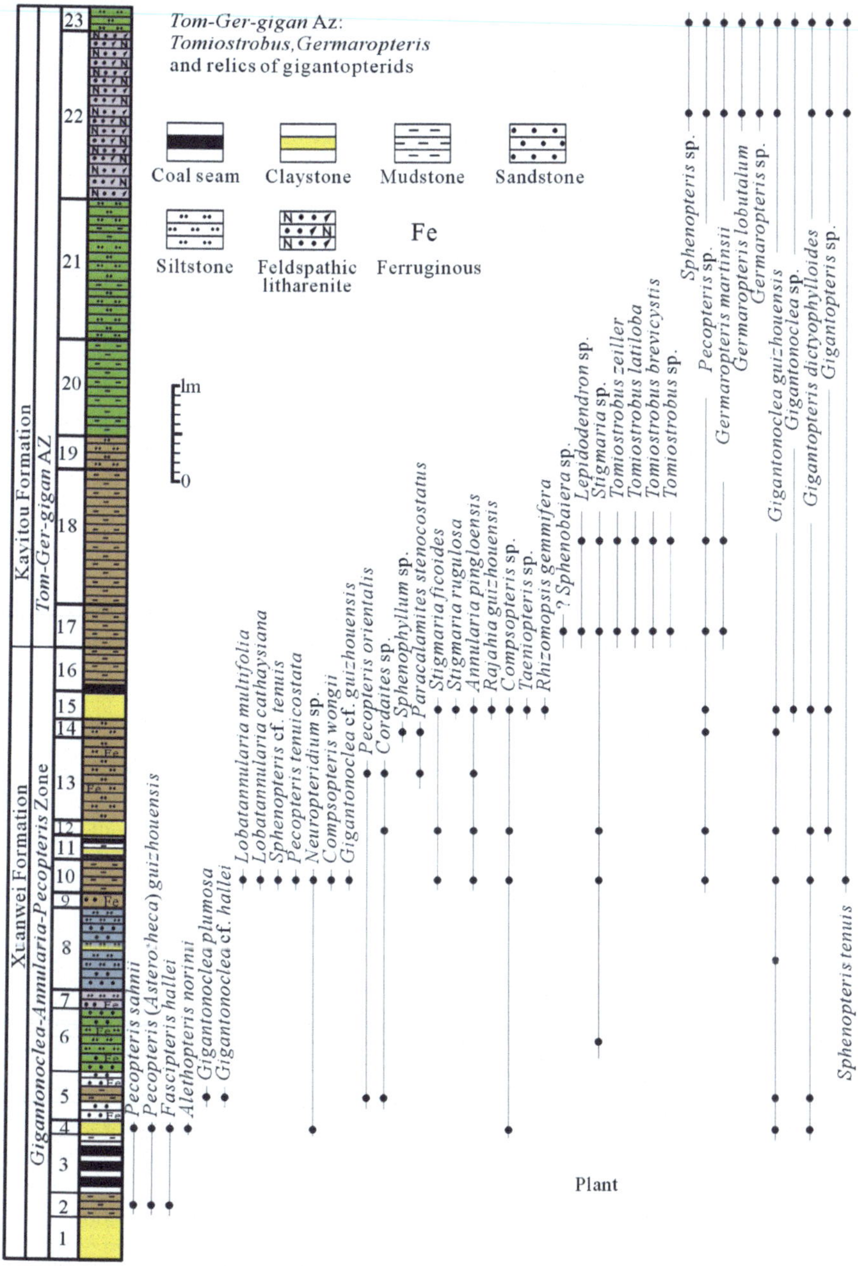

Fig. 3.27 Stratigraphic column and occurrences of plant fossils at the Tucheng section (plant fossils referred to Yu et al. 2022)

22. Grayish red to brownish red feldspathic litharenite, with plant fossils *Sphenopteris* sp., *Pecopteris* sp., *Germaropteris martinsii*, *Germaropteris lobutalum*, *Germaropteris* sp., *Gigantonoclea guizhouensis*, *Gigantopteris dictyophylloides*, *Gigantopteris* sp. and *Sphenopteris tenuis*. 175 cm
21. Grayish green to yellowish green thin-bedded siltstone intercalated with mudstone, with fragments of plant fossils. 145 cm
20. Yellow to grayish green thin-bedded mudstone. 100 cm
19. Yellow thin-bedded siltstone with laminations, with abundant bivalves and fragments of plant fossils at the top part. 35 cm
18. Light yellow mudstone with laminations. Bivalves and brachiopods were found. Plant fossils *Lepidodendron* sp., *Stigmaria* sp., *Tomiostrobus zeiller*, *Tomiostrobus latiloba*, *Tomiostrobus brevicystis*, *Tomiostrobus* sp., *Pecopteris* sp. and *Germaropteris martinsii* were identified. 140 cm
17. Light yellow mudstone with laminations, with marine bivalves and plant fossils ?*Sphenobaiera* sp., *Lepidodendron* sp., *Stigmaria* sp., *Tomiostrobus zeiller*, *Tomiostrobus latiloba*, *Tomiostrobus brevicystis*, *Tomiostrobus* sp., *Pecopteris* sp. and *Germaropteris martinsii*. 45 cm

Conformity
Xuanwei Formation in the Upper Permian

16. Brown to yellow mudstone overlying black coal seams. 45 cm
15. Yellow to grayish brown claystone, with plant fossils *Stigmaria ficoides*, *Stigmaria rugulosta*, *Annularia pingloensis*, *Rajahia guizhouensis*, *Compsopteris* sp., *Taeniopteris* sp., *Rhizomopsis gemmofera*, *Pecopteris* sp., *Gigantonoclea guizhuensis*, *Gigantopteris dictyophylloides*, *Gigantonoclea* sp. and *Gigantopteris* sp.. 30 cm
14. Yellow to grayish red siltstone, with plant fossils *Sphenophyllum* sp., *Paracalamites stenocostatus*, *Pecopteris* sp. and *Gigantonoclea guizhouensis*. 20 cm
13. Yellow medium- to thin-bedded siltstone intercalated with a layer (6 cm) of ferruginous siltstone, with plant fossils *Paracalamites stenocostatus*, *Annularia pingloensis*, *Pecopteris orientalis* and *Cordaites* sp.. 85 cm
12. Pale to grayish red claystone, with plant fossils *Cordaites* sp., *Stigmaria ficoides*, *Annularia pingloensis*, *Compsopteris* sp., *Stigmaria* sp., *Pecopteris* sp., *Gigantonoclea guizhuensis*, *Gigantopteris dictyophylloides* and *Gigantopteris* sp.. 15 cm
11. Black coal seams intercalated with brown mudstone and pale claystone. 25 cm
10. Yellow to gray mudstone, with plant fossils *Lobatannularia multifolia*, *Lobatannularia cathaysiana*, *Sphenopteris* cf. *tenuis*, *Pecopteris tenuicostata*, *Neuropteridium* sp., *Compsopteris wongii*, *Gigantonoclea* cf. *guizhouensis*, *Stigmaria ficoides*, *Annularia pingloensis*, *Compsopteris* sp., *Stigmaria* sp.,

Pecopteris sp., *Gigantonoclea guizhuensis*, *Gigantopteris dictyophylloides* and *Sphenopteris tenuis*. 35 cm
9. Yellow fine-grained ferruginous sandstone intercalated with siltstone. 15 cm
8. Gray to bluish gray thin-bedded fine-grained sandstone and siltstone, intercalated with pale claystone, with the plant fossil *Gigantonoclea guizhuensis*. 85 cm
7. Grayish red to brown fine-grained ferruginous sandstone intercalated with pale siltstone. 20 cm
6. Grayish green thin-bedded ferruginous fine-grained sandstone and siltstone, with the plant fossil *Stigmaria* sp.. 65 cm
5. Brown ferruginous fine-grained sandstone intercalated with yellow to grayish red thin-bedded mudstone, with plant fossils *Gigantonoclea plumosa*, *Gigantonoclea* cf. *hallei*, *Pecopteris orientalis*, *Cordaites* sp., *Gigantonoclea guizhouensis* and *Gigantopteris dictyophylloides*. 50 cm
4. Yellow to pale claystone, with plant fossils *Pecopteris sahnii*, *Asterotheca guizhouensis*, *Fascipteris hallei*, *Alethopteris norinii*, *Neuropteridium* sp., *Compsopteris* sp., *Gigantonoclea guizhouensis* and *Gigantopteris dictyophylloides*. 15 cm
3. Black coal seams with a layer (5 cm) of purplish brown mudstone at the top part. 62 cm
2. Yellow to pale mudstone with plant fossils *Pecopteris sahnii*, *Asterotheca guizhouensis* and *Fascipteris hallei*. 25 cm
1. Pale to light gray claystone. 43 cm

Unexposed downwards

Note: Plant fossils are referred to Yu et al. (2022) (updated in taxonomy).

3.12 Zhongzhai Section Nearby Zhongzhai Village, Liuzhi County

The Zhongzhai section (26° 09′ 00″ N, 105° 17′ 24″ E) is situated about 1 km northeast of Zhongzhai Village, Liuzhi County, southwestern Guizhou Province, South China (Fig. 3.28).

The sequence of the Zhongzhai section is composed of the Lungtan Formation and the conformably overlying Feixianguan Formation (Fig. 3.29). The Lungtan Formation is dominated by grayish yellow siltstones and grayish green mudstones, intercalated with gray limestones. The Feixianguan Formation is characterized by grayish green, yellowish green and purplish red silt-bearing calcareous mudstones and calcareous siltstones, intercalated with calcareous mudstones. Abundant brachiopods in the Lungtan Formation, a few conodonts in the Permian–Triassic interval and some bivalves throughout the section were discovered (Figs. 3.29, 3.30).

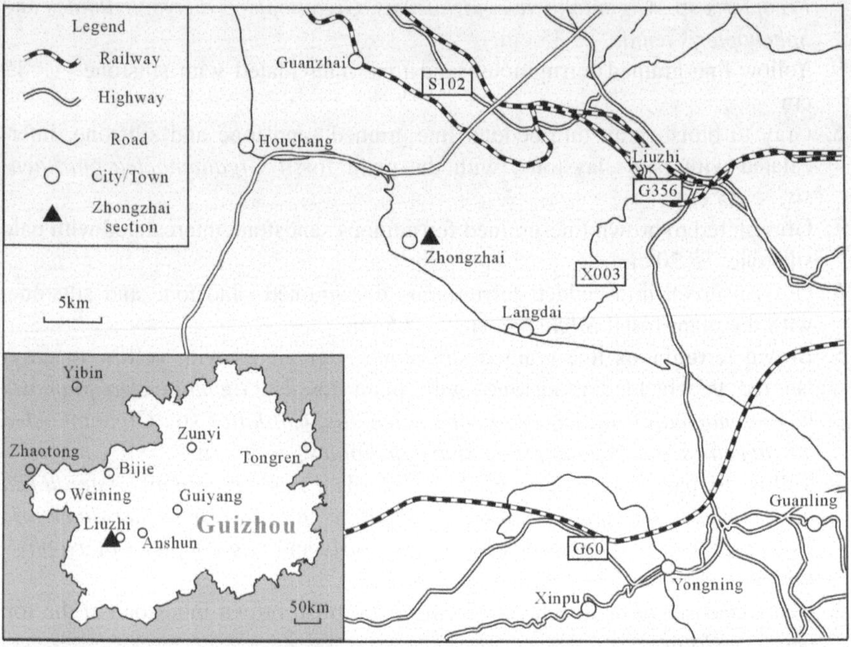

Fig. 3.28 Geographic location of the Zhongzhai section

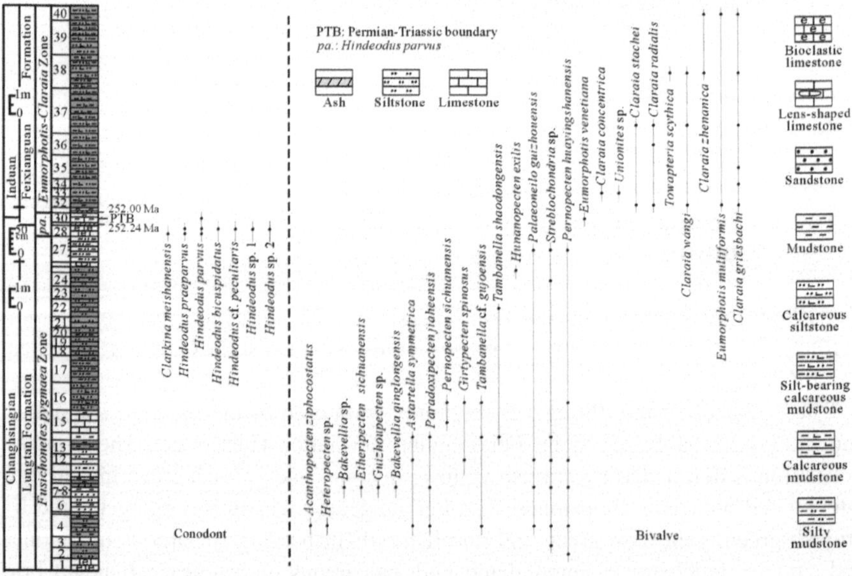

Fig. 3.29 Stratigraphic column and fossil (conodont, bivalve) occurrences at the Zhongzhai section (conodonts after Zhang et al. 2014a)

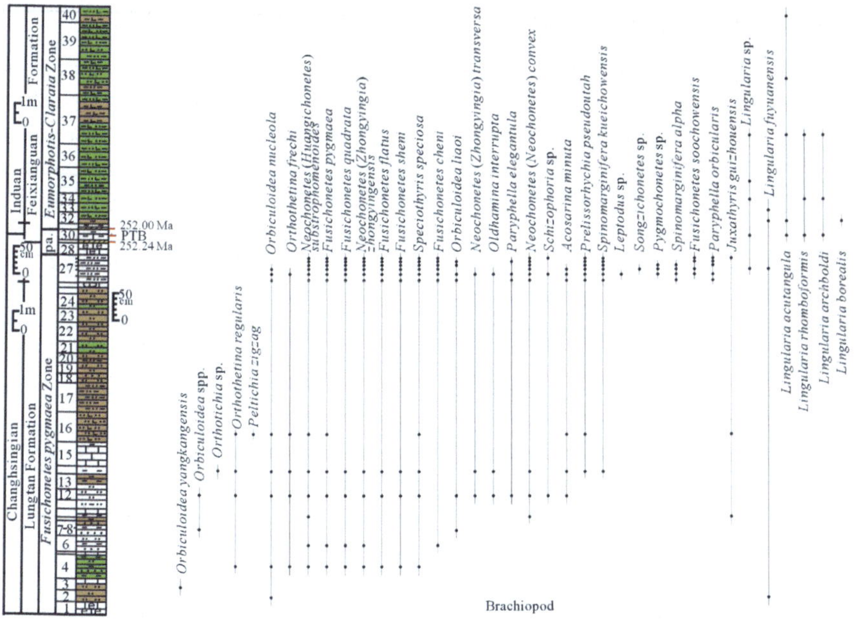

Fig. 3.30 Stratigraphic column and fossil (brachiopod) occurrences at the Zhongzhai section (brachiopods referred to Zhang et al. 2013, 2014b, 2015; U–Pb age after Shen et al. 2011). Legends same to Fig. 3.29

Feixianguan Formation in the Lower Triassic
Not described upwards

40. Yellowish green thin-bedded silt-bearing calcareous mudstone with laminations in the upper part (60 cm thick); yellow thin-bedded calcareous mudstone with laminations in the lower part (20 cm thick). Bivalves *Claraia zhenanica*, *Claraia griesbachi* and *Eumorphotis multiformis* were identified in the bed. 80 cm
39. Yellowish green thin-bedded silt-bearing calcareous mudstone with laminations in the upper part (92 cm thick); gray medium-bedded calcareous siltstone with laminations in the middle part (40 cm thick); brown thin-bedded calcareous mudstone in the lower part (3 cm thick); yellowish green thin-bedded silt-bearing mudstone in the basal part (60 cm thick). 195 cm
38. Yellow thin-bedded calcareous mudstone at the top (10 cm thick); yellowish green medium-bedded silt-bearing calcareous mudstone in the upper part (50 cm thick); gray thin-bedded calcareous siltstone in the middle part (50 cm thick); yellowish green thin-bedded silt-bearing calcareous mudstone in the lower part (50 cm thick); yellow thin-bedded calcareous mudstone at the base (20 cm thick). Laminations were prevalent throughout the bed. Bivalves *Towapteria scythica*, *Claraia wangi*, *Claraia zhenanica* and *Claraia*

griesbachi and the brachiopod *Lingularia acutangula* were examined in the bed. 180 cm

37. Yellowish green thin-bedded silt-bearing calcareous mudstone in the upper part (80 cm); yellow thin-bedded calcareous mudstone intercalated with brown thin-bedded calcareous mudstone in the middle part (80 cm); yellowish green thin-bedded silt-bearing calcareous mudstone in the lower part (90 cm). Laminations were commonly found in the bed. Bivalves *Claraia stachei*, *Claraia radialis*, *Claraia wangi* and *Claraia griesbachi* and brachiopods *Lingularia* sp., *Lingularia acutangula*, *Lingularia rhomboformis* and *Lingularia archboldi* were recovered. 250 cm
36. Yellow thin-bedded silt-bearing calcareous mudstone in the upper part (35 cm); yellowish green thin-bedded silt-bearing calcareous mudstone with laminations in the lower part (85 cm). The bivalve *Claraia radialis* was examined. 120 cm
35. Yellowish green thin-bedded silt-bearing calcareous mudstone in the upper part (88 cm); yellow thin-bedded silt-bearing mudstone with laminations in the middle part (12 cm); yellowish green thin-bedded silt-bearing calcareous mudstone in the lower part (35 cm). Laminations were recorded in the middle and lower parts. The bivalve *Claraia griesbachi* and brachiopod *Lingularia* sp. were discovered in the bed. 135 cm
34. Yellowish green thin-bedded silt-bearing calcareous mudstone with the bivalve *Claraia griesbachi* and brachiopods *Lingularia* sp., *Lingularia rhomboformis* and *Lingularia archboldi*. 50 cm
33. Yellowish green thin-bedded silt-bearing calcareous mudstone with bivalves *Claraia concentrica* and *Unionites* sp. and the brachiopod *Lingularia fuyuanensis*. 40 cm
32. Grayish green thin-bedded silty and calcareous mudstone altered with yellow thin-bedded mudstone. Abundant bivalves *Claraia wangi*, *Claraia griesbachi*, *Claraia stachei*, *Claraia radialis* and *Eumorphotis multiformis* and brachiopods *Lingularia fuyuanensis*, *Lingularia acutangula* and *Lingularia borealis* were collected. 55 cm
31. Yellow thin-bedded mudstone with a layer of pale volcanic ash (3 cm thick) at the base. 10 cm
30. Yellow thin-bedded mudstone with abundant brachiopods *Lingularia* sp., *Lingularia acutangula*, *Lingularia rhomboformis* and *Lingularia archboldi* in the upper part (8 cm); gray thin-bedded argillaceous limestone with the bivalve *Eumorphotis venetiana* in the lower part (10 cm). The conodont *Hindeodus parvus* was found in the bed. 18 cm
29. Pale volcanic ash. 7 cm

Conformity
Lungtan Formation in the Upper Permian

28. Light gray medium-bedded bioclastic limestone. The conodons *Hindeodus praeparvus*, *Hindeodus parvus*, *Hindeodus bicuspidatus*, *Hindeodus* cf.

peculiaris and *Hindeodus* spp. were recorded in the upper part (10 cm); conodonts *Clarkina meishanensis*, *Hindeodus praeparvus* and *Hindeodus parvus* were recovered in the lower part (10 cm). 20 cm

27. Grayish black thin-bedded silty mudstone. A few bivalves *Pernopecten huayingshanensis*, *Streblochondria* sp. and *Palaeoneilo guizhouensis* and abundant brachiopods *Orbiculoidea nucleoli*, *Orbiculoidea liaoi*, *Orthothetina frechi*, *Neochonetes* (*Huangichonetes*) *substrophomenoides*, *Neochonetes* (*Zhongyingia*) *zhongyingensis*, *Neochonetes* (*Zhongyingia*) *transversa*, *Neochonetes* (*Neochonetes*) *convex*, *Fusichonetes pygmaea*, *Fusichonetes quadrata*, *Fusichonetes flatus*, *Fusichonetes sheni*, *Fusichonetes cheni*, *Fusichonetes soochowensis*, *Speciothyris speciosa*, *Oldhamina interrupta*, *Paryphella elegantula*, *Paryphella orbicularis*, *Schizophoria* sp., *Acosarina minuta*, *Prelissorhychia pseudoutah*, *Spinomarginifera kueichowensis*, *Spinomarginifera alpha*, *Leptodus* sp., *Songzichonetes* sp., *Pygmochonetes* sp., *Juxathyris guizhouensis*, *Lingularia fuyuanensis* and *Lingularia* sp. were found in the bed. 50 cm
26. Gray lens-shaped limestone. 18 cm
25. Yellow thin-bedded mudstone in the upper part (10 cm); grayish yellow thin-bedded siltstone in the lower part (20 cm). The bivalve *Hunanopecten exilis* was collected in the bed. 30 cm
24. Grayish yellow thin-bedded silty mudstone in the upper part (40 cm); grayish green thin-bedded siltstone in the lower part (30 cm). The bivalve *Streblochondria* sp. was discovered in the bed. 70 cm
23. Grayish yellow thin-bedded siltstone. 70 cm
22. Grayish yellow thin-bedded siltstone with the bivalve *Tambanella shaodongensis*. 100 cm
21. Grayish green thin-bedded siltstone, laminations evident. 60 cm
20. Grayish yellow thin-bedded siltstone with laminations in the upper part (24 cm); grayish black thin-bedded silty mudstone in the middle part (6 cm); grayish yellow thin-bedded siltstone in the lower part (20 cm). 50 cm
19. Grayish yellow thin-bedded siltstone, laminations prevalent. 55 cm
18. Grayish yellow thin-bedded fine-grained sandstone intercalated with laminated mudstone in the upper part (30 cm); grayish yellow thin-bedded siltstone intercalated with lens-shaped limestone in the lower part (20 cm). 50 cm
17. Grayish black thin-bedded silty mudstone with laminations (covered by the Quaternary at the base). 150 cm
16. Grayish yellow thin-bedded calcareous siltstone. Abundant brachiopods *Orthothetina regularis*, *Orthothetina frechi*, *Peltichia zigzag*, *Orbiculoidea nucleoli*, *Neochonetes* (*Huangichonetes*) *substrophomenoides*, *Fusichonetes pygmaea*, *Speciothyris speciosa*, *Acosarina minuta*, *Prelissorhychia pseudoutah* and *Juxathyris guizhouensis* and bivalves *Pernopecten sichuanensis*, *Pernopecten huayingshanensis*, *Girtypecten spinosus* and *Tambanella* cf. *gujoensis* were collected in the lower part of the bed. 180 cm

15. Light gray thick-bedded limestone, with grayish black thin-bedded mudstone at the top (3 cm thick). The bivalve *Pernopecten sichuanensis* was discovered in the limestone. 123 cm
14. Grayish black thin-bedded mudstone, laminations evident. Abundant brachiopods *Orthotichia* sp., *Orthothetina regularis*, *Orthothetina frechi*, *Orbiculoidea nucleoli*, *Neochonetes* (*Huangichonetes*) *substrophomenoides*, *Neochonetes* (*Zhongyingia*) *zhongyingensis*, *Neochonetes* (*Zhongyingia*) *transversa*, *Neochonetes* (*Neochonetes*) *convex*, *Fusichonetes pygmaea*, *Fusichonetes quadrata*, *Fusichonetes flatus*, *Fusichonetes sheni*, *Speciothyris speciosa*, *Oldhamina interrupta*, *Acosarina minuta*, *Prelissorhychia pseudoutah* and *Spinomarginifera kueichowensis* and bivalves *Paradoxipecten jiaheensis*, *Streblochondria* sp., *Tambanella shaodongensis*, *Tambanella* cf. *gujoensis* and *Astartella symmetrica* were recorded. 45 cm
13. Grayish yellow thin-bedded mudstone in the upper part (40 cm); grayish black thin-bedded siltstone with laminations in the lower part (35 cm). 75 cm
12. Grayish black thin-bedded siltstone intercalated with laminated mudstone. Abundant brachiopods *Orbiculoidea nucleoli*, *Orbiculoidea liaoi*, *Orbiculoidea* spp., *Orthothetina regularis*, *Orthothetina frechi*, *Neochonetes* (*Huangichonetes*) *substrophomenoides*, *Neochonetes* (*Zhongyingia*) *zhongyingensis*, *Neochonetes* (*Zhongyingia*) *transversa*, *Neochonetes* (*Neochonetes*) *convex*, *Fusichonetes pygmaea*, *Fusichonetes quadrata*, *Fusichonetes flatus*, *Fusichonetes sheni*, *Speciothyris speciosa*, *Oldhamina interrupta*, *Oldhamia squamosa*, *Paryphella elegantula*, *Schizophoria* sp. and *Acosarina minuta* and bivalves *Pernopecten huayingshanensis*, *Astartella symmetrica*, *Streblochondria* sp., *Palaeoneilo guizhouensis* and *Girtypecten spinosus* were recovered. 47 cm
11. Grayish yellow thin-bedded argillaceous siltstone in the upper part (10 cm); grayish black thin-bedded silty mudstone in the lower part (28 cm). 38 cm
10. Gray thick-bedded limestone, with thin-bedded mudstone at the base. The brachiopods *Neochonetes* (*Huangichonetes*) *substrophomenoides*, *Neochonetes* (*Neochonetes*) *convex* and *Juxathyris guizhouensis* were recorded in the mudstone. 40 cm
9. Grayish yellow siltstone intercalated with mudstone, laminations prevalent at the top (5 cm); grayish yellow thin-bedded mudstone in the upper part (4 cm); grayish yellow thin-bedded siltstone in the lower part (8 cm); grayish black thin-bedded mudstone at the base (5 cm). 22 cm
8. Grayish yellow thin-bedded fine-grained sandstone intercalated with laminated mudstone in the upper part (35 cm); grayish black thin-bedded mudstone at the base (5 cm). Abundant bivalves *Pernopecten huayingshanensis*, *Etheripecten sichuanensis*, *Paradoxipecten jiaheensis*, *Bakevellia qinglongensis*, *Bakevellia* sp., *Guizhoupecten* sp., *Palaeoneilo guizhouensis* and *Streblochondria* sp. and a few plant-fossil fragments were respectively found in the sandstone and mudstone. 40 cm

7. Grayish black thin-bedded fine-grained sandstone intercalated with laminated mudstone. A few brachiopods *Orbiculoidea liaoi* and *Orbiculoidea* spp. and plant-fossil fragments were preserved in the sandstone. 40 cm
6. Gray thin-bedded siltstone at the top (23 cm); gray thin-bedded mudstone in the upper part (40 cm); gray thin-bedded fine-grained sandstone in the lower part (10 cm); gray thin-bedded mudstone at the base (7 cm). Laminations were commonly observed in the bed. Abundant brachiopods *Neochonetes* (*Huangichonetes*) *substrophomenoides*, *Fusichonetes pygmaea*, *Fusichonetes quadrata*, *Fusichonetes cheni* and *Neochonetes* (*Zhongyingia*) *zhongyingensis* were recovered. 80 cm
5. Grayish yellow thin-bedded limestone. 7 cm
4. Grayish green thin-bedded mudstone intercalated with siltstone in the upper part (116 cm); gray thin-bedded fine-grained sandstone in the middle part (4 cm); grayish green thin-bedded mudstone in the lower part (40 cm). Abundant brachiopods *Orthothetina regularis*, *Orthothetina frechi*, *Orbiculoidea nucleola*, *Neochonetes* (*Huangichonetes*) *substrophomenoides*, *Neochonetes* (*Zhongyingia*) *zhongyingensis*, *Fusichonetes pygmaea*, *Fusichonetes quadrata*, *Fusichonetes flatus*, *Fusichonetes sheni* and *Speciothyris speciosa* and bivalves *Pernopecten huayingshanensis*, *Streblochondria* sp., *Acanthopecten ziphocostatus*, *Palaeoneilo guizhouensis*, *Heteropecten* sp., *Paradoxipecten jiaheensis* and *Tambanella* cf. *gujoensis* were recorded in the bed. 200 cm
3. Light gray medium-bedded limestone in the upper part; grayish yellow thin-bedded silstone with laminations in the lower part. The bivalve *Streblochondria* sp. and the brachiopod *Orbiculoidea yangkangensis* were examined. 60 cm
2. Grayish yellow thin-bedded siltstone with laminations in the upper part (55 cm); grayish black silty mudstone in the lower part (2 cm); grayish yellow siltstone at the base (3 cm). The bivalve *Pernopecten huayingshanensis* and brachiopods *Orbiculoidea nucleoli* and *Lingularia fuyuanensis* were identified. 60 cm
1. Light gray thick-bedded bioclastic limestone. > 50 cm

The conodonts are based on Zhang et al. (2014a); brachiopods are re-examinded in the book, based on the plates of Zhang et al. (2013, 2014b, 2015).

3.13 Meishan D Section Nearby Meishan Town, Changxing County

The Meishan section (D section) (31° 04′ 50″ N, 119° 42′ 22″ E) is located about 2 km southwest of the Meishan Town, Changxing County, Zhejiang Province, China (Fig. 3.31).

The sequence of the Meishan section is composed of the Changhsing Formation and the Yinkeng Formation (Fig. 3.32). The Changhsing Formation is

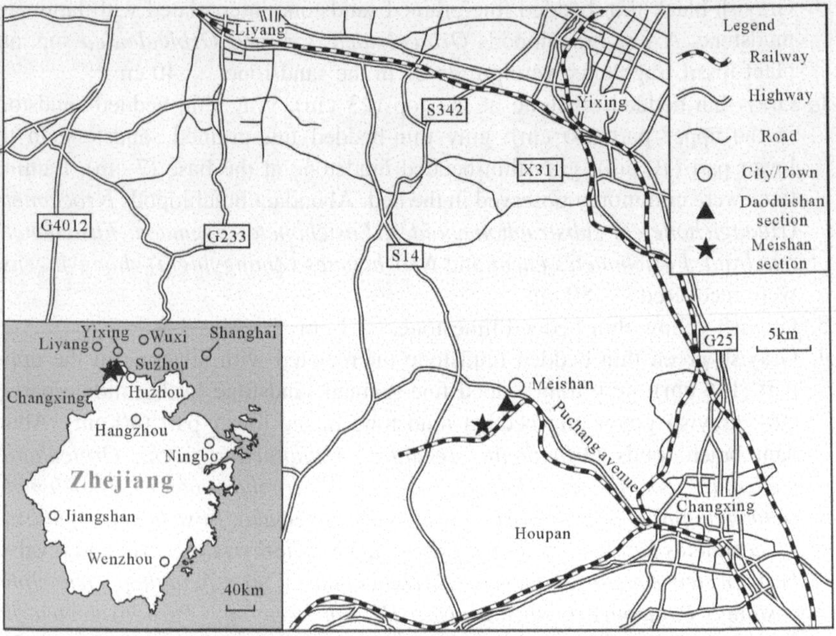

Fig. 3.31 Geographic location of Meishan and Daoduishan sections

dominated by gray medium-bedded micritic limestones and bioclastic limestones. The Yinkeng Formation is dominated by grayish black thin-bedded carbonaceous and calcareous mudstones and gray to grayish green thin-bedded calcareous mudstones, and intercalated with gray thin-bedded argillaceous limestones at the basal part and with gray thin-bedded micritic limestones in the upper of the formation. Laminations are commonly recorded in the section. Abundant foraminifers (not listed in the figure) and conodonts have been yielded and a few radiolarians, brachiopods, ammonoids and bivalves were found in the section (Fig. 3.32).

Yinkeng Formation in the Lower Triassic

30. Grayish green thin-bedded argillaceous limestone with laminations. The **ammonoids** ophiceratids, **bivalves** *Claraia griesbachi*, *C. concentrica* and *C. wangi* and **conodonts** *Neogondollella carinata*, *N. planata*, *Hindeodus inflatus*, *H. typicalis*, *H. latidentatus*, *H. turgidus*, *H. changxingensis*, *H. eurypyge*, *H. parvus* and *H. praeparvus* were found. 48 cm
29. Gray medium-bedded dolomitic and argillaceous limestone. The **ammonoid** *Ophiceras* sp., **bivalves** *Claraia wangi* and *C. dieneri*, **conodonts** *Hindeodus parvus*, *H. turgidus*, *H. inflatus*, *H. latidentatus*, *H. changxingensis*, *H. eurypyge* and *Isarcicella staeschei*, and the **foraminifer** *Nodosinelloides aequiampla* were examined. 18 cm

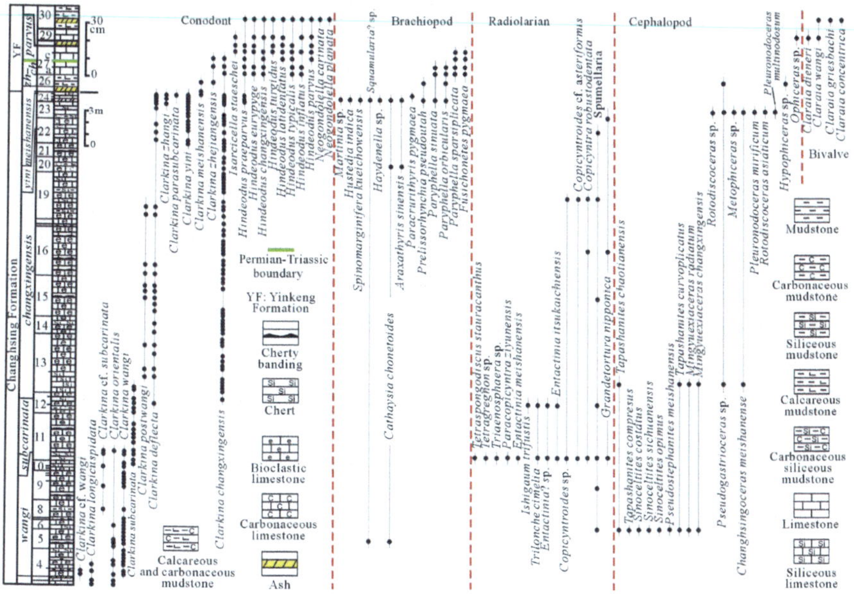

Fig. 3.32 Stratigraphic column and fossil occurrences at the Meishan section. *Note wangi*: *Clarkina wangi* Zone, *subcarinata*: *Clarkina subcarinata* Zone, *changxingensis*: *Clarkina changxingensis* Zone, *yini*: *Clarkina yini* Zone, *meishanensis*: *Clarkina meishanensis* Zone, *zhch*: *Clarkina zhejiangensis–Hindeodus changxingensis* Zone, *parvus*: *Hindeodus parvus*. Conodonts mostly referred to Zhang et al. (2007) and Yuan et al. (2014); radiolarians based on He et al. (2005); cephalopods referred to Zhao et al. (1981) and Yang et al. (1987)

28. Pale volcanic ash with **conodonts** *Hindeodus parvus*, *H. praeparvus*, *H. eurypyge*, *H. typicalis*, *H. turgidus*, *H. latidentatus*, *Isarcicella staeschei* and *C. zhejiangensis*. 4 cm

27. Gray medium-bedded dolomitic and argillaceous limestone. 16 cm

27d. **Brachiopods** *Fusichonetes pygmaea*, *Paryphella sparsiplicata* and *Paracrurithyris pygmaea*, **conodonts** *Hindeodus parvus*, *H. praeparvus*, *H. inflatus*, *H. turgidus*, *H. latidentatus*, *H. typicalis*, *H. changxingensis*, *H. eurypyge* and *Isarcicella staeschei* and **foraminifers** *Ammodiscus* sp., *Diplosphaerina inaequalis*, *Frondina* sp., *Geinitzina araxensis*, *Globivalvulina bulloides*, *Robuloides lens* and *Nodosinelloides aequiampla* were recorded. 4 cm

27c. **Brachiopods** *Fusichonetes pygmaea* and *Paryphella sparsiplicata*, **Conodonts** *Hindeodus parvus*, *H. praeparvus*, *H. latidentatus*, *H. typicalis*, *H. changxingensis*, *Clarkina changxingensis* and *C. zhejiangensis*, and **foraminifers** *Frondina permica*, *Ichthyofrondina palmata*, *Nodosinelloides netschajewi*, *N. aequiampla*, *N.* sp., *Rectostipulina quadrata*, *Tuberitina* sp., *Diplosphaerina inaequalis*, *Geinitzina araxensis*, *Globivalvulina bulloides* and *Robuloides lens* were found. 4 cm

27b. **Brachiopods** *Fusichonetes pygmaea*, *Paryphella orbicularis*, *Paryphella sinuata* and *Paryphella sparsiplicata*, **conodonts** *Clarkina changxingensis*,

C. zhejiangensis, *Hindeodus changxingensis*, *H. eurypyge*, *H. typicalis*, *H. inflatus* and *H. praeparvus* were recovered. 4 cm

27a. **Brachiopods** *Fusichonetes pygmaea*, *Paryphella orbicularis*, *Paryphella sinuata* and *Paryphella sparsiplicata*, **conodonts** *Hindeodus inflatus*, *H. changxingensis*, *H. typicalis*, *H. eurypyge*, *H. praeparvus* and *Clarkina changxingensis* were collected. 4 cm

Foraminifers *Glomospira* spp., *Cryptoseptida* sp., *Frondina permica*, *'Nodosaria' skyphica*, *Robuloides* sp., *Cryptoseptida anatoliensis*, *Hemigordius* sp., *Neoendothyra* sp., *Nodosinelloides netschajewi*, *Rectostipulina quadrata*, *Diplosphaerina inaequalis*, *Geinitzina araxensis*, *Globivalvulina bulloides*, *Robuloides lens* and *Nodosinelloides aequiampla* were recorded in beds 27a and 27b.

26. Grayish black calcareous mudstone with **ammonoids** *Hypophiceras* sp. and *Pseudogastrioceras* sp., **conodonts** *Clarkina meishanensis* and *C. zhejiangensis*, and **brachiopods** *Paryphella orbicularis*, *Paracrurithyris pygmaea* and *Prelissorhynchia pseudoutah.* 6 cm

25. Pale volcanic ash with the conodont *Clarkina meishanensis.* 4 cm

Conformity

Changhsing Formation in the Upper Permian

24e. Gray thin-bedded bioclastic limestone intercalated with micritic limestone, locally intercalated with chert. The **brachiopod** *Paracrurithyris pygmaea*, **conodonts** *Clarkina yini*, *C. meishanensis*, *C. parasubcarinata*, *C. zhangi*, *C. deflecta*, *C. changxingensis* and *Hindeodus praeparvus* were recorded. 10 cm

24d. Dark gray thin-bedded micritic limestone intercalated with bioclastic limestone, with **conodonts** *Clarkina yini*, *C. deflecta* and *C. changxingensis.* 23 cm

24c. Dark gray thin-bedded dolomitic micritic limestone. **Conodonts** *Clarkina yini*, *C. deflecta*, *C. changxingensis*, *C. zhangi* and *Hindeodus praeparvus* and **brachiopods** *Araxathyris sinensis*, *Cathaysia chonetoides*, *Haydenella* sp., *Squamularia*? sp., *Prelissorhynchia pseudoutah*, *Spinomarginifera kueichowensis*, *Paryphella orbicularis*, *P. sinuata*, *Hustedia indica* and *Martinia* sp. were found. 17 cm

24b. Dark gray thin-bedded dolomitic micritic limestone with **conodonts** *Clarkina yini*, *C. deflecta*, *C. changxingensis* and *Hindeodus praeparvus.* 11 cm

24a. Dark gray thin-bedded micritic limestone with **conodonts** *Clarkina yini*, *C. deflecta*, *C. changxingensis* and *C. zhangi.* 10 cm

Foraminifers *Hemigordius rotundus*, *Pachyphloia robusta*, *Ammovertella inversus*, *Colaniella nana*, *Hemigordius parvus*, *H. zaninettiae*, *H. bronnimanni*, *H. regularis*, *Multidiscus padangensis*, *Neoendothyra permica*, *N. parva*, *Neotuberitina maljavkini*, *Nodosinelloides camerata*, *N. Aequiampla*, *N. netschajewi*, *N. aceraeformis*, *N. ronda*, *N. sagitta*, *Pachyphloia ovata*,

Palaeotextularia spp., *Rectostipulina pentamerata*, *Aeolisaccus dunningtoni*, *Cryptoseptida fragilis*, *C. anatoliensis*, *Dagmarita altilis*, *Geinitzina spandeli*, *G. araxensis*, *G. uralica*, *Paraglobivalvulina* sp., *Pseudoammodiscus parvus*, *'Pseudoglandulina' conica*, *Reichelina* spp., *Sengoerina argandi*, *Vervilleina bradyi*, *Frondina permica*, *Ichthyofrondina palmata*, *Rectostipulina quadrata*, *Tuberitina* sp., *Diplosphaerina inaequalis*, *Globivalvulina bulloides* and *Robuloides lens* were recorded in bed 24.

23b. Dark gray thin-bedded micritic limestone intercalated with carbonaceous and siliceous mudstone and chert (3 cycles formed). 105 cm

23a. Dark gray medium-bedded limestone, with carbonaceous and siliceous mudstone at the base. 25 cm
Conodonts *Clarkina changxingensis*, *Clarkina yini* and *C. deflecta*, **foraminifers** *Dagmarita chanakchiensis*, *Tournayella*? spp., *Hemigordius guangdongensis*, *H. rotundus*, *H. parvus*, *H. zaninettiae*, *H. bronnimanni*, *Ammovertella inversus*, *Neotuberitina maljavkini*, *Nodosinelloides camerata*, *N. ronda*, *N. sagitta*, *N. aceraeformis*, *N. aequiampla*, *Pachyphloia ovata*, *Aeolisaccus dunningtoni*, *Colaniella* spp., *Cryptoseptida fragilis*, *C. anatoliensis*, *Dagmarita altilis*, *Geinitzina uralica*, *G. araxensis*, *Paraglobivalvulina* sp., *Reichelina* spp., *Vervilleina bradyi*, *Frondina permica*, *Ichthyofrondina palmata*, *Rectostipulina quadrata*, *Diplosphaerina inaequalis*, *Globivalvulina bulloides* and *Glomospira regularis*, and the **radiolarian** *Grandetortura nipponica* were reported in bed 23.

22. Dark gray medium-bedded carbonaceous and bioclastic limestone intercalated with thin-bedded carbonaceous micritic limestone, with normal graded beddings and hummocky cross stratifications. **Conodonts** *Clarkina changxingensis*, *C. yini* and *C. deflecta*, **foraminifers** *Millerella* sp., *Tetrataxis angusta*, *Rectostipulina pentamerata*, *'Pseudoglandulina' inflataeforma*, *'Pseudoglandulina' conica*, *Cryptomorphina limonitica*, *Palaeofusulina* sp., *Biorbis* spp., *Pachyphloia schwageri*, *P. ovata*, *P. robusta*, *Postendothyra guangxiensis*, *Dagmarita chanakchiensis*, *D. altilis*, *Tournayella*? spp., *Hemigordius rotundus*, *H. parvus*, *H. zaninettiae*, *H. bronnimanni*, *H. regularis*, *Ammovertella inversus*, *Colaniella nana*, *Multidiscus padangensis*, *Neoendothyra permica*, *Neotuberitina maljavkini*, *Nodosinelloides camerata*, *N. ronda*, *N. sagitta*, *N. netschajewi*, *Palaeotextularia* spp., *Aeolisaccus dunningtoni*, *Cryptoseptida fragilis*, *C. anatoliensis*, *Geinitzina spandeli*, *G. uralica*, *G. araxensis*, *Neoendothyra parva*, *Nodosinelloides aceraeformis*, *N. aequiampla*, *Paraglobivalvulina* sp., *Pseudoammodiscus parvus*, *Reichelina* spp., *Sengoerina argandi*, *Vervilleina bradyi*, *Glomospira regularis*, *Frondina permica*, *Ichthyofrondina palmata*, *Rectostipulina quadrata*, *Diplosphaerina inaequalis*, *Globivalvulina bulloides* and *Robuloides lens*, and **radiolarian** Spumellaria forms were recorded in the bed. **Ammonoids** *Pseudogastrioceras* sp., *Changhsingoceras meishanense*, *Pachydiscoceras changhsingense*, *Pleuronodoceras mirificum*, *Pl. multinodosum*, *Rotodiscoceras asiaticum*, *Stacheoceras pachydiscum* and *Trigonogastrioceras changxingense* were reported in bed 24 to the top 80 cm of bed 22 (Zhao et al. 1981). 207 cm

21. Dark gray medium-bedded bioclastic micritic limestone intercalated with cherty banding, with laminations. The **conodont** *Clarkina changxingensis* and **foraminifers** *Rectostipulina pentamerata, 'Pseudoglandulina' inflataeforma, 'Pseudoglandulina' conica, Pachyphloia schwageri, P. robusta, P. ovata, Dagmarita chanakchiensis, Neotuberitina maljavkini, Aeolisaccus dunningtoni, Cryptoseptida fragilis, Geinitzina araxensis, Hemigordius zaninettiae, H. bronnimanni, H. regularis, Paraglobivalvulina* sp., *Reichelina* spp., *Glomospira* spp., *Frondina permica, Nodosinelloides netschajewi, N. aequiampla, Rectostipulina quadrata, Diplosphaerina inaequalis* and *Globivalvulina bulloides* were collected. 117 cm
20. Gray medium-bedded bioclastic limestone intercalated with dark gray thin-bedded carbonaceous micritic limestone and calcareous and carbonaceous mudstone. The **conodont** *Clarkina changxingensis*, **foraminifers** *Rectostipulina pentamerata, Rectostipulina quadrata, Pachyphloia robusta, Neotuberitina maljavkini, Nodosinelloides camerata, N. aceraeformis, N. netschajewi, N. aequiampla, Cryptoseptida fragilis, Dagmarita altilis, Geinitzina spandeli, G. uralica, G. araxensis, Paraglobivalvulina* sp., *Reichelina* spp., *Glomospira regularis, Hemigordius* sp., *Frondina permica, Ichthyofrondina palmata, Diplosphaerina inaequalis* and *Globivalvulina bulloides*, and **brachiopods** *Araxathyris sinensis* and *Cathaysia chonetoides* were recorded. 120 cm
19. Gray medium-bedded bioclastic limestone altered with grayish black thin-bedded siliceous limestone or siliceous mudstone, with laminations in the mudstone. **Conodonts** *Clarkina changxingensis, C. postwangi* and *C. deflecta*, **foraminifers** *Protonodosaria* spp., *'Pseudoglandulina' inflataeforma, 'Pseudoglandulina' conica, Dagmarita chanakchiensis, Dagmarita altilis, Hemigordius rotundus, H. bronnimanni, H. regularis, Pachyphloia robusta, Neotuberitina maljavkini, Nodosinelloides camerata, N. sagitta, N. aceraeformis, N. netschajewi, N. aequiampla, Rectostipulina pentamerata, Rectostipulina quadrata, Aeolisaccus dunningtoni, Cryptoseptida fragilis, Geinitzina spandeli, G. uralica, G. araxensis, Paraglobivalvulina* sp., *Pseudoammodiscus parvus, Reichelina* spp., *Glomospira* spp., *Frondina permica, Ichthyofrondina palmata, Tuberitina* sp., *Diplosphaerina inaequalis, Globivalvulina bulloides* and *Robuloides lens*, and **radiolarians** *Copicyntra robustodentata, Copicyntroides* cf. *asteriformis, Copicyntroides* sp. and abundant other Spumellaria forms were recovered. 400 cm
18. Dark gray medium-bedded bioclastic limestone intercalated with grayish black thin-bedded siliceous limestone, with laminations in the siliceous limestone. The **conodont** *Clarkina changxingensis* and **foraminifers** *'Pseudoglandulina' inflataeforma, Frondina permica, Nodosinelloides* sp., *Diplosphaerina inaequalis* and *Globivalvulina bulloides* were obtained. 40 cm
17. Grayish black silt-bearing calcareous mudstone. 9 cm
16. Gray medium-bedded bioclastic micritic limestone intercalated with black nodule- and banding-shaped siliceous limestone. **Conodonts** *Clarkina*

changxingensis, *C. postwangi* and *C. deflecta*, **foraminifers** *Protonodosaria* spp., *Biorbis* spp., *Pachyphloia schwageri*, *P. robusta*, *Dagmarita chanakchiensis*, *D. altilis*, *Hemigordius rotundus*, *H.* sp., *Multidiscus padangensis*, *Neoendothyra permica*, *Neotuberitina maljavkini*, *Nodosinelloides camerata*, *N. ronda*, *N. sagitta*, *N. aceraeformis*, *N. netschajewi*, *N. aequiampla*, *Rectostipulina pentamerata*, *R. quadrata*, *Aeolisaccus dunningtoni*, *Cryptoseptida fragilis*, *Geinitzina araxensis*, *G. spandeli*, *G. uralica*, *Paraglobivalvulina* sp., '*Pseudoglandulina*' *conica*, *Reichelina* spp., *Sengoerina argandi*, *Glomospira* spp., *Frondina permica*, *Ichthyofrondina palmata*, *Tuberitina* sp., *Diplosphaerina inaequalis* and *Globivalvulina bulloides*, and **radiolarians** *Copicyntra robustodentata* and *Grandetortura nipponica* were recorded. 410 cm

15. Dark gray medium-bedded bioclastic micritic limestone intercalated with cherty banding. **Conodonts** *Clarkina changxingensis*, *C. postwangi*, *C. subcarinata* and *C. deflecta*, **foraminifers** *Protonodosaria* spp., *Hemigordius rotundus*, *H. bronnimanni*, *H.* sp., *Pachyphloia robusta*, *Multidiscus padangensis*, *Neotuberitina maljavkini*, *Nodosinelloides camerata*, *N. netschajewi*, *N. aequiampla*, *Dagmarita* spp., *Geinitzina spandeli*, *G. araxensis*, *Pseudoammodiscus parvus*, '*Pseudoglandulina*' *conica*, *Reichelina* spp., *Glomospira regularis*, *Frondina permica*, *Diplosphaerina inaequalis* and *Globivalvulina bulloides*, and **radiolarian** Spumellaria forms were reported. 350 cm
14. Dark gray thin- to medium-bedded micritic limestone intercalated with black thin-bedded chert. **Conodonts** *Clarkina changxingensis*, *C. subcarinata* and *C. deflecta* and **foraminifers** *Protonodosaria* spp., *Millerella* sp., *Tetrataxis angusta*, *Postendothyra guangxiensis*, *Dagmarita chanakchiensis*, *Hemigordius guangdongensis*, *H. rotundus*, *H. parvus*, *H. zaninettiae*, *H. bronnimanni*, *Colaniella nana*, *Neoendothyra permica*, *Neotuberitina maljavkini*, *Nodosinelloides camerata*, *N. ronda*, *N. netschajewi*, *N. aequiampla*, *Palaeotextularia* spp., *Aeolisaccus dunningtoni*, *Cryptoseptida fragilis*, *Geinitzina spandeli*, *G. araxensis*, *Paraglobivalvulina* sp., *Pseudoammodiscus parvus*, *Reichelina* spp., *Glomospira regularis*, *Cryptoseptida anatoliensis*, *Frondina permica*, *Rectostipulina quadrata*, *Diplosphaerina inaequalis* and *Globivalvulina bulloides* were discovered. 153 cm

13b. Grayish black medium-bedded bioclastic micritic limestone with laminations. 320 cm

13a. Grayish black sponge-spicule limestone intercalated with thin-bedded siliceous limestone, with laminations. 183 cm

Conodonts *Clarkina changxingensis*, *C. subcarinata*, *C. postwangi* and *C. deflecta* and **foraminifers** *Protonodosaria* spp., *Millerella* sp., *Postendothyra guangxiensis*, *Hemigordius rotundus*, *H. zaninettiae*, *H. bronnimanni*, *H. regularis*, *Colaniella nana*, *Multidiscus padangensis*, *Neotuberitina maljavkini*, *Nodosinelloides camerata*, *N. ronda*, *N. aceraeformis*, *N. aequiampla*, *N. netschajewi*, *Aeolisaccus dunningtoni*, *Cryptoseptida fragilis*, *Dagmarita altilis*, *Dagmarita* spp., *Geinitzina araxensis*, *Reichelina* spp.,

Glomospira spp., *Frondina permica, Diplosphaerina inaequalis* and *Globivalvulina bulloides* were collected in bed 13. **Ammonoids** *Pseudogastrioceras* sp., *Changhsingoceras meishanense, Liuchengoceras* sp., *Mingyuexiaceras changxingense, M. radiatum, Pseudostephanites* sp., *Rongjiangoceras* cf. *lenticulare, Sinoceltites* sp., *Tapashanites changxingense, T. chaotianensis, T. costatus* and *T. curvoplicatus* were recorded in the lower part of bed 13 (Zhao et al. 1981).

12c. Grayish black thin- to medium-bedded bioclastic micritic limestone with laminations. 38 cm

12b. Grayish black medium-bedded siliceous micritic limestone with laminations. 75 cm

12a. Grayish black thin- to medium-bedded bioclastic micritic limestone intercalated with grayish black thin-bedded carbonaceous mudstone with laminations. 80 cm

Conodonts *Clarkina subcarinata, C. changxingensis* and *C. deflecta* and **radiolarians** *Entactinia itsukaichiensis, Entactinia*? sp., *Trilonche cimelia* and *Ishigaum trifustis* and Spumellaria forms were recovered in bed 12.

11. Dark gray medium-bedded bioclastic micritic limestone intercalated with black cherty banding, laminations prevalent. **Conodonts** *Clarkina orientalis, C. wangi, C. subcarinata* and *Clarkina* cf. *subcarinata* and abundant **radiolarians** *Entactinia itsukaichiensis, E. meishanensis, Entactinia*? sp., *Trilonche cimelia, Paracopicyntra ziyunensis, Copicyntroides* cf. *asteriformis, Copicyntroides* sp., *Copicyntra robustodentata, Grandetortura nipponica, Triaenosphaera* sp., *Tetragregnon* sp., *Tetraspongodiscus stauracanthus* and other Spumellaria forms were recorded. 400 cm

10c. Light gray medium-bedded bioclastic micritic limestone altered with thin-bedded micritic limestone (two cycles formed). 48 cm

10b. Light gray thin-bedded bioclastic micritic limestone altered with thin-bedded siliceous limestone. 13 cm

10a. Light gray medium-bedded bioclastic micritic limestone with cross beddings. 32 cm

Conodonts *Clarkina wangi* and *Clarkina* cf. *subcarinata* were found in bed 10.

9d. Gray bioclastic micritic limestone. 30 cm

9c. Light gray thin-bedded bioclastic micritic limestone altered with dark gray thin-bedded bioclastic micritic limestone (seven cycles formed). 29 cm

9b. Light gray thin-bedded bioclastic micritic limestone altered with dark gray thin-bedded bioclastic micritic limestone (five cycles formed). 50 cm

9a. Dark gray thin-bedded bioclastic limestone with cherty nodules and bandings. 162 cm

Conodonts *Clarkina wangi* and *Clarkina* cf. *subcarinata* and **radiolarian** Spumellaria forms were recorded in bed 9.

8. Gray medium-bedded bioclastic limestone intercalated with grayish black thin-bedded siliceous mudstone, with **conodonts** *Clarkina wangi* and *C.* cf. *subcarinata* in the limestone. 167 cm

7. Grayish yellow calcareous and carbonaceous mudstone. 5 cm
6. Light gray medium-bedded bioclastic limestone altered with dark gray thin-bedded carbonaceous and siliceous limestone with laminations. **Conodonts** *Clarkina orientalis* and *C. wangi* and **radiolarian** Spumellaria forms were recorded in the bed. **Ammonoids** *Mingyuexiaceras changxingense, M. radiatum, Tapashanites chaotianensis, T. curvoplicatus, T. compresus, Pseudogastrioceras gigantum, Pseudostephanites meishanensis, Sinoceltites costatus, S. opimus* and *S. sichuanensis* were found in the basal part of bed 6 (Yang et al. 1987). 85 cm
5. Dark gray thin- to medium-bedded bioclastic micritic limestone with cherty bandings, with **conodonts** *Clarkina orientalis* and *C. wangi* and brachiopods *Cathaysia chonetoides* and *Squamularia*? sp.. 180 cm

4b5. Bioclastic micritic limestone, with grayish black thin-bedded chert at the top. 55 cm

4b4. Medium-bedded bioclastic limestone with hummocky cross stratifications. 53 cm

4b3. Grayish brown mudstone with laminations. 3 cm

4b2. Medium-bedded bioclastic limestone with hummocky cross stratifications. 47 cm

4b1. Gray thin- to medium-bedded bioclastic micritic limestone with laminations. 22 cm

Conodonts *Clarkina orientalis*, C. *longicuspidata* and *C. wangi* were recorded in bed 4b.

4a. Grayish black thick-bedded bioclastic micritic limestone, with dark gray thin-bedded silty and calcareous mudstone at the top (10 cm thick). **Conodonts** *Clarkina orientalis*, C. *longicuspidata*, *C. wangi* and *C.* cf. *wangi* were discovered in bed 4a. 77.5 cm

3. Grayish black carbonaceous and argillaceous limestone, with pale volcanic ash at the top (3 cm thick). **Foraminifers** *Collaniella* sp., *Eacristellaria* sp. and *Geinitzina postcarbonica, Pseudoglandulina conica* and **conodonts** *Clarkina longicuspidata* and *C. orientalis* were recorded. 24.5 cm
2. Dark gray thick-bedded silt-bearing micritic limestone, with **foraminifers** *Collaniella* sp., *Eacristellaria* sp., *Geinitzina postcarbonica* and *Pseudoglandulina conica* and **conodonts** *Clarkina longicuspidata* and *C. orientalis.* 56 cm

Note: Ammonoids are based on Zhao et al. (1981) and Yang et al. (1987); conodonts are based on Mei et al. (1998), Jiang et al. (2007), Zhang et al. (2007) and Yuan et al. (2014) (updated and summarized here); foraminifers are based on Song et al. (2009).

3.14 Daoduishan Section Nearby Meishan Town, Changxing County

As a supplement to the materials of fauna and biodiversity of Meishan D section, the Daoduishan section was selected to study, because the Daoduishan section is very close to the Meishan D section in geography and can be regarded as an auxiliary section for the latter whose rock is fresh and preserved in a national geopark (not easy for excavating abundant macrofossils). The Daoduishan section is located about 2.5 km northeast of the Meishan D section, 20 km northwest of Changxing County, Zhejiang Province, China (Fig. 3.31).

The sequence of the Daoduishan section is composed of the upper part of the Changhsing Formation and the basal part of the Yinkeng Formation (Fig. 3.33). The upper part of the Changhsing Formation is dominated by gray medium-bedded bioclastic limestones. The basal part of the Yinkeng Formation is dominated by grayish yellow thin-bedded argillaceous limestones (Fig. 3.34).

Yinkeng Formation in the Lower Triassic
Covered by the Quaternary at the Top

30-2. Gray medium-bedded micritic limestone, with **brachiopods** lingulids, *Speciothyris flabilifomis* and *Paracruithyris pygmaea*. 23 cm

30-1. Yellowish green calcareous mudstone. 4 cm

29. Gray medium-bedded dolomitic and argillaceous limestone, with laminations in the lower part of bed 29. The **conodont** *Hindeodus parvus* and **brachiopods** *Fusichonetes pigmaea*, *Prelissorhynchia pseudoutah*, *Paracruithyris pygmaea*, *Spinomarginifera kueichowensis*, *Speciothyris flabilifomis*, *S. speciosa*, *Paryphella orbicularis*, *P. sparsiplicata* and *Neochonetes* sp. were collected. 16 cm

28. Pale claystone. 1 cm

27. Gray thin-bedded dolomitic and argillaceous limestone, with **brachiopods** *Fusichonetes pigmaea*, *Speciothyris flabilifomis*, *Paryphella orbicularis*, *P. sparsiplicata* and *P. sinuata* and **conodonts** *Hindeodus parvus*, *Hindeodus praeparvus*, *Hindeodus latidentatus*, *Clarkina zhejiangensis* and *Hindeodus typicalis*. 8 cm

26. Dark gray thin-bedded calcareous mudstone. **Brachiopods** *Fusichonetes pigmaea*, *Prelissorhynchia pseudoutah*, *Speciothyris flabilifomis*, *Spinomarginifera kueichowensis*, *Paracruithyris pygmaea*, *Paryphella orbicularis*, *P. sinuata*, *P. sparsiplicata*, *P. undata*, *P.* sp., *Neochonetes* sp. and lingulids and **conodonts** *Clarkina changxingensis*, *C. yini* and *C. meishanensis* were recovered. 12 cm

25. Yellowish brown claystone. 4 cm

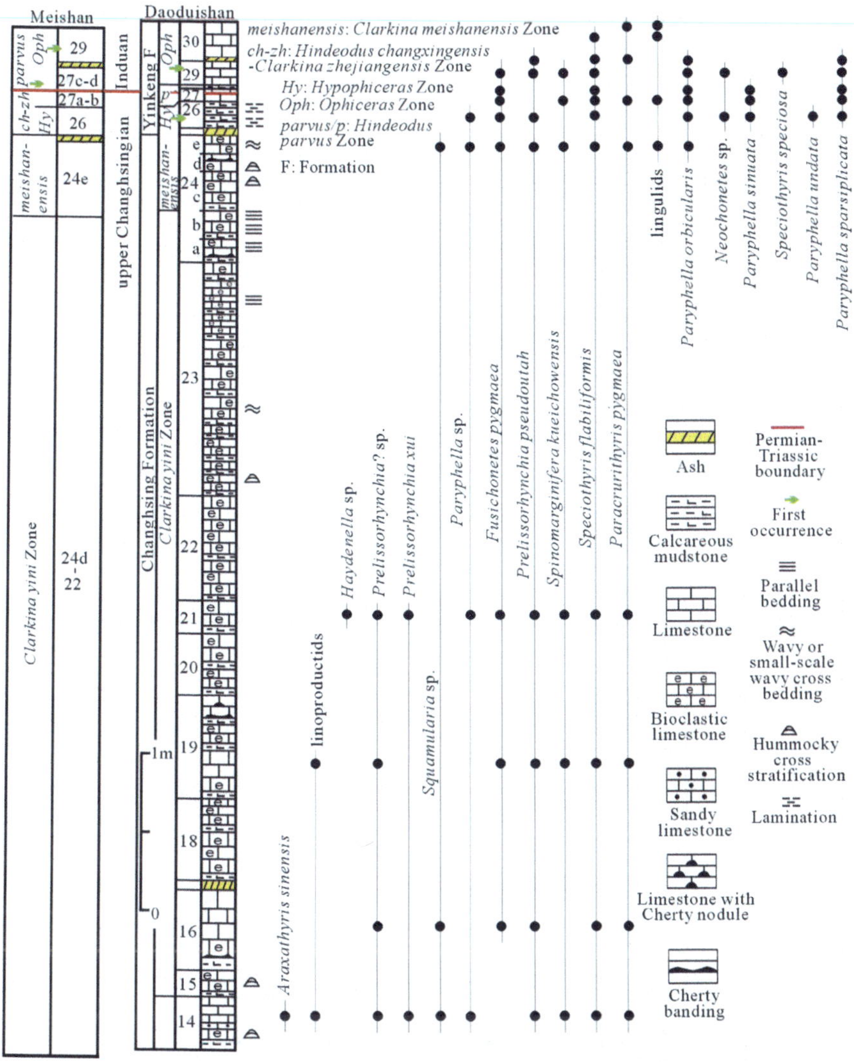

Fig. 3.33 Stratigraphic column and fossil occurrences (brachiopod) at the Daoduishan section (brachiopods referred to He et al. 2016)

Conformity
Changhsing Formation in the Upper Permian

24e-2. Light gray thin-bedded bioclastic limestone, with grayish yellow thin-bedded calcareous mudstone (0.5 cm) at the base. 6 cm

24e-1. Light gray medium-bedded bioclastic limestone, with grayish yellow thin-bedded calcareous mudstone (1 cm thick) at the base. **Brachiopods**

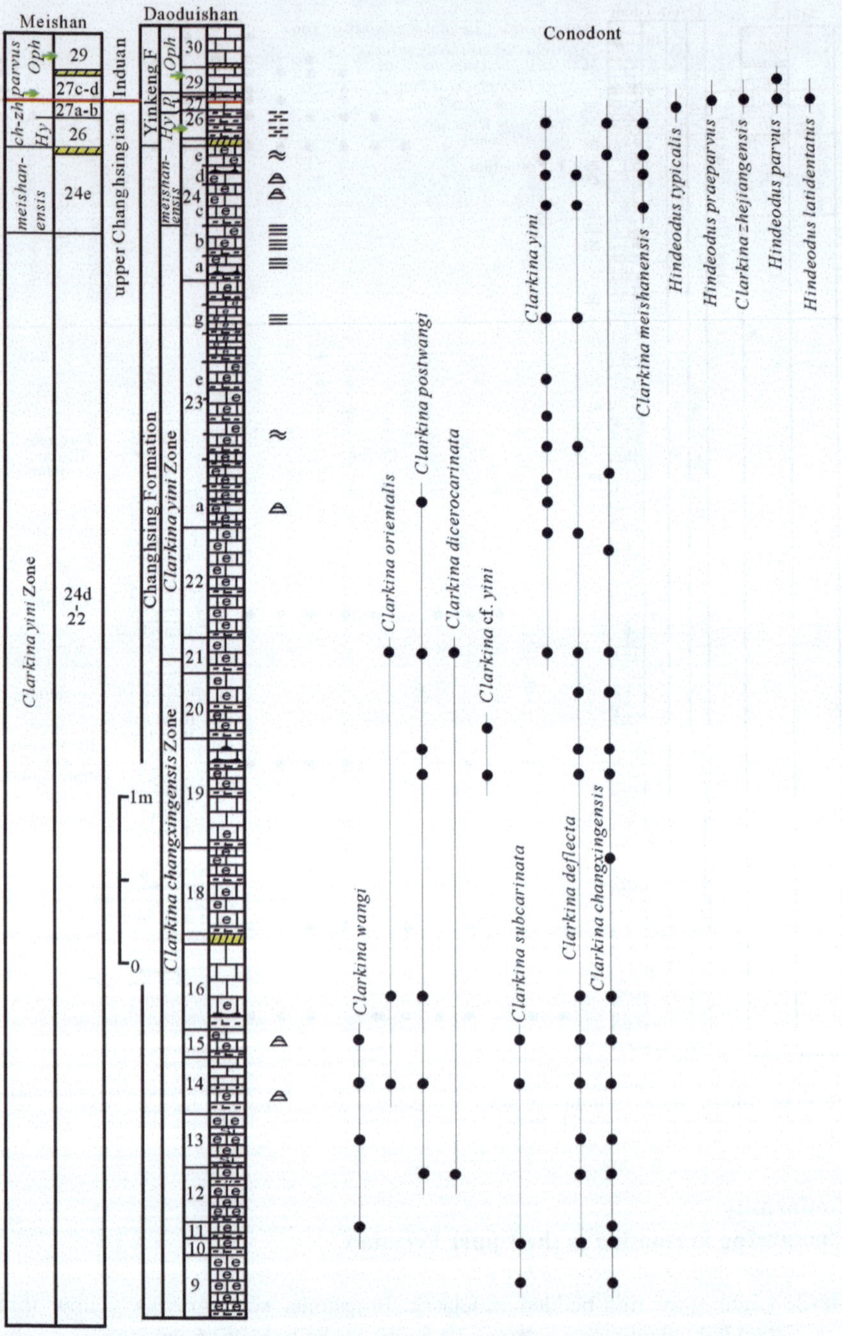

Fig. 3.34 Stratigraphic column and fossil occurrences (conodont) at the Daoduishan section. Legends and abbreviations same to Fig. 3.33

Squamularia sp., *Paryphella orbicularis*, *P.* sp., *Fusichonetes pigmaea*, *Prelissorhynchia pseudoutah*, *Spinomarginifera kueichowensis*, *Speciothyris flabilifomis*, *Paracruithyris pygmaea* and lingulids were found in the limestone. 5 cm

Additionally, **conodonts** *Clarkina changxingensis*, *C. yini* and *C. meishanensis* were examined in bed 24e.

24d. Light gray medium-bedded bioclastic limestone intercalated with cherty nodule and banding, with grayish yellow calcareous mudstone (2 cm thick) at the base. **Conodonts** *Clarkina deflecta*, *C. meishanensis* and *C. yini* were collected. 12 cm

24c. Gray thin- to medium-bedded bioclastic limestone, with grayish yellow calcareous mudstone (3 cm thick) at the base and with laminations at the top. **Conodonts** *Clarkina deflecta*, *C. meishanensis* and *C. yini* were examined. 24 cm

24b-2. Gray thin- to medium-bedded bioclastic limestone intercalated with cherty banding; grayish yellow thin-bedded calcareous mudstone at the base. 9 cm

24b-1. Dark gray medium-bedded bioclastic limestone intercalated with cherty banding; grayish yellow thin-bedded calcareous mudstone at the base. 7.5 cm

24a. Dark gray medium-bedded bioclastic limestone with cherty nodule; grayish yellow calcareous mudstone (4 cm thick) at the base. 16 cm

23h-2. Light gray bioclastic limestone, with grayish yellow mudstone (1 cm thick) at the base. 11 cm

23h-1. Gray bioclastic limestone with laminations, with grayish yellow calcareous mudstone at the base. 4 cm

23g-2. Light gray bioclastic limestone in the upper part (8 cm thick); grayish yellow calcareous mudstone altered with cherty banding in the lower part (4 cm thick). 12 cm

23g-1. Light gray bioclastic limestone, with yellow mudstone (0.5 cm thick) at the base. **Conodonts** *Clarkina deflecta* and *C. yini* were discovered in the limestone. 4.5 cm

23f-2. Gray bioclastic limestone, with grayish yellow calcareous mudstone (1.5 cm thick) at the base. 9.5 cm

23f-1. Gray bioclastic limestone, with grayish yellow calcareous mudstone at the base. 6 cm

23e-3. Gray bioclastic limestone, with grayish yellow calcareous mudstone at the base. The **conodont** *Clarkina yini* was collected in the limestone. 10 cm

23e-2. Gray bioclastic limestone intercalated with cherty banding, with grayish yellow calcareous mudstone (3 cm thick) at the base. 6 cm

23e-1. Grayish black siliceous limestone, with grayish yellow calcareous mudstone at the base. 3 cm

23d. Light gray bioclastic limestone, with grayish yellow calcareous mudstone (5 cm thick) at the base. The **conodont** *Clarkina yini* was recovered in the limestone. 19 cm

23c. Light gray bioclastic limestone with parallel beddings; grayish yellow calcareous mudstone at the base (1 cm thick). **Conodonts** *Clarkina yini* and *C. deflecta* were found. 16 cm

23b-3. Light gray bioclastic limestone, with grayish yellow calcareous mudstone at the base. The **conodont** *Clarkina changxingensis* was examined. 7 cm

23b-2. Light gray bioclastic limestone with laminations; grayish yellow calcareous mudstone at the base. The **conodont** *Clarkina yini* was identified. 4 cm

23b-1. Light gray bioclastic limestone; grayish yellow calcareous mudstone at the base. 6 cm

23a-3. Light gray bioclastic limestone; grayish yellow calcareous mudstone at the base. **Conodonts** *Clarkina postwangi* and *C. yini* were discovered in the limestone. 12 cm

23a-2. Light gray bioclastic limestone; grayish yellow calcareous mudstone at the base. 3 cm

23a-1. Gray bioclastic limestone; grayish yellow calcareous mudstone at the base (3 cm thick). 6 cm

22b-3. Light gray medium-bedded bioclastic limestone with a few cherty nodules; grayish yellow calcareous mudstone at the base (2 cm thick). **Conodonts** *Clarkina deflecta* and *C. yini* were examined in the limestone. 17 cm

22b-2. Light gray bioclastic limestone intercalated with cherty banding; yellow calcareous mudstone at the base. The **conodont** *Clarkina changxingensis* was recovered in the limestone. 5 cm

22b-1. Gray bioclastic limestone intercalated with cherty banding; pale claystone at the base (2 cm thick). 12 cm

22a-2. Gray medium-bedded bioclastic limestone; grayish yellow calcareous mudstone at the base (1 cm thick). 14 cm

22a-1. Gray medium-bedded bioclastic limestone; grayish yellow calcareous mudstone at the base (1 cm thick). 13 cm

21. Light gray medium-bedded bioclastic limestone intercalated with cherty banding, cross beddings evident; grayish yellow calcareous mudstone at the base (1 cm thick). **Conodonts** *Clarkina changxingensis, C. deflecta, C. postwangi, C. dicerocarinata, C. orientalis* and *C. yini* and **brachiopods** *Haydenella* sp., *Pressorhynchia xui, P.*? sp., *Paryphella* sp., *Fusichonetes pigmaea, Prelissorhynchia pseudoutah, Spinomarginifera kueichowensis, Speciothyris flabilifomis* and *Paracrurithyris pygmaea* were found. 22 cm

20-2. Gray medium-bedded bioclastic limestone; yellow calcareous mudstone (1 cm thick) with laminations at the base. **Conodonts** *Clarkina changxingensis* and *C. deflecta* were identified in the limestone. 24 cm

20-1. Dark gray bioclastic limestone in the upper part (8 cm thick); calcareous mudstone intercalated with cherty banding in the lower part (5 cm thick). The **conodont** *C.* cf. *yini* was examined in the limestone. 13 cm

19b. Dark gray medium-bedded bioclastic limestone with cherty nodule; grayish yellow thin-bedded calcareous mudstone at the base. **Conodonts** *Clarkina changxingensis*, *C. deflecta* and *C. postwangi* were obtained in the limestone. 19 cm

19a-3. Dark gray medium-bedded bioclastic limestone intercalated with cherty banding; grayish yellow calcareous mudstone at the base (0.5 cm thick). Conodonts *Clarkina changxingensis*, *C. deflecta*, *C. postwangi* and *C.* cf. *yini* were discovered in the limestone. 12.5 cm

19a-2. Gray medium-bedded bioclastic limestone; grayish yellow calcareous mudstone at the base. Conodonts and brachiopods were collected in the limestone. 20 cm

19a-1. Gray medium-bedded bioclastic limestone, with yellow calcareous mudstone at the base. Conodonts were obtained in the limestone. 11 cm

Brachiopods *Prelissorhynchia pseudoutah*, *Pressorhynchia*? sp., *Fusichonetes pigmaea*, *Spinomarginifera kueichowensis*, *Speciothyris flabilifomis*, *Paracrurithyris pygmaea* and linoproductids were recorded in bed 19.

18-2. Dark gray bioclastic limestone with cherty nodule; yellow calcareous mudstone and black cherty banding at the base (1 cm thick). The **conodont** *Clarkina changxingensis* was examined in the limestone. 7 cm

18-1. Dark gray medium- to thick-bedded bioclastic limestone, with cherty banding at the base. 40 cm

17. Pale claystone. 5 cm

16-2. Black chert. 8 cm

16-1. Dark gray thin- to medium-bedded bioclastic limestone intercalated with cherty banding in the upper part; yellow calcareous mudstone with laminations in the lower part. **Brachiopods** *Prelissorhynchia pseudoutah*, *Pressorhynchia*? sp., *Squamularia* sp., *Fusichonetes pigmaea*, *Speciothyris flabilifomis* and *Paracrurithyris pygmaea* and **conodonts** *Clarkina changxingensis*, *C. deflecta*, *C. postwangi* and *C. orientalis* were collected in the bed. 33 cm

15. Dark gray bioclastic limestone, with yellow calcareous mudstone at the base and with hummocky cross stratifications in the basal part of the limestone. **Conodonts** *Clarkina changxingensis*, *C. subcarinata*, *C. deflecta* and *C. wangi* were recovered in the limestone. 11 cm

14-2. Dark gray medium-bedded bioclastic limestone with cherty nodule in the upper part (17 cm thick); dark gray thin-bedded bioclastic limestone with cherty nodule in the lower part (17 cm thick). Hummocky cross stratifications and wavy cross beddings are commonly found in the basal part of the bed. **Brachiopods** linoproductids, *Orbicoelia flabiliformis*, *Paracrurithyris pygmaea*, *Araxathyris sinensis*, *Prelissorhynchia* *pseudoutah*, *Pressorhynchia*? sp., *Squamularia* sp., *Paryphella* sp., *Spinomarginifera kueichowensis* and *Speciothyris flabilifomis* and **conodonts** *Clarkina changxingensis*, *C. subcarinata*, *C. deflecta*, *C. wangi*, *C. postwangi* and *C. orientalis* were found. 34 cm

14-1. Yellow calcareous mudstone. 2 cm

13-2. Dark gray bioclastic limestone intercalated with cherty banding, with wavy cross beddings and laminations in the upper part (7 cm thick); dark gray medium-bedded bioclastic limestone intercalated with mudstone in the middle part (13 cm thick), with laminations; dark gray bioclastic limestone intercalated with cherty banding in the lower part (10 cm thick). **Conodonts** *Clarkina changxingensis*, *C. deflecta* and *C. wangi* were collected in bed 13–2. 30 cm

13-1. Yellow calcareous mudstone. 1.5 cm

12-2. Dark gray medium-bedded bioclastic limestone in the upper part; grayish yellow thin-bedded silty mudstone with wedge-shaped cross beddings in the lower part. **Conodonts** *Clarkina changxingensis*, *C. deflecta*, *C. postwangi* and *C. dicerocarinata* were examined in the limestone. 12 cm

12-1. Dark gray medium-bedded bioclastic limestone intercalated with light yellow sandy limestone, with hummocky cross stratifications in the basal part of the bioclastic limestone; yellow calcareous mudstone (1.5 cm thick) at the base. 23 cm

11. Dark gray bioclastic limestone with laminations; yellow calcareous mudstone with a scour channel at the base. **Conodonts** *Clarkina changxingensis* and *C. wangi* were discovered in the limestone. 10 cm

10. Dark gray bioclastic limestone, with grayish yellow calcareous mudstone (1 mm thick) at the base. **Conodonts** *Clarkina changxingensis*, *C. subcarinata* and *C. deflecta* were collected. 10 cm

9. Dark gray medium-bedded bioclastic limestone, with yellow calcareous mudstone (2 cm thick) at the base. **Conodonts** *Clarkina changxingensis* and *C. subcarinata* were examined in the limestone. 38 cm

8. Light gray medium-bedded bioclastic limestone intercalated with cherty banding in the upper part (11 cm thick); dark gray medium-bedded micritic limestone with chery nodule in the middle part (13 cm thick); yellow calcareous mudstone at the base (1 cm thick). 25 cm

7. Dark gray medium-bedded limestone with parallel beddings; black cherty bandings at the base. 19 cm

Unexposed Downwards

Note: Brachiopods are based on He et al. (2016), but updated in taxonomy according to He et al. (2019).

3.15 Huangzhishan Section at Huangzhishan Village, Wuxing Town

The Huangzhishan section (30° 55′ 18″ N, 119° 59′ 21″ E) is located nearby the Huangzhishan Village, Huzhou City, Zhejiang Province, South China (Fig. 3.35).

The sequence of the Huangzhishan section is composed of the upper part of Changhsing Formation and the conformably overlying Yinkeng Formation (Fig. 3.36). The upper part of the Changhsing Formation is dominated by light gray bioclastic limestones intercalated with carbonaceous mudstones. The Yinkeng Formation is characterized by calcareous mudstones (basal part of the formation) and argillaceous limestones.

Yinkeng Formation in the Uppermost Permian to the Lower Triassic Covered by the Quaternary at the top

47. Grayish green thick-bedded argillaceous limestone, with brachiopods *Lingularia* sp., Discinidae, *Fusichonetes quadrata, Speciothyris flabelliformis* and *Acosarina minuta.* > 60 cm
46. Grayish green thick-bedded argillaceous limestone intercalated with yellowish green thin-bedded calcareous mudstone, with brachiopods *Lingularia* sp. and *Speciothyris flabelliformis.* 87 cm

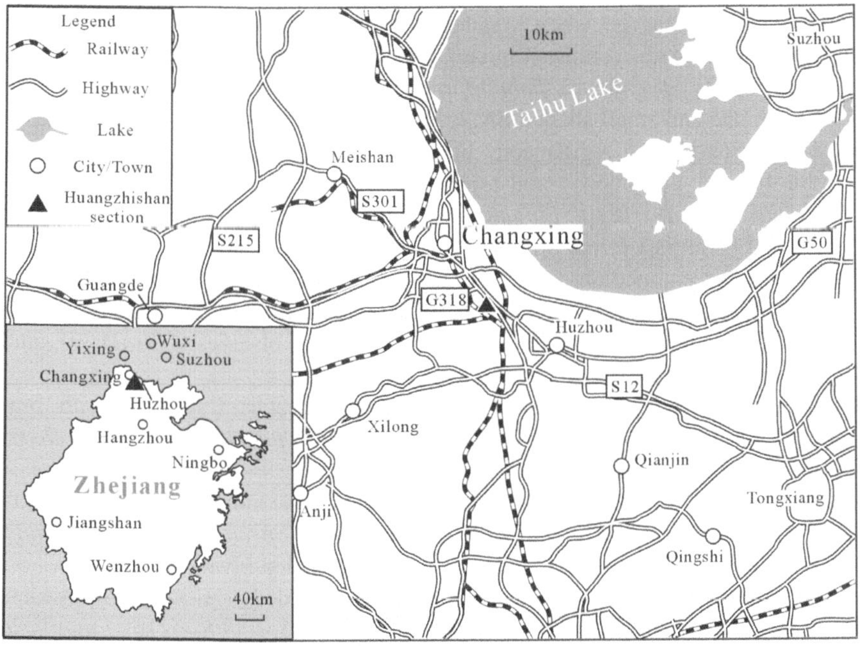

Fig. 3.35 Geographic location of the Huangzhishan section

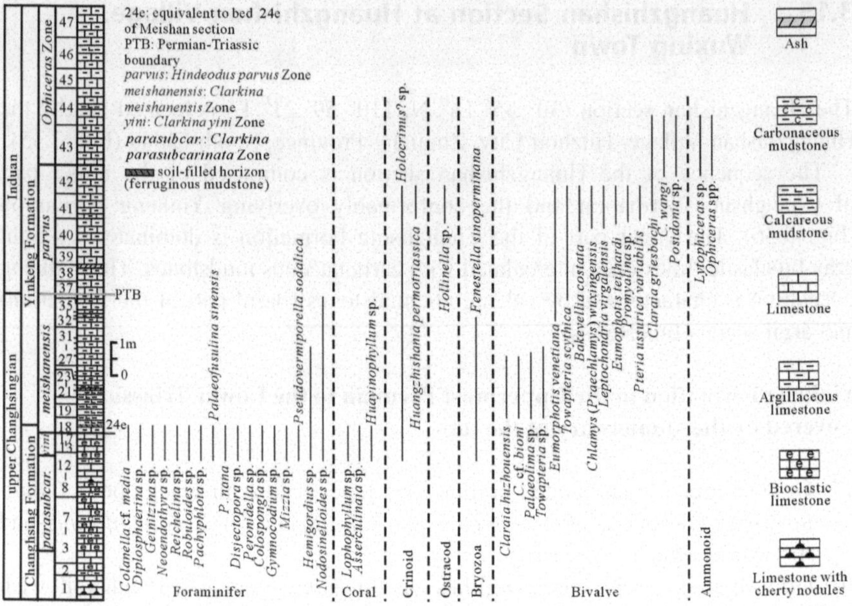

Fig. 3.36 Stratigraphic column and fossil occurrences (foraminifer, coral, crinoid, ostracod, bryozoa, bivalve) at the Huangzhishan section (fossils based on Chen et al. 2009)

45. Gray (brownish red when weathered) thick-bedded argillaceous limestone intercalated with yellowish green thin-bedded calcareous mudstone, with brachiopods *Speciothyris flabelliformis*, *Speciothyris* sp., *Lingularia* sp. and Discinidae and small-shelled bivalves. 76 cm
44. Gray thick-bedded argillaceous limestone intercalated with yellowish green thin-bedded calcareous mudstone, with brachiopods *Speciothyris flabelliformis*, *Speciothyris* sp., *Lingularia* sp. and Discinidae, ammonoid *Ophiceras* sp. and bivalves *Claraia* sp. and *Eumorphotis* sp.. 106 cm
43. Grayish green medium-bedded argillaceous limestone with laminations. A few brachiopods *Speciothyris flabelliformis*, *Speciothyris* sp., *Lingularia* sp. and Discinidae, ammonoid *Ophiceras* sp., and bivalves *Claraia* sp. and *Eumorphotis* sp. were recorded. 48 cm
42. Grayish green medium- to thin-bedded argillaceous limestone with brachiopods *Fusichonetes soochowensis*, *Speciothyris flabelliformis*, *Speciothyris* sp. and *Lingularia* sp. and bivalves. 87 cm
41. Dark gray medium-bedded argillaceous limestone altered with light gray argillaceous limestone, with a few brachiopods *Speciothyris flabelliformis*, *Speciothyris* sp. and *Lingularia* sp. and small-shelled bivalves. 78 cm
40. Gray (brownish red when weathered) medium-bedded argillaceous limestone, with a few brachiopods *Speciothyris flabelliformis*, *Speciothyris* sp., *Lingularia* sp. and *Fusichonetes quadrata.* 77 cm

39. Yellowish green (brown when weathered) medium-bedded argillaceous limestone, with brachiopods *Speciothyris flabelliformis*, *Speciothyris* sp. and *Lingularia* sp.. 40 cm
38. Gray medium-bedded argillaceous limestone altered with yellowish green argillaceous limestone, with grayish green ash at the top. A few brachiopods *Speciothyris flabelliformis*, *Speciothyris* sp., *Lingularia* sp. and *Fusichonetes quadrata*, small ammonoids and crinoids were yielded. 47 cm
37. Yellowish green medium-bedded argillaceous limestone intercalated with thin-bedded mudstone, with brachiopods *Meekella* sp., *Prelissorhynchia pseudoutah*, *Speciothyris flabelliformis*, *Speciothyris* sp., *Acosarina minuta*, *Semilingula*? sp. and *Lingularia* sp.. 35 cm
36. Yellowish green medium-bedded argillaceous limestone, with brachiopods *Prelissorhynchia pseudoutah*, *Orthothetina* sp., *Speciothyris flabelliformis*, *Speciothyris* sp., *Lingularia* sp., Discinidae, *Acosarina minuta* and *Fusichonetes quadrata.* 32 cm
35. Yellowish green medium- to thin-bedded argillaceous limestone, with brachiopods *Araxathyris* sp., *Spinomarginifera kueichowensis*, *Fusichonetes pygmaea*, *Prelissorhynchia pseudoutah*, *Speciothyris flabelliformis*, *Speciothyris* sp., *Lingularia* sp., *Fusichonetes quadrata* and *Acosarina minuta.* 28 cm
34. Yellowish green thin-bedded argillaceous limestone. Brachiopods *Spinomarginifera kueichowensis*, *Prelissorhynchia pseudoutah*, *Orthothetina* sp., *Lingularia* sp., *Fusichonetes quadrata*, *Speciothyris flabelliformis* and *Acosarina minuta* were examined. 10 cm
33. Yellowish green thin-bedded argillaceous limestone, with brachiopods *Paryphella* sp., *Spinomarginifera kueichowensis*, *Fusichonetes pygmaea*, *Fusichonetes quadrata*, *Speciothyris flabelliformis*, *Speciothyris* sp. and *Acosarina minuta.* 6 cm
32. Gray (brownish red when weathered) thin-bedded argillaceous limestone altered with yellowish green thin-bedded argillaceous limestone. Brachiopods *Haydenella kiangsiensis*, *Paryphella* sp., *Araxathyris* sp., *Spinomarginifera kueichowensis*, *Fusichonetes pygmaea*, *Speciothyris* sp., *Lingularia* sp., *Fusichonetes quadrata*, *Speciothyris flabelliformis* and *Acosarina minuta* were recovered. 23 cm
31. Yellowish green medium-bedded argillaceous limestone with laminations. Brachiopods *Uncinunellina* sp., *Paryphella* sp., *Spinomarginifera kueichowensis*, *Orthothetina* sp., *Speciothyris* sp., *Lingularia* sp., *Fusichonetes quadrata*, *Speciothyris flabelliformis* and *Acosarina minuta* were recorded. 40 cm
30. Yellowish green thin-bedded argillaceous limestone with laminations. Brachiopods *Spinomarginifera kueichowensis*, *Fusichonetes pygmaea*, *Orthothetina* sp., *Speciothyris* sp., *Fusichonetes quadrata*, *Speciothyris flabelliformis* and *Acosarina minuta* were discovered. 15 cm
29. Yellowish green medium-bedded argillaceous limestone with laminations. Brachiopods *Leptodus nobilis*, *Paryphella* sp., *Spinomarginifera kueichowensis*, *Fusichonetes pygmaea*, *Prelissorhynchia pseudoutah*, *Orthothetina* sp.,

Speciothyris sp., *Fusichonetes quadrata*, *Speciothyris flabelliformis* and *Acosarina minuta* were collected. 25 cm

28. Yellowish green medium-bedded argillaceous limestone, with laminations in the lower part. Moderately abundant brachiopods *Paryphella sinuata*, *Uncinunellina* sp., *Haydenella kiangsiensis*, *Paryphella* sp., *Spinomarginifera kueichowensis*, *Fusichonetes pygmaea*, *Prelissorhynchia pseudoutah*, *Orthothetina* sp., *Speciothyris* sp., *Lingularia* sp., Discinidae, *Fusichonetes quadrata*, *Speciothyris flabelliformis* and *Acosarina minuta* were recorded. 15 cm
27. Yellowish green medium-bedded argillaceous limestone, with moderately abundant brachiopods *Neochonetes* sp., *Derbyia schellwieni*, *Uncinunellina* sp., *Haydenella kiangsiensis*, *Paryphella* sp., *Araxathyris* sp., *Spinomarginifera kueichowensis*, *Fusichonetes pygmaea*, *Prelissorhynchia pseudoutah*, *Fusichonetes quadrata*, *Speciothyris flabelliformis* and *Acosarina minuta.* 20 cm
26. Yellowish green medium-bedded argillaceous limestone, with laminations in the lower part. 12 cm
25. Yellowish green medium-bedded argillaceous limestone, with moderately abundant brachiopods *Paryphella orbicularis*, *Leptodus nobilis*, *Uncinunellina* sp., *Haydenella kiangsiensis*, *Paryphella* sp., *Spinomarginifera kueichowensis*, *Prelissorhynchia pseudoutah*, *Fusichonetes soochowensis*, *Lingularia* sp., Discinidae, *Fusichonetes quadrata*, *Speciothyris flabelliformis* and *Acosarina minuta.* 12 cm
24. Yellowish green medium- to thin-bedded argillaceous limestone with a few spongy spicules, bivalves and algaes. 20 cm
23. Yellowish green medium-bedded argillaceous limestone, with a few bivalves and moderately abundant brachiopods *Paryphella orbicularis*, *Leptodus nobilis*, *Uncinunellina* sp., *Neochonetes* sp., *Derbyia schellwieni*, *Paryphella* sp., *Spinomarginifera kueichowensis*, *Fusichonetes pygmaea*, *Prelissorhynchia pseudoutah*, *Fusichonetes quadrata*, *Speciothyris flabelliformis* and *Acosarina minuta.* 20 cm
22. Yellowish green medium-bedded calcareous mudstone, with pale volcanic ash at the base. A few small-shelled bivalves were yielded. 15 cm
21. Yellowish green thin-bedded calcareous mudstone with a few small-shelled bivalves. 15 cm
20. Yellowish green (purplish red when weathered) medium- to thin-bedded calcareous mudstone. Moderately abundant brachiopods *Paryphella orbicularis*, *Derbyia schellwieni*, *Leptodus nobilis*, *Paryphella* sp., *Spinomarginifera kueichowensis*, *Fusichonetes pygmaea*, *Prelissorhynchia pseudoutah*, *Fusichonetes soochowensis*, *Lingularia* sp., Discinidae, *Fusichonetes quadrata*, *Speciothyris flabelliformis* and *Acosarina minuta* were recovered. 17 cm
19. Yellowish green medium-bedded calcareous mudstone with abundant brachiopods *Fusichonetes flatus*, *Fusichonetes quadrata*, *Fusichonetes pygmaea*, *Fusichonetes soochowensis*, *Fusichonetes* sp., *Keyserlingina guizhouensis*, *Paryphella transversa*, *Oldhamina* sp., *Speciothyris speciosa*, *Speciothyris flabelliformis*, *Leptodus cancriniformis*, *Spinomarginifera alpha*, *Paryphella*

sinuata, Paryphella orbicularis, Paryphella sp., *Neochonetes* sp., *Derbyia schellwieni, Leptodus nobilis, Haydenella kiangsiensis, Araxathyris* sp., *Spinomarginifera kueichowensis, Meekella* sp., *Prelissorhynchia pseudoutah, Orthothetina* sp., *Lingularia* sp. and *Acosarina minuta.* 25 cm

18. Yellowish green thin-bedded calcareous mudstone in the upper part; purplish red ferruginous mudstone (2 cm thick, soil-filled) in the middle part; light gray medium-bedded argillaceous limestone in the lower part. Abundant bivalves, ammonoids, corals and brachiopods *Spiriferellina fastigata, Alphaneospirifer* sp., *Paryphella undata, Paryphella transversa, Paryphella orbicularis, Paryphella sinuata, Martinia* sp., *Terebratuloidea depressa, Fusichonetes pygmaea, Fusichonetes flatus, Fusichonetes soochowensis, Fusichonetes quadrata, Keyserlingina guizhouensis, Derbyia schellwieni, Leptodus nobilis, Uncinunellina* sp., *Araxathyris* sp., *Spinomarginifera kueichowensis, Meekella* sp., *Prelissorhynchia pseudoutah, Lingularia* sp., *Acosarina minuta* and *Speciothyris flabelliformis* were recorded in the upper part. 28 cm

Conformity
Changhsing Formation in the Upper Permian

17. Light gray (purplish red on the top surface of bed when weathered) thin-bedded limestone; black carbonaceous mudstone (0.5 cm thick) at the base. 11 cm
16. Light gray (purplish red on the top surface of bed when weathered) thin-bedded bioclastic limestone; black carbonaceous mudstone (0.5 cm thick) at the base. 7 cm
15. Light gray (purplish red when weathered) thin-bedded bioclastic limestone; black carbonaceous mudstone (0.3 cm thick) at the base. 10 cm
14. Light gray (purplish red when weathered) thin-bedded limestone; black carbonaceous mudstone (0.3 cm thick) at the base. 7 cm
13. Light gray thin-bedded bioclastic limestone intercalated with two layers of black carbonaceous mudstone; black carbonaceous mudstone (0.3 cm thick) at the base. 21 cm
12. Light gray medium-bedded bioclastic limestone with sponges, algaes and corals; black carbonaceous mudstone (0.3 cm thick) at the base. 13 cm
11. Light gray thin-bedded limestone; black carbonaceous mudstone (0.3 cm thick) at the base. 21 cm
10. Gray medium-bedded limestone with cherty nodules; black carbonaceous mudstone at the base. 31 cm
9. Light gray medium- to thin-bedded bioclastic limestone with cherty nodules; black carbonaceous mudstone (0.3 cm thick) at the base. 40 cm
8. Light gray thin-bedded bioclastic limestone; black carbonaceous mudstone (0.3 cm thick) at the base. 37 cm
7. Light gray thick-bedded bioclastic limestone with sponges, bryozoans, corals, crinoids and cephalopods; black carbonaceous mudstone at the base. 54 cm

6. Light gray (brownish red on the top surface of bed when weathered) medium-bedded bioclastic limestone with abundant bryozoans, sponges, corals and fusulinids; black carbonaceous mudstone (0.5 cm thick) at the base. 57 cm
5. Light gray medium-bedded bioclastic limestone with a few cherty nodules, corals found; black carbonaceous mudstone (0.5 cm thick) at the base. 40 cm
4. Light gray medium-bedded bioclastic limestone with cherty nodules; black carbonaceous mudstone (0.5 cm thick) at the base. 20 cm
3. Light gray medium- to thin-bedded bioclastic limestone; black carbonaceous mudstone (0.5 cm thick) at the base. 60 cm
2. Light gray medium-bedded bioclastic limestone with black cherty nodules, abundant brachiopods and corals discovered; black carbonaceous mudstone (0.5 cm thick) at the base. 37 cm
1. Light gray medium-bedded limestone with cherty nodules. > 10 cm

Note: Brachiopods referred to He et al. (2015), but updated in taxonomy; additionally, conodonts, foraminifers, corals, crinoids, bryozoans, ostracods and bivalves have been recorded in the upper part of the Changhsing Formation (Chen et al. 2008, 2009; Figs. 3.36, 3.37).

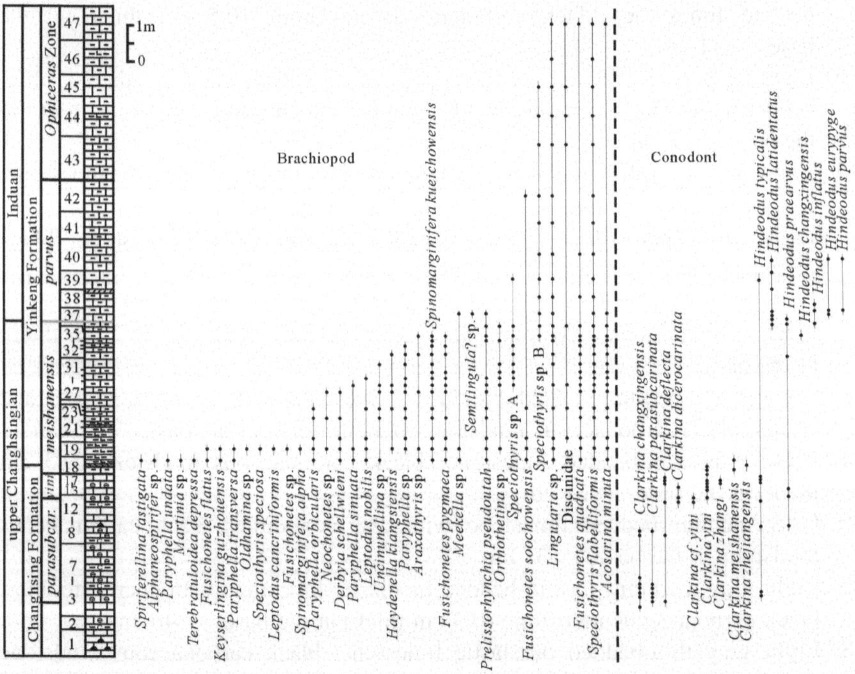

Fig. 3.37 Stratigraphic column and fossil occurrences (brachiopod, conodont) at the Huangzhishan section. *Note* brachiopods referred to He et al. (2015); conodonts referred to Chen et al. (2008, 2009); legends and abbreviations same to Fig. 3.36

3.16 Yangou Section in Leping City

The Yangou section is located at a quarry nearby a reservoir, about 1.5 km northeast to the Yangou Coal Mine (or about 5 km northeast to the Yanshan Town), Leping City, Jiangxi Province, China (Fig. 3.38).

The sequence of the Yangou section is composed of the Changhsing Formation and the conformably overlying Taye Formation (Fig. 3.39). The Changhsing Formation is dominated by dark gray thick-bedded bioclastic limestones. The Taye Formation is characterized by dark gray to grayish green argillaceous limestones. Abundant foraminifers, conodonts and gastropods were discovered in the section (Figs. 3.39, 3.40).

Taye Formation in the Lower Triassic (Induan)

47. Dark gray medium-bedded argillaceous limestone intercalated with mudstone. 80 cm
46. Dark gray medium-bedded argillaceous limestone intercalated with mudstone, laminations evident. 120 cm
45. Grayish green medium- to thin-bedded argillaceous limestone with laminations and pyrites. 100 cm
44. Grayish green thin-bedded argillaceous limestone intercalated with mudstone, laminations prevalent. 40 cm

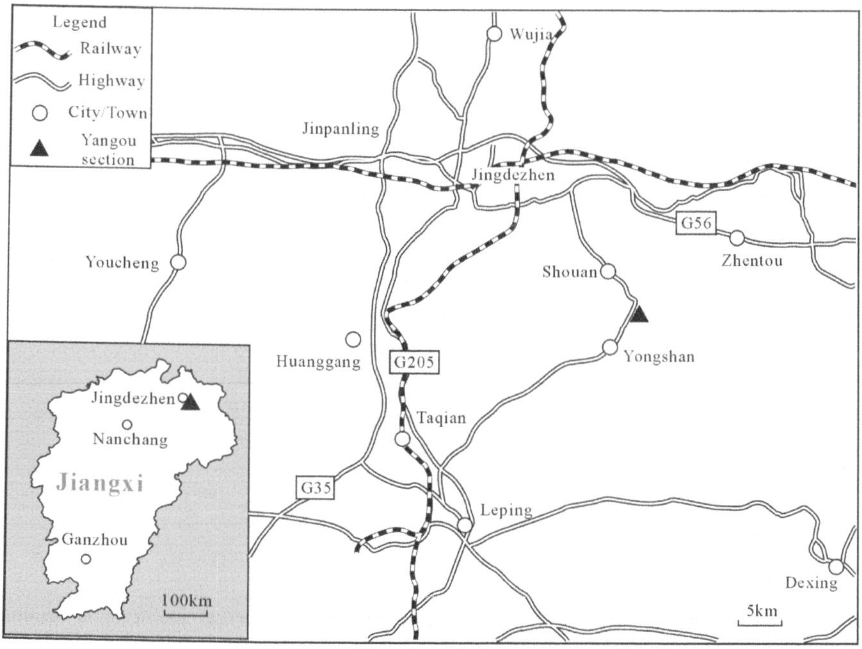

Fig. 3.38 Geographic location of the Yangou section

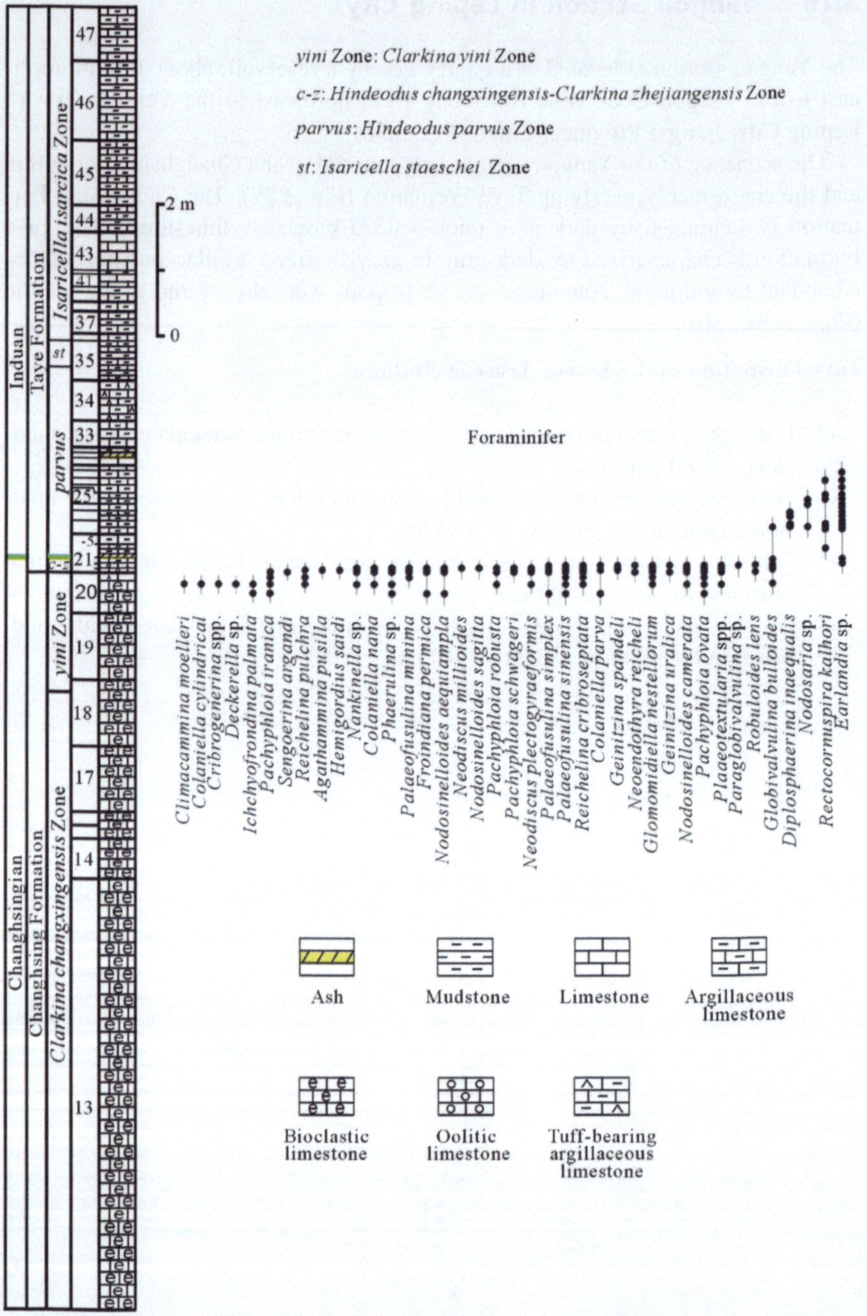

Fig. 3.39 Stratigraphic column and fossil occurrences (foraminifer) at the Yangou section (foraminifers referred to Tian et al. 2014a)

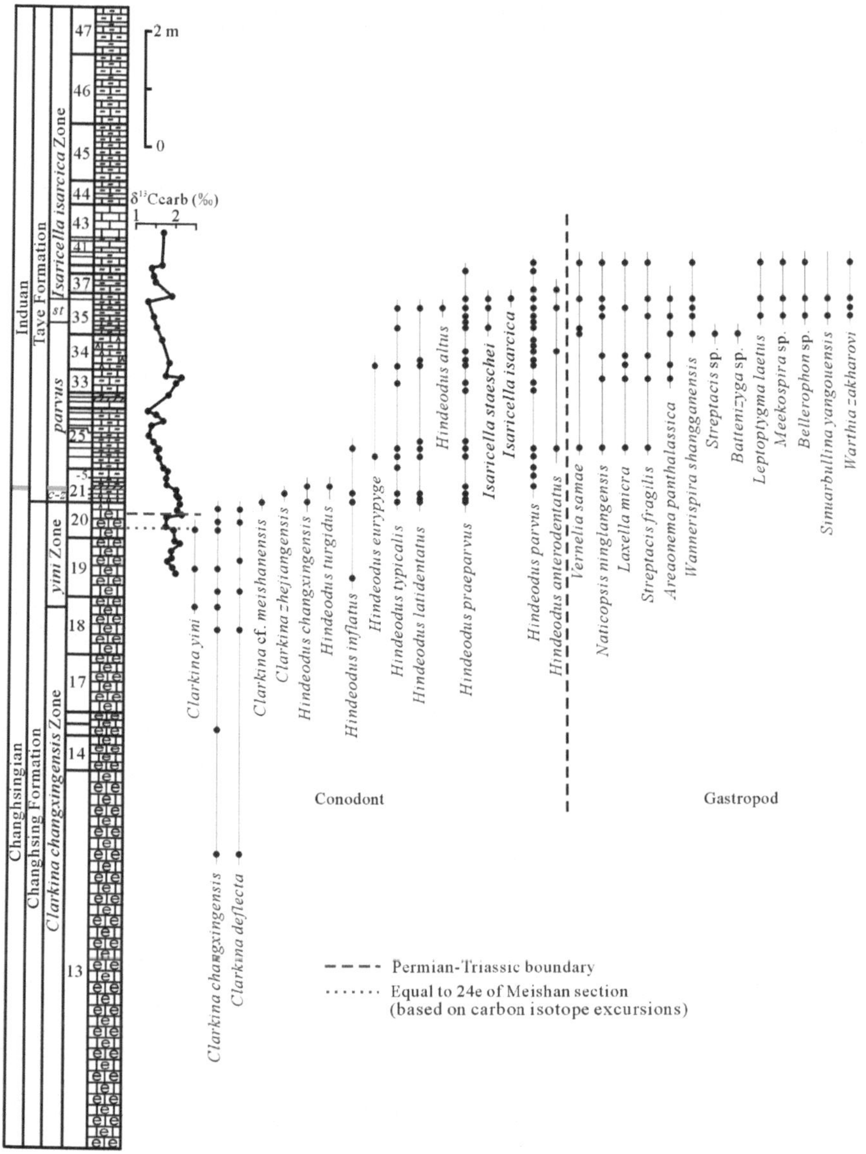

Fig. 3.40 Stratigraphic column and fossil occurrences (conodont, gastropod) at the Yangou section. *Note* Data of conodonts referred to Sun et al. (2012), gastropods after Sun et al. (2021), carbon isotope after Tian et al. (2014b). Legends and other abbreviations same to Fig. 3.39

43. Grayish green limestone with laminations, rare foraminifers found. 57 cm
42. Yellow mudstone. 2 cm
41. Dark gray medium-bedded limestone with abundant pyrites. 19 cm
40. Dark gray mudstone with a few crinoids. 4 cm
39. Dark gray thin-bedded argillaceous limestone, with mudstone (1 cm) at the top part. The conodont *Hindeodus parvus* and gastropods *Vernelia samae, Naticopsis ninglangensis, Laxella micra, Streptacis fragilis, Wannerispira shangganensis, Leptoptygma laetus, Meekospira* sp., *Bellerophon* sp. and *Warthia zakharovi* were identified in the limestone. 12 cm
38. Dark gray thin-bedded argillaceous limestone intercalated with mudstone, with conodonts *Hindeodus parvus* and *Hindeodus praeparvus* and rare gastropods. 12 cm
37. Gray medium-bedded argillaceous limestone, intercalated with mudstone in the upper part. Conodonts *Hindeodus parvus* and *Hindeodus anterodentatus* were collected in the limestone. 22 cm
36. Grayish green argillaceous limestone, intercalated with mudstone in the upper part. Abundant gastropods *Vernelia samae, Naticopsis ninglangensis, Streptacis fragilis, Areaonema panthalassica, Wannerispira shangganensis, Leptoptygma laetus, Meekospira* sp., *Bellerophon* sp., *Sinuarbullina yangouensis* and *Warthia zakharovi*, ostracods and conodonts *Hindeodus praeparvus, Hindeodus parvus, Isarcicella staeschei* and *Isaricella isarcica* were found in the limestone. 10 cm
35. Dark gray thin-bedded argillaceous limestone intercalated with mudstone, with abundant gastropods *Vernelia samae, Naticopsis ninglangensis, Laxella micra, Streptacis fragilis, Areaonema panthalassica, Wannerispira shangganensis, Streptacis* sp., *Battenizyga* sp., *Leptoptygma laetus, Meekospira* sp., *Bellerophon* sp., *Sinuarbullina yangouensis* and *Warthia zakharovi*, ostracods, and conodonts *Hindeodus praeparvus, Hindeodus typicalis, Hindeodus parvus, Hindeodus latidentatus, Hindeodus anterodentatus, Hindeodus altus* and *Isarcicella staeschei.* 60 cm
34. Dark gray tuff-bearing argillaceous limestone with pyrites, abundantly with gastropods *Naticopsis ninglangensis, Laxella micra, Streptacis fragilis* and *Areaonema panthalassica*, crinoids, and conodonts *Hindeodus praparvus, Hindeodus typicalis, Hindeodus latidentatus, Hindeodus parvus, Hindeodus eurypyge* and *Hindeodus anterodentatus.* 60 cm
33. Dark gray to grayish green medium- to thin-bedded argillaceous limestone, abundantly with gastropods *Naticopsis ninglangensis, Laxella micra, Streptacis fragilis* and *Areaonema panthalassica*, and conodonts *Hindeodus praeparvus, H. parvus* and *H. typicalis.* 40 cm
32. Yellow mudstone. 3 cm
31. Pale to grayish green volcanic ash. 4 cm
30. Grayish green to yellow thin-bedded micritic limestone, with a few bivalves. 11 cm
29. Grayish green micritic limestone. 5 cm

28. Gray argillaceous limestone with a few gastropods and foraminifer *Earlandia* sp.. 10 cm
27. Gray thin-bedded argillaceous limestone with laminations. Foraminifers *Rectocornuspira kalhori* and *Earlandia* sp. were recovered. 11 cm
26. Dark gray thin-bedded argillaceous limestone with abundant pyrites. Gastropods, conodont *Hindeodus altus*, and foraminifer *Earlandia* sp. were collected. 12 cm
25. Dark gray thin-bedded argillaceous limestone with abundant pyrite framboids, with conodonts *Hindeodus praeparvus* and *H. latidentatus* and foraminifers *Nodosaria* sp., *Rectocornuspira kalhori* and *Earlandia* sp.. 15 cm
24. Dark gray argillaceous limestone with pyrite framboids. Gastropods *Vernelia samae*, *Naticopsis ninglangensis*, *Laxella micra* and *Streptacis fragilis*, conodonts *Hindeodus inflatus*, *H. praeparvus*, *H. typicalis*, *H. latidentatus*, *H. parvus* and *H. anterodentatus* and rare foraminifers *Nodosaria* sp. and *Earlandia* sp. were discovered. 10 cm
23. Dark gray thin-bedded argillaceous limestone with pyrite framboids. A few gastropods, some conodonts *Hindeodus praeparvus*, *H. typicalis*, *H. latidentatus*, *H. parvus* and *H. eurypyge* and rare foraminifers *Diplosphaerina inaequalis* and *Earlandia* sp. were collected. 15 cm
22. Dark gray thin-bedded argillaceous limestone with laminations. A few gastropods, conodonts *Hindeodus parvus* and *H. typicalis*, and foraminifers *Globivalvulina bulloides*, *Diplosphaerina inaequalis*, *Nodosaria* sp., *Rectocornuspira kalhori* and *Earlandia* sp. were found. 20 cm

21-5. Dark gray medium-bedded argillaceous limestone with a few pyrites; volcanic ash at the base. The conodont *Hindeodus parvus* was examined. 14 cm

21-4. Dark gray medium- to thin-bedded limestone, with volcanic ash at the base. Conodonts *Hindeodus parvus*, *H. changxingensis*, *H. praeparvus* and *H. turgidus* and the foraminifer *Rectocornuspira kalhori* were recorded. 20 cm

21-3. Dark gray argillaceous limestone with a few pyrites. Conodonts *Hindeodus inflatus*, *H. praeparvus*, *H. typicalis*, *H. latidentatus* and *Clarkina zhejiangensis* and a few foraminifers *Globivalvulina bulloides* and *Earlandia* sp. were examined. 10 cm

21-2. Dark gray tuff-bearing oolitic limestone with laminations. A few algaes, conodonts *Hindeodus praeparvus* and *H. latidentatus*, and abundant foraminifers *Nankinella* sp., *Colaniella nana*, *Phaerulina* sp., *Palaeofusulina minima*, *Froindiana permica*, *Nodosinelloides aequiampla*, *Neodiscus millioides*, *Nodosinelloides sagitta*, *Pachyphloia robusta*, *Pachyphloia schwageri*, *Neodiscus plectogyraeformis*, *Palaeofusulina simplex*, *Palaeofusulina sinensis*, *Reichelina cribroseptata*, *Colaniella parva*, *Geinitzina spandeli*, *Neoendothyra reicheli*, *Glomomidiella nestellorum*, *Geinitzina uralica*, *Nodosinelloides camerata*, *Pachyphloia ovata*, *Plaaeotextularia* spp., *Paraglobivalvulina* sp., *Robuloides lens*, *Globivalvulina bulloides* and *Earlandia* sp. were recovered. 8 cm

21-1. Dark gray oolitic limestone with laminations. A few algaes, conodonts *Clarkina* cf. *meishanensis*, *Hindeodus changxingensis*, *H. inflatus*, *H. praeparvus*, *H. typicalis*, *H. latidentatus*, *Clarkina changxingensis*, *C. yini* and *C. deflecta*, and abundant foraminifers *Pachyphloia iranica*, *Sengoerina argandi*, *Reichelina pulchra*, *Agathammina pusilla*, *Hemigordius saidi*, *Nankinella* sp., *Colaniella nana*, *Phaerulina* sp., *Palaeofusulina minima*, *Froindiana permica*, *Nodosinelloides aequiampla*, *Pachyphloia robusta*, *Pachyphloia schwageri*, *Neodiscus plectogyraeformis*, *Palaeofusulina simplex*, *Palaeofusulina sinensis*, *Reichelina cribroseptata*, *Colaniella parva*, *Geinitzina spandeli*, *Neoendothyra reicheli*, *Glomomidiella nestellorum*, *Geinitzina uralica*, *Nodosinelloides camerata*, *Pachyphloia ovata*, *Plaaeotextularia* spp., *Robuloides lens* and *Globivalvulina bulloides* were discovered. 2.5 cm

Conformity (with sedimentation hiatus)
Changhsing Formation in the Upper Permian (Changhsingian)

20. Gray thick-bedded bioclastic limestone, with dark gray tuffaceous limestone at the top part and pyrites in the lower part. Abundant algaes, ostracods, sponges, gastropods, crinoids, bryozoans, brachiopods, bivalves, conodonts *Clarkina changxingensis*, *C. yini* and *C. deflecta*, and foraminifers *Climacammina moelleri*, *Colaniella cylindrical*, *Cribrogenerina* spp., *Deckerella* sp., *Ichchyofrondina palmata*, *Pachyphloia iranica*, *Reichelina pulchra*, *Nankinella* sp., *Colaniella nana*, *Phaerulina* sp., *Palaeofusulina minima*, *Froindiana permica*, *Nodosinelloides aequiampla*, *Pachyphloia robusta*, *Neodiscus plectogyraeformis*, *Palaeofusulina simplex*, *Palaeofusulina sinensis*, *Reichelina cribroseptata*, *Colaniella parva*, *Glomomidiella nestellorum*, *Nodosinelloides camerata*, *Pachyphloia ovata*, *Plaaeotextularia* spp. and *Globivalvulina bulloides* were collected. 60 cm
19. Gray thick-bedded bioclastic limestone, with pyrites in the upper part. Abundant algaes, fusulinids, gastropods, crinoids, brachiopods and conodonts *Clarkina changxingensis*, *C. deflecta*, *C. yini* and *Hindeodus inflatus* were found. 100 cm
18. Gray thick-bedded bioclastic limestone. Abundant algaes, fusulinids, gastropods, crinoids, brachiopods and conodonts *Clarkina changxingensis*, *C. yini* and *C. deflecta* were obtained. 100 cm
17. Gray thick-bedded bioclastic limestone with foraminifers, brachiopods, crinoids, gastropods and algaes. 100 cm
16. Gray medium-bedded bioclastic limestone with foraminifers, brachiopods, crinoids and algaes. 20 cm
15. Gray medium-bedded bioclastic limestone with foraminifers, brachiopods, crinoids, algaes and conodont *Clarkina changxingensis*. 20 cm

14. Gray medium-bedded bioclastic limestone with algaes, crinoids, bryozoans, gastropods, brachiopods, foraminifers and bivalves. 60 cm
13. Gray thick-bedded bioclastic limestone with algaes, crinoids, bryozoans, gastropods, brachiopods, bivalves, foraminifers and conodonts *Clarkina changxingensis* and *C. deflecta.* 650 cm

Note: Conodonts are referred to Sun et al. (2012) (but updated in taxonomy in this book), foraminifers are based on Tian et al. (2014a), gastropods are based on Sun et al. (2021).

3.17 Chibi Section in Chibi County

The Chibi section is located about 4 km west to the Chibi County, Hubei Province, South China (Fig. 3.41).

The sequence of the Chibi section is composed of the Changhsing and Taye formations in an ascending order (Fig. 3.42). The Changhsing Formation is mainly composed of medium- to thin-bedded chert-nodule limestones. The Taye Formation is characterized by thin-bedded argillaceous limestones intercalated with

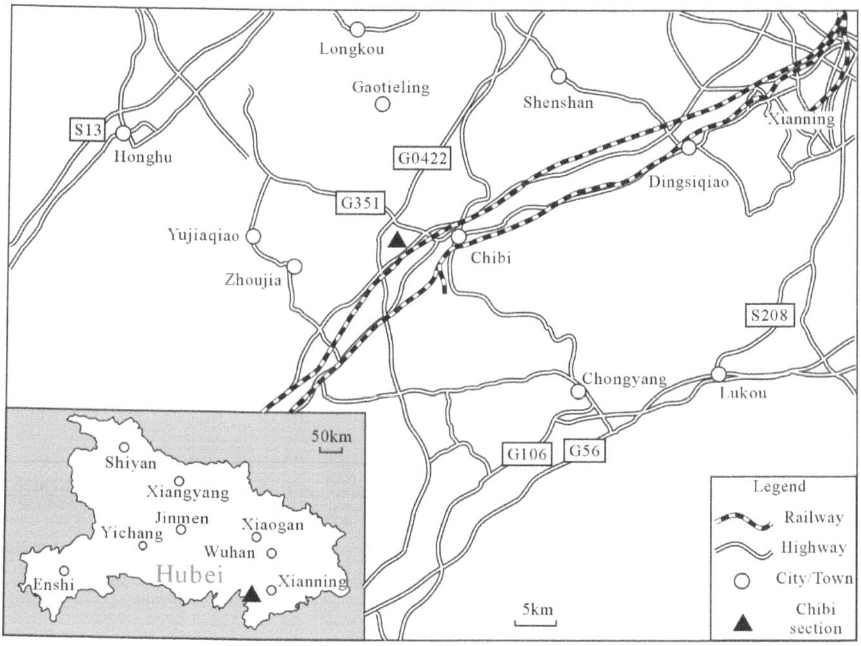

Fig. 3.41 Geographic location of the Chibi section

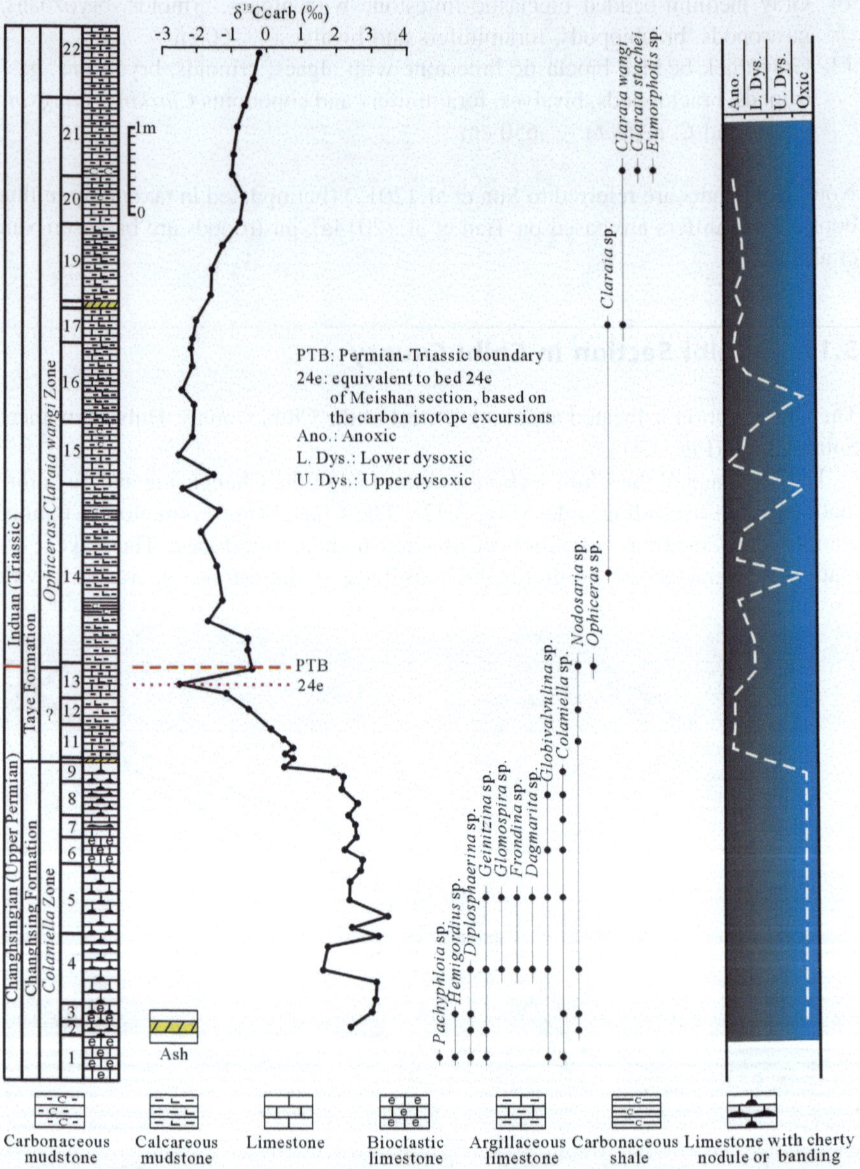

Fig. 3.42 Stratigraphic column and fossil occurrences at the Chibi section. *Note* Data of fossils, redox conditions based on framboid pyrites and carbon isotopes are referred to Li (2016)

calcareous mudstones. A few foraminifers were found in the Changhsing Formation and the basal part of Taye Formation, and a few bivalves and ammonoids were discovered in the Taye Formation (Fig. 3.42).

Taye Formation in the Lower Triassic (Induan)

22. Gray medium- to thin-bedded argillaceous limestone. 82 cm
21. Gray medium-bedded argillaceous limestone, with black carbonaceous mudstone at the basal part (13 cm). Bivalves *Claraia wangi* and *Claraia stachei* (abundant for both genera) and *Eumorphotis* sp. (a few) were found at the basal part. 89 cm
20. Gray medium-bedded argillaceous limestone intercalated with a layer of grayish black carbonaceous mudstone (2 cm). 56 cm
19. Grayish black calcareous mudstone. 87 cm
18. Grayish green volcanic ash. 6 cm
17. Gray thin-bedded argillaceous limestone intercalated with grayish black calcareous mudstone with laminations. Bivalves *Claraia wangi* and *Claraia* spp. were abundantly yielded in the mudstone. 40 cm
16. Gray thin-bedded calcareous mudstone with three layers (4 cm for each layer) of argillaceous limestone in the middle and upper parts. Bivalves and microgastropods were discovered. 92 cm
15. Gray thin-bedded argillaceous limestone, with two layers of carbonaceous mudstone in the middle and upper parts, and with nodular limestone at the top part (20 cm). Microgastropods and ostracods were found. 71 cm
14. Grayish black calcareous mudstone intercalated with black carbonaceous shale, with a few bivalve *Claraia* sp. and brachiopods. 204 cm
13. Grayish black thin-bedded argillaceous limestone intercalated with calcareous mudstone, with ammonoid *Ophiceras* sp. and non-fusulinid foraminiferal *Nodosaria* sp.. 40 cm
12. Grayish black calcareous mudstone, laminations evident. The non-fusulinid foraminiferal *Nodosaria* sp. was examined. 30 cm
11. Grayish black thin-bedded argillaceous limestone, with ostracods and non-fusulinid foraminiferal *Nodosaria* sp.. 38 cm
10. Pale volcanic ash (yellowish brown when weathered). 2 cm

Conformity
Changhsing Formation in the Upper Permian (Changhsingian)

9. Gray black medium-bedded limestone with abundant chery nodules. The non-fusulinid foraminiferal *Colaniella* sp. was examined. 22 cm
8. Grayish black thin-bedded limestone with cherty banding. The non-fusulinid foraminiferal *Colaniella* sp. and *Globivalvulina* sp. were discovered. 38 cm
7. Grayish black thin-bedded limestone intercalated with cherty banding. The non-fusulinid foraminiferal *Colaniella* sp. was obtained. 26 cm
6. Gray medium-bedded bioclastic limestone. The non-fusulinid foraminiferal *Colaniella* sp. and *Globivalvulina* sp. were recovered. 30 cm

5. Grayish black medium-bedded limestone with chery nodules, with calcareous algaes and crinoids. Additionally, the foraminifers *Colaniella* sp., *Globivalvulina* sp., *Dagmarita* sp., *Glomospira* sp., *Geinitzina* sp. and *Frondina* sp. were examined. 80 cm
4. Grayish black thin-bedded limestone with cherty nodules and bandings. Ostracods and foraminifers *Diplosphaerina* sp., *Globivalvulina* sp., *Glomospira* sp., *Geinitzina* sp., *Nodosaria* sp., *Frondina* sp. and *Dagmarita* sp. were found. 78 cm
3. Gray medium-bedded bioclastic limestone with cherty nodules. Calcareous algaes, crinoids, and foraminfers *Diplosphaerina* sp., *Colaniella* sp., *Hemigordius* sp. and *Nodosaria* sp. were collected. 21 cm
2. Grayish black thin-bedded limestone with cherty nodules. Ostracods and foraminfers *Diplosphaerina* sp., *Geinitzina* sp. and *Nodosaria* sp. were discovered. 14 cm
1. Gray medium-bedded bioclastic limestone, with a few crinoids and abundant foraminifers *Globivalvulina* sp., *Diplosphaerina* sp., *Hemigordius* sp., *Geinitzina* sp., *Colaniella* sp. and *Pachyphloia* sp.. 50 cm

3.18 Liangfengya Section in Chongqing City

The Liangfengya section (29° 30′ 14″ N, 106° 24′ 15″ E) is situated about 1 km southwest of Liangfengya Village, Chongqing City, South China (Fig. 3.43).

The sequence of the Liangfengya section is composed of the Changhsing Formation and the conformably overlying Feixianguan Formation (Fig. 3.44). The Changhsing Formation is dominated by light gray to gray thick-bedded bioclastic limestones and micritic limestones, occasionally intercalated with thin-bedded calcareous mudstones and argillaceous limestones. The basal part of Feixianguan Formation is characterized by grayish green thin-bedded argillaceous limestones interbedded with calcareous limestones.

Feixianguan Formation in the Lower Triassic

36. Grayish green thin-bedded calcareous mudstone. 10 cm
35. Grayish green (brown when weathered) thin-bedded argillaceous limestone. 19 cm
34. Grayish green (brown when weathered) thin-bedded argillaceous limestone. Abundant small brachiopods *Acosarina minuta*, *Speciothyris speciosa* and *Lingularia* sp., bivalves *Towapteria scythium* and *Pteria ussurica variabilis*, and conodont *Hindeodus parvus* were recorded. 14 cm
33. Pale volcanic ash. 2 cm

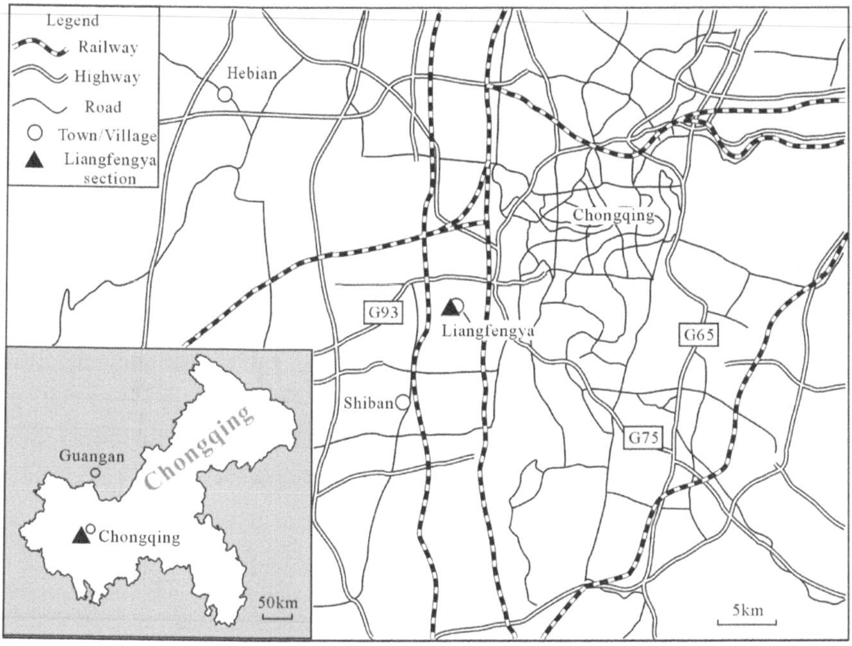

Fig. 3.43 Geographic location of the Liangfengya section

Conformity
Changhsing Formation in the Upper Permian

32. Grayish green argillaceous limestone with abundant pyrite crystals. Brachiopods ?*Lingularia fuyuanensis*, *Acosarina minuta*, *Cathaysia chonetoides*, *Fusichonetes pygmaea*, *Neochonetes* sp., *Haydenella kiangsiensis*, *Spinomarginifera kueichowensis*, *S. chenyaoyenensis*, *S. sulcata*, *Prelissorhynchia pseudoutah* and *Crurithyris flabelliformis* were identified. 13 cm
31. Pale volcanic ash, with a few brachiopods *Acosarina minuta*, *Orthothetina regularis*, *Meekella* sp., *Fusichonetes pygmaea*, *Cathaysia* sp., *Spinomarginifera* sp., *Leptodus* sp., *Uncinunellina* sp. and *Crurithyris* sp.. 5cm
30. Grayish green argillaceous limestone, abundantly with brachiopods ?*Lingularia* sp., *Fusichonetes soochowensis*, *Orbiculoidea nucleola*, *Meekella* cf. *deltoides*, *Perigeyerella* sp., *Spinomarginifera chenyaoyenensis*, *S. alpha*, S. *pseudosintanensis*, *Compressoproductus* sp., ?*Parapulchratia nitens*, *Richthofenia sinensis*, *Stenoscisma mutabilis*, *Martinia corculum*, *Juxathyris araxensis*, *Hustedia remota*, *Notothyris crassa*, *Permianella typica* and *Dicystoconcha* (*Neodicystoconcha*) *cuneiformis* and the coral *Lophophyllidium* sp. The conodont *Hindeodus praeparvus* was collected in the upper part and conodonts *Clarkina meishanensis* and *Hindeodus eurypyge* were found in the lower part of the bed. 13 cm

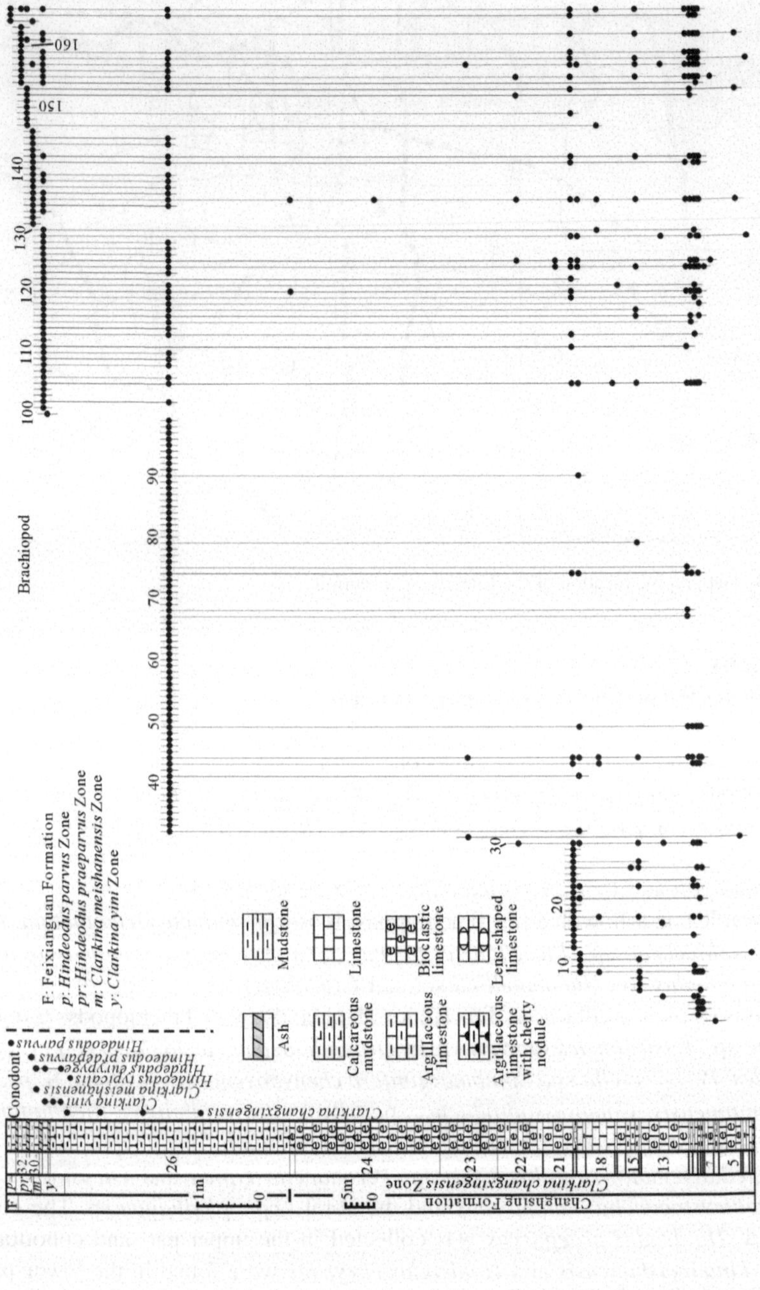
Brachiopod
Conodont
Hindeodus parvus
Hindeodus praeparvus
Hindeodus eurypyge
Hindeodus typicalis
Clarkina meishanensis
Clarkina yini
Clarkina changxingensis
F: Feixianguan Formation
p: Hindeodus parvus Zone
pr: Hindeodus praeparvus Zone
m: Clarkina meishanensis Zone
y: Clarkina yini Zone
Mudstone
Limestone
Bioclastic limestone
Lens-shaped limestone
Ash
Calcareous mudstone
Argillaceous limestone
Argillaceous limestone with cherty nodule
Changhsing Formation
Clarkina changxingensis Zone
1m
5m

◂**Fig. 3.44** Stratigraphic column and fossil occurrences at the Liangfengya section 1—*Orthothetina* sp., 2—*Derbyia schellwieni*, 3—*Derbyia guangdongensis*, 4—*Orthothetina eusarkos*, 5—*Oldhamina grandis*, 6—*Juxathyris yuananensis*, 7—*Spinomarginifera intercedens*, 8—*Streptorhynchus* sp., 9—*Qinglongia ovalis*, 10—*Juxathyris zhijingensis*, 11—*Wellerellina opima*, 12—*Acosarina* sp., 13—*Qinglongia praepinguis*, 14—*Q. costalliformis*, 15—*Juxathyris undulata*, 16—*J. protea*, 17—*J. jiangyouensis*, 18—*Crurithyris cordata*, 19—*Anchorhynchia subpentagona*, 20—*Oldhamina interrupta*, 21—*O. anshunensis*, 22—*O. longirostrata*, 23—*O. squamosa*, 24—*Alatoproductus* sp., 25—*Haydenella crispescens*, 26—*Cathaysia beifengjingensis*, 27—*C. parvulia*, 28—*C. yongdingensis*, 29—*Meekella eximia*, 30—*Oldhamina decipiens*, 31—*Qinglongia* sp., 32—*Q. gigantea*, 33—*Q. exilis*, 34—*Q. difforma*, 35—*Q. elegans*, 36—*Zhongliangshania zhongliangshanensis*, 37—*Notothyris triplicata*, 38—*Hustedia pseudocardia*, 39—*H. indica*, 40—*Composita* sp., 41—*Juxathyris rectimarginalis*, 42—*J. quadrilobata*, 43—*J. dilatatus*, 44—*J. bisulcata*, 45—*J. excavata*, 46—*J.* aff. *sulcata*, 47—*Spiriferillina nasuta*, 48—*S.* cf. *pandri*, 49—*Paraspiriferina multiplicata*, 50—*Martinia abrupta*, 51—*M. orbicularis*, 52—*Squamularia grandis*, 53—*S. elongantula*, 54—*S. subrostrata*, 55—*Phricodothyris asiatica*, 56—*P. echinata*, 57—*Stenoscisma inclusus*, 58—*S. amoenum*, 59—*S. zhongliangshanensis*, 60—*Uncinunellina jabensis*, 61—*Anchorhynchia ignobilis*, 62—*Neorichthofenia mabutii*, 63—*Richthofenia guangdeensis*, 64—*Leptodus richthofeni*, 65—*L. nobilis*, 66—*Spinomarginifera sichuanensis*, 67—*Compressoproductus compressus*, 68—*C. mongolicus*, 69—*Tyloplecta yangtzeensis*, 70—*Anidanthus guichiensis*, 71—*A. sinosus*, 72—?*Parapulchratia intercedens*, 73—*Pygmochonetes qinglongensis*, 74—*Haydenella qinglongensis*, 75—*H. elongata*, 76—*H. deminutus*, 77—*H. chianensis*, 78—*Paryphella orbicularis*, 79—*Tschernyschewia sinensis*, 80—*Edriosteges kayseri*, 81—*Cathaysia speciosa*, 82—*Streptorhynchus longyangensis*, 83—*S. pelargonatus*, 84—*Perigeyerella fastigata*, 85—*P. costellata*, 86—*Derbyia yangtzeensis*, 87—*Meekella versiformis*, 88—*M. perigeyerelloides*, 89—*M. arakeljani*, 90—*M. dorsisulcata*, 91—*Orthothetina ellipsoides*, 92—*O. planoconvexa*, 93—*O. rara*, 94—*O. exquisita*, 95—*O. shuangtangensis*, 96—*Enteletes* sp., 97—*Peltichia zigzag*, 98—*Peltichia sinensis*, 99—*P. traversa*, 100—*Hustedia* sp., 101—*Juxathyris plicata*, 102—*J. triangularis*, 103—*Notothyris subnucleolus*, 104—*Juxathyris* aff. *hunanensis*, 105—*J. guizhouensis*, 106—*Martinia semiplana*, 107—*Squamularia extensiformis*, 108—*S. indica*, 109—*S. rostrata*, 110—*S. squamularioides*, 111—*Crurithyris extumida*, 112—*Stenoscisma* sp., 113—*Anchorhynchia sarciniformis*, 114—*Richthofenia anshunensis*, 115—*Keyserlingina dzulfensis*, 116—*Leptodus tenuis*, 117—*L. deminutus*, 118—*Haydenella paradoxica*, 119—*H. nusata*, 120—*H. wenganensis*, 121—*Tschernyschewia pseudoirginae*, 122—*T. typica*, 123—*Edriosteges poyangensis*, 124—*Neochonetes* (*Sommeriella*) *strophomenoides*, 125—*Neochonetes* (*Huangichonetes*) *substrophomenoides*, 126—*Tropidelasma zhongliangshanensis*, 127—*Derbyia acutangula*, 128—*Meekella sichuanensis*, 129—*Orthothetina ruber*, 130—*Peltichia* sp., 131—*Dicystoconcha* (*Neodicystoconcha*) *cuneiformis*, 132—*Permianella typica*, 133—*Notothyris crassa*, 134—*Hustedia remota*, 135—*Juxathyris araxensis*, 136—*Martinia corculum*, 137—*Stenoscisma mutabilis*, 138—*Richthofenia sinensis*, 139—?*Parapulchratia nitens*, 140—*Compressoproductus* sp., 141—*Spinomarginifera pseudosintanensis*, 142—*S. alpha*, 143—*Perigeyerella* sp., 144—*Meekella* cf. *deltoides*, 145—*Orbiculoidea nucleola*, 146—*Fusichonetes soochowensis*, 147—*Crurithyris* sp., 148—*Uncinunellina* sp., 149—*Leptodus* sp., 150—*Spinomarginifera* sp., 151—*Cathaysia* sp., 152—*Meekella* sp., 153—*Orthothetina regularis*, 154—*Crurithyris flabelliformis*, 155—*Prelissorhynchia pseudoutah*, 156—*Spinomarginifera sulcata*, 157—*S. chenyaoyenensis*, 158—*S. kueichowensis*, 159—*Haydenella kiangsiensis*, 160—*Neochonetes* sp., 161—*Fusichonetes pygmaea*, 162—*Cathaysia chonetoides*, 163—?*Lingularia fuyuanensis*, 164—?*Lingularia* sp., 165—*Speciothyris speciosa*, 166—*Acosarina minuta*. *Note* Brachiopods referred to Shen and He (1991) and conodonts after Yuan and Shen (2011)

29. Pale volcanic ash. 6 cm
28. Light gray argillaceous limestone with abundant pyrite crystals. Abundant brachiopods *Acosarina minuta, Peltichia* sp., *Orthothetina ruber, Meekella sichuanensis, Derbyia acutangula, Tropidelasma zhongliangshanensis, Fusichonetes pygmaea, Neochonetes (Huangichonetes) substrophomenoides, Neochonetes (Sommeriella) strophomenoides, Edriosteges poyangensis, Tschernyschewia typica, T. pseudoirginae, Cathaysia chonetoides, Haydenella kiangsiensis, H. wenganensis, H. nusata, H.? paradoxica, ?Parapulchratia nitens, Spinomarginifera kueichowensis, S. chenyaoyenensis, S. alpha, S. sulcata, Leptodus deminutus, L. tenuis, Keyserlingina dzulfensis, Richthofenia anshunensis, R. sinensis, Anchorhynchia sarciniformis, Prelissorhynchia pseudoutah, Stenoscisma* sp., *Crurithyris flabelliformis, C. extumida, Squamularia squamularioides, S. rostrata, S. indica, S. extensiformis, Martinia corculum, M. semiplana, Juxathyris araxensis, J. guizhouensis, J.* aff. *hunanensis, J. triangularis, J. plicata, Hustedia remota, Notothyris crassa* and *N. subnucleolus*, bivalves *Guizhoupecten regularis* and *Permoperna trapezoides*, trilobite *Pseudophillipsia obtusicauda*, and fusulinid *Palaeofusulina* sp. were recorded in the bed. Conodonts *Clarkina yini* and *Hindeodus eurypyge* were found in the upper part of the bed. 9 cm
27. Yellow volcanic ash with the brachiopod *Hustedia* sp.. 3 cm
26. Gray argillaceus limestone with a number of cherty nodules. Abundant brachiopods *Orbiculoidea nucleola, Peltichia traversa, P. sinensis, P. zigzag, Enteletes* sp., *Acosarina minuta, Orthothetina regularis, O. shuangtangensis, O. exquisita, O. rara, O. planoconvexa, O. ellipsoides, Meekella dorsisulcata, M. sichuanensis, M.* cf. *deltoides, M. arakeljani, M. perigeyerelloides, M. versiformis, Derbyia acutangula, D. yangtzeensis, Perigeyerella costellata, P. fastigata, Streptorhynchus pelargonatus, S. longyangensis, Tropidelasma zhongliangshanensis, Neochonetes (Huangichonetes) substrophomenoides, Neochonetes (Sommeriella) strophomenoides, Cathaysia speciosa, Edriosteges poyangensis, E. kayseri, Tschernyschewia sinensis, T. pseudoirginae, Paryphella orbicularis, Haydenella kiangsiensis, H. wenganensis, H. nusata, H.? paradoxica, H. chianensis, H. deminutus, H. elongata, H. qinglongensis, Pygmochonetes qinglongensis, ?Parapulchratia nitens, ?P. intercedens, Anidanthus sinosus, A. guichiensis, Tyloplecta yangtzeensis, Compressoproductus mongolicus, C. compressus, Spinomarginifera kueichowensis, S. chenyaoyenensis, S. alpha, S. sulcata, S. pseudosintanensis, S. sichuanensis, Leptodus nobilis, L. richthofeni, L. tenuis, L. deminutus, Keyserlingina dzulfensis, Richthofenia sinensis, R. anshunensis, R. guangdeensis, Neorichthofenia mabutii, Anchorhynchia sarciniformis, A. ignobilis, Uncinunellina jabensis, Prelissorhynchia pseudoutah, Stenoscisma mutabilis, S. zhongliangshanensis, S. amoenum, S. inclusus, Speciothyris speciosa, Crurithyris flabelliformis, Phricodothyris echinata, P. asiatica, Squamularia squamularioides, S. indica, S. rostrata, S. subrostrata, S. elongantula, S. grandis, Martinia corculum, M. semiplana, M. orbicularis, M. abrupta, Paraspiriferina multiplicata, Spiriferillina* cf. *pandri, S. nasuta, Juxathyris araxensis, J.* aff. *sulcata, J. excavata,*

J. triangularis, J. bisulcata, J. guizhouensis, J. dilatatus, J. quadrilobata, J. rectimarginalis, Composita sp., *Hustedia remota, H. indica, H. pseudocardia, Notothyris triplicata, Zhongliangshania zhongliangshanensis, Qinglongia elegans, Q. difforma, Q. exilis* and *Q. gigantea*, a few bivalves *Promytilus* sp., *Permoperna trapzoidelis* and *Guizhoupecten* sp., fusulinids *Palaeofusulina sinensis, P. nana, P. wangi* and *Nankinella* sp., and rare coral *Lophophylidium* sp. were identified. The conodont *Clarkina yini* was found in the upper part and conodonts *Clarkina yini* and *Hindeodus eurypyge* were discovered in the lower part of the bed. 392 cm

25. Light gray thick-bedded bioclastic limestone with abundant cherty nodules. Brachiopods *Haydenella wenganensis* and *Juxathyris araxensis*, fusulinids *Palaeofusulina simplicata, P. wong, P.* cf. *nata* and *Nankinella* sp. and conodont *Clarkina changxingensis* were examined. 96 cm
24. Gray thick-bedded bioclastic limestone with abundant cherty nodules. The brachiopod *Juxathyris araxensis*, fusulinid *Nankinella* sp. and conodonts *Clarkina changxingensis* and *Hindeodus typicalis* were recovered. 2618 cm
23. Gray thick-bedded bioclastic limestone, with brachiopods *Spinomarginifera chenyaoyenensis, Juxathyris bisulcata* and *Qinglongia* sp. and fusulinid *Nankinella* sp.. 1050 cm
22. Gray thick-bedded bioclastic limestone intercalated with banding-shaped mudstone, with brachiopods *Meekella* sp., *Neochonetes* (*Huangichonetes*) *substrophomenoides, Oldhamina decipiens, Prelissorhynchia pseudoutah* and *Juxathyris araxensis* and fusulinids *Palaeofusulina sinensis, P. minima, P. longa, Codonofusiella bicornis* and *C. lini.* 831 cm
21. Light gray thick-bedded bioclastic limestone, with brachiopods *Neochonetes* (*Huangichonetes*) *substrophomenoides* and *Neochonetes* (*Sommeriella*) *strophomenoides*, and fusulinids *Palaeofusulina simplicata* and *Nankinella* sp.. 569 cm
20. Thin-bedded calcareous mudstone intercalated with lens-shaped limestone. Abundant brachiopods *Acosarina minuta, Orthothetina ruber, O. regularis, Meekella eximia, Neochonetes* (*Huangichonetes*) *substrophomenoides* and *Neochonetes* (*Sommeriella*) *strophomenoides, Cathaysia chonetoides, C. yongdingensis, C. parvulia, C. beifengjingensis, Haydenella kiangsiensis, H. wenganensis, H. qinglongensis, H. nasuta, H. crispescens, Alatoproductus* sp., *Tschernyschewia typica, Spinomarginifera alpha, S. kueichowensis, S. chenyaoyenenis, S. pseudosintanensis, Leptodus* sp., *Oldhamina decipiens, O. squamosa, O. longirostrata, O. anshunensis, O. interrupta, Anchorhynchia sarciniformis, A. subpentagona, Prelissorhynchia pseudoutah, Speciothyris speciosa, Crurithyris cordata, C. extumida, Juxathyris araxensis, J. bisulcata, J. jiangyouensis, J. protea, J. dilatatus, J. undulata, J. guizhouensis, Qinglongia costalliformis* and *Q. praepinguis*, and bivalves *Aviculopecten fasciculicostatus, Etheripecten sichuanensis, Girtypecten chongqingensis* and *Euchondria longtangensis* were examined. 48 cm
19. Gray thick-bedded limestone intercalated with banding-shaped mudstone. Brachiopods *Acosarina* sp., *Orthothetina ruber, Oldhamina decipiens, O.*

squamosa, O. anshunensis, Wellerellina opima, Paraspiriferina multiplicata, Meekella dorsisulcata, Neochonetes (*Huangichonetes*) *substrophomenoides* and *Neochonetes* (*Sommeriella*) *strophomenoides, Haydenella qinglongensis, Spinomarginifera chenyaoyenenis, Juxathyris bisulcata, J. araxensis, J. guizhouensis, J. zhijingensis, J. rectimarginalis, Qinglongia gigantea* and *Q. ovalis*, and fusulinid *Palaeofusulina simplex* were identified. 169 cm

18. Gray thick-bedded micritic limestones, with brachiopods *Acosarina minuta, Peltichia* sp., *Crurithyris* sp., *Juxathyris araxensis, J. bisulcata, J. dilatatus* and *Qinglongia ovalis.* 567 cm
17. Light gray argillaceous limestone with abundant cherty nodules, with the brachiopod *Juxathyris guizhouensis* and fusulinids *Codonofusiella* cf. *kueichowensis* and *Palaeofusulina* cf. *sinensis.* 54 cm
16. Gray thick-bedded bioclastic limestone, with the brachiopod *Tschernyschewia pseudoirginae* and fusulinids *Chenia* sp., *Palaeofusulina simplex* and *Nankinella* sp.. 169 cm
15. Gray argillaceous limestone with the brachiopod *Cathaysia beifengjingensis.* 246 cm
14. Thin-bedded lens-shaped argillaceous limestone with banding-shaped mudstone. Brachiopods *Acosarina minuta, Streptorhynchus* sp., *Neochonetes* (*Sommeriella*) *strophomenoides, Cathaysia chonetoides, C. parvulia, C. beifengjingensis, Tschernyschewia sinensis, Spinomarginifera kueichowensis, S. chenyaoyenensis, S. alpha, S. intercedens, L. tenuis, L. deminutus, Oldhamina decipiens, O. squamosa, Prelissorhynchia pseudoutah, Wellerellina opima, Juxathyris araxensis, J. bisulcata, J. guizhouensis, J. yuananensis, Crurithyris cordata* and *Qinglongia ovalis*, trilobite *Pseudophillipsia obtusicauda*, and fusulinids *Palaeofusulina wongi, P. minima, P.* cf. *sinensis, Reichelina cribroseptata* were discovered. 36 cm
13. Gray thick-bedded bioclastic limestone with brachiopods *Orthothetina ruber, Spinomarginifera chenyaoyenensis, Oldhamina decipiens, O. grandis* and *Juxathyris jiangyouensis.* 883 cm
12. Gray thick-bedded micritic limestone intercalated with banding-shaped mudstone. Brachiopods *Acosarina minuta, Neochonetes* (*Sommeriella*) *strophomenoides, Cathaysia chonetoides, Haydenella kiangsiensis, H. wenganensis, H. elongata, H. qinglongensis, Compressoproductus compressus, C. mongolicus, Spinomarginifera chenyaoyenensis, S. pseudosintanensis, Crurithyris extumida, Prelissorhynchia pseudoutah, Paraspiriferina multiplicata, Juxathyris araxensis, J. bisulcata* and *J. guizhouensis*, bivalves *Aviculopecten fasciculicostatus, Towapteria guizhouensis, Palaeolima hunanensis, Etheripecten sichuanensis* and *Girtypecten beipeiensis*, and trilobite *Pseudophillipsia chongqingensis* were identified. 83 cm
11. Gray thick-bedded limestone intercalated with banding-shaped mudstone. Brachiopods *Acosarina minuta, Orthothetina ruber, O. regularis, Meekella* sp., *Neochonetes* (*Huangichonetes*) *substrophomenoides, Neochonetes* (*Sommeriella*) *strophomenoides, Cathaysia chonetoides, Haydenella kiangsiensis, H. wenganensis, H. qinglongensis, Spinomarginifera kueichowensis, S.*

chenyaoyenensis, *S. alpha*, *L. tenuis*, *Keyserlingina dzulfensis*, *Oldhamina decipiens*, *O. grandis*, *O. squamosa*, *Prelissorhynchia pseudoutah*, *Crurithyris cordata*, *Speciothyris speciosa*, *Paraspiriferina multiplicata*, *Juxathyris araxensis*, *J. bisulcata*, *J. guizhouensis*, *J. dilatatus*, *J. zhijingensis*, *J. undulata*, *Qinglongia ovalis* and *Alatoproductus* sp., bivalves *Etheripecten sichuanensis*, *Limipecten* sp., *Pernopecten huayingshanensis* and *Euchondria longtangensis*, and trilobite *Pseudophillipsia chongqingensis* were discovered. 100 cm

10. Dark gray argillaceous limestone. Brachiopods *Acosarina minuta*, *Peltichia* sp., *Orthothetina eusarkos*, *Derbyia guangdongensis*, *D. schellwieni*, *Neochonetes* (*Huangichonetes*) *substrophomenoides*, *Neochonetes* (*Sommeriella*) *strophomenoides*, *Cathaysia chonetoides*, *C. beifengjingensis*, *Haydenella kiangsiensis*, *H. wenganensis*, *H. elongata*, *H. nasuta*, *Tschernyschewia pseudoirginae*, *Spinomarginifera kueichowensis*, *S. chenyaoyenensis*, *S. alpha*, *S. pseudosintanensis*, *Oldhamina decipiens*, *O. grandis*, *O. squamosa*, *Anchorhynchia sarciniformis*, *Prelissorhynchia pseudoutah*, *Speciothyris speciosa*, *Paraspiriferina multiplicata*, *Juxathyris araxensis*, *J. guizhouensis*, *J. dilatatus* and *Qinglongia ovalis* were examined. 50 cm
9. Gray calcareous mudstone with brachiopods *Acosarina minuta*, *Orthothetina eusarkos*, *O. ruber*, *Derbyia guangdongensis*, *Neochonetes* (*Huangichonetes*) *substrophomenoides*, *Neochonetes* (*Sommeriella*) *strophomenoides*, *Cathaysia chonetoides*, *Haydenella kiangsiensis*, *H. wenganensis*, *H. qinglongensis*, *Spinomarginifera kueichowensis*, *S. chenyaoyenensis*, *S. alpha*, *Oldhamina decipiens*, *O. anshunensis*, *Prelissorhynchia pseudoutah*, *Crurithyris flabelliformis*, *Crurithyris cordata*, *Speciothyris speciosa*, *Paraspiriferina multiplicata*, *Juxathyris araxensis*, *J. guizhouensis*, *J. dilatatus*, *J. bisulcata*, *J. yuananensis* and *Qinglongia ovalis*, and trilobite *Pseudophillipsia* sp.. 42 cm
8. Dark gray limestone with brachiopods *Acosarina minuta*, *Orthothetina eusarkos*, *O. ruber*, *Derbyia guangdongensis*, *Neochonetes* (*Sommeriella*) *strophomenoides*, *Cathaysia chonetoides*, *C. beifengjingensis*, *Haydenella kiangsiensis*, *H. wenganensis*, *H. ? paradoxica*, *Spinomarginifera kueichowensis*, *S. chenyaoyenensis*, *S. alpha*, *S. pseudosintanensis*, *Leptodus tenuis*, *Prelissorhynchia pseudoutah*, *Paraspiriferina multiplicata*, *Juxathyris araxensis*, *J. guizhouensis*, *J. bisulcata* and *Qinglongia ovalis.* 37 cm
7. Dark gray thick-bedded bioclastic limestone intercalated with banding-shaped mudstone, with the brachiopod *Neochonetes* (*Sommeriella*) *strophomenoides* and fusulinids *Palaeofusulina sinensis*, *P. rhomboides*, *P. prisca* and *Dunbarula* sp.. 179 cm
6. Gray to light gray shale intercalated with lens-shaped argillaceous limestone, with brachiopods *Orthothetina* sp., *Neochonetes* (*Huangichonetes*) *substrophomenoides* and *Prelissorhynchia pseudoutah.* 100 cm
5. Dark gray thick-bedded bioclastic limestone. 491 cm
4. Dark gray limestone intercalated with shale, with brachiopods *Orthothetina regularis*, *Cathaysia chonetoides*, *Juxathyris araxensis* and *Qinglongia* sp.. 21 cm.

3. Dark gray thick-bedded limestone intercalated with banding-shaped mudstone, with the fusulinid *Palaeofusulina* sp.. 142 cm
2. Dark gray thick-bedded micritic limestone. 50 cm
1. Grayish black argillaceous limestone with brachiopods *Orthothetina ruber*, *Spinomarginifera kueichowensis* and *Acosarina* sp.. 38 cm

Conformity

Lungtan Formation in the Upper Permian
Gray black shale (not described downwards).

Note: Conodonts are after Yuan and Shen (2011); other fossils are based on Shen and He (1991), but updated in taxonomy.

3.19 Zhaixia Section in Chongyang County

The Zhaixia section is located at the Zhaixia Village, Chongyang County, Xianning City, Hubei Province, South China (Fig. 3.45).

The sequence of the Zhaixia section is composed of the Changhsing and Taye formations in an ascending order (Fig. 3.46). The Changhsing Formation

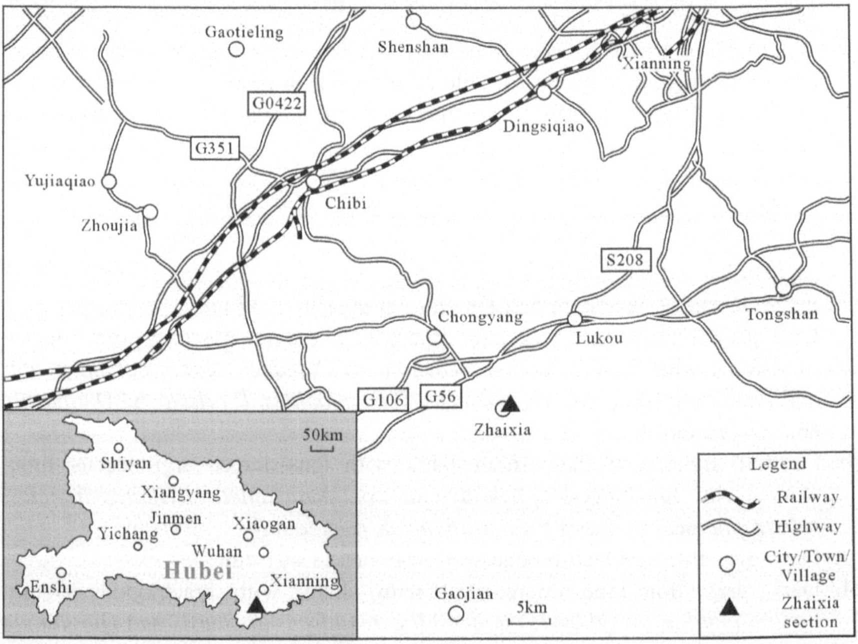

Fig. 3.45 Geographic location of the Zhaixia section

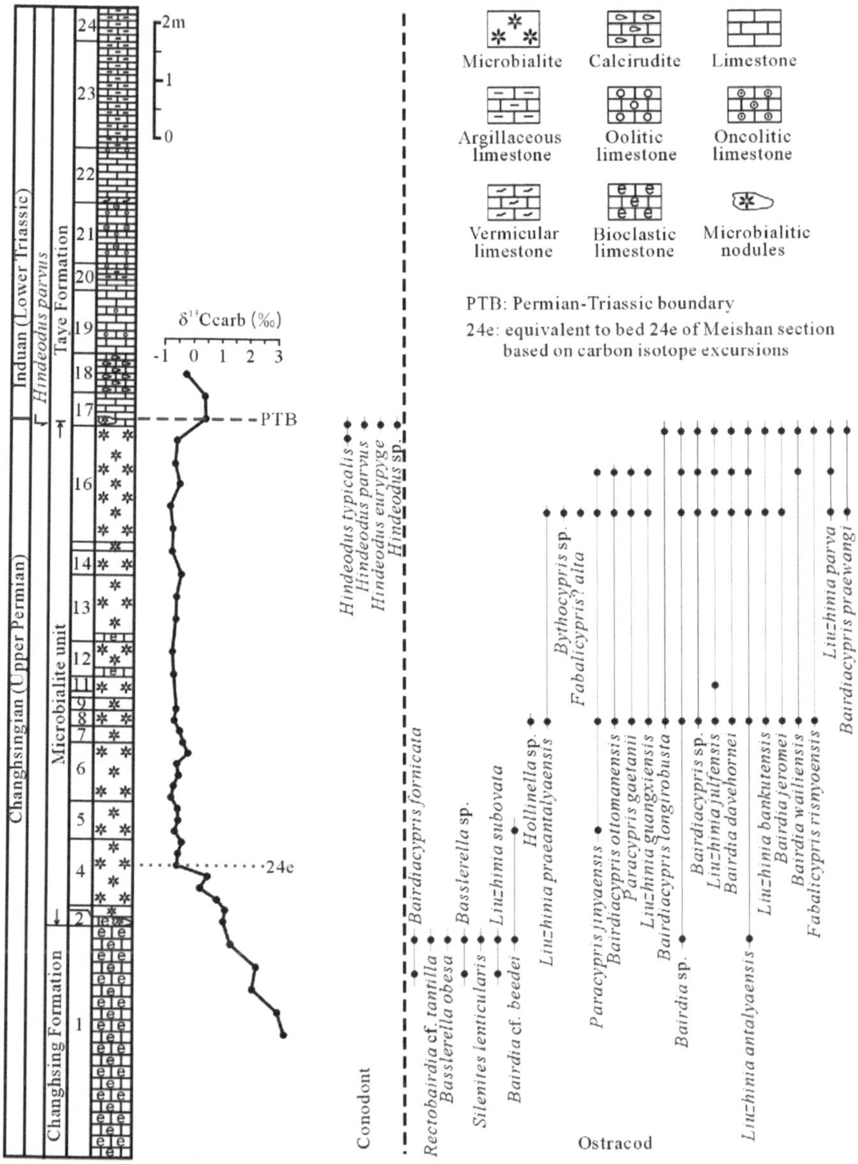

Fig. 3.46 Stratigraphic column and fossil occurrences at the Zhaixia section. *Note* Data of carbon isotope are after Wang et al. (2019), Lower Triassic ostracods after Wan (2021), Upper Permian ostracods after Liu (2010), and conodonts after Yang et al. (2006)

is mainly composed of bioclastic limestones. The Taye Formation is characterized by micritic limestones and oolitic limestones, with microbialites at the basal part. Abundant ostracods and a few conodonts were yielded in the Changhsing Formation and the basal part of the Taye Formation (Fig. 3.46).

Unexposed Upwards
Taye Formation in the Lower Triassic

24. Gray thin-bedded argillaceous limestone. 59 cm
23. Dark gray thin-bedded argillaceous limestone. 185 cm
22. Gray thin-bedded micritic limestone, with dark gray oolitic limestone at the top part (1 cm) and dark gray vermicular limestone at the basal part (8 cm). 97 cm
21. Three cycles were deposited and each cycle includes: oncolitic limestone at the top part; oolitic limestone in the upper part; micritic limestone in the lower part. 104 cm
20. Dark gray oolitic limestone at the top part (3 cm); dark gray thin-bedded argillaceous limestone in the upper part; dark gray thin-bedded micritic limestone in the lower part. 45 cm
19. Dark gray thin-bedded micritic limestone intercalated with two layers of oolitic limestone. 110 cm
18. Dark gray thin-bedded calcirudite. 67 cm
17. Dark gray thin-bedded limestone with microbialitic nodules. The conodonts *Hindeodus typicalis*, *Hindeodus parvus*, *Hindeodus eurypyge* and *Hindeodus* sp., and ostracods *Bairdiacypris longirobusta*, *Bairdiacypris* sp., *Bairdia davehornei*, *Bairdia jeromei*, *Bairdia wailiensis*, *Bairdia* sp., *Liuzhinia julfensis*, *Liuzhinia antalyaensis*, *Liuzhinia bankutensis*, *Liuzhinia parva*, *Fabalicypris risnyoensis* and *Bairdiacypris praewangi* were examined at the basal part of the bed. 59 cm

Microbialite Unit (assigned to basal part of the Taye Formation)

16. Dark gray medium-bedded microbialite. Ostracods *Liuzhinia praeantalyaensis*, *Liuzhinia guangxiensis*, *Liuzhinia julfensis*, *Liuzhinia antalyaensis*, *Liuzhinia bankutensis*, *Liuzhinia parva*, *Bythocypris* sp., *Fabalicypris? alta*, *Paracypris jinyaensis*, *Paracypris gaetanii*, *Bairdiacypris ottomanensis*, *Bairdiacypris praewangi*, *Bairdiacypris* sp., *Bairdia davehornei*, *Bairdia jeromei*, *Bairdia wailiensis* and *Bairdia* sp. were discovered in the bed. The conodont *Hindeodus typicalis* was identified at the top part of the bed. 200 cm
15. Gray medium-bedded microbialite. 15 cm
14. Gray thick-bedded thrombolite (a kind of microbialite). 42 cm
13. Gray thick-bedded thrombolite with bioclastic limestone at the base. 115 cm

12. Gray thick-bedded thrombolite with bioclastic limestone and stromatolite (regarded as microbialite) at the base. 62 cm
11. Gray medium-bedded thrombolite with the ostracod *Liuzhinia julfensis.* 25 cm
10. Gray stromatolite intercalated with thrombolite. 10 cm
9. Gray medium-bedded thrombolite. 22 cm
8. Gray medium-bedded thrombolite in the upper part (15 cm); gray stromatolite in the lower part (12 cm). Ostracods *Hollinella* sp., *Liuzhinia praeantalyaensis*, *Liuzhinia guangxiensis*, *Liuzhinia julfensis*, *Liuzhinia bankutensis*, *Liuzhinia antalyaensis*, *Paracypris jinyaensis*, *Paracypris gaetanii*, *Bairdiacypris ottomanensis*, *Bairdiacypris longirobusta*, *Bairdiacypris* sp., *Bairdia davehornei*, *Bairdia jeromei*, *Bairdia wailiensis*, *Bairdia* sp. and *Fabalicypris risnyoensis* were recorded. 27 cm
7. Gray stromatolite in the upper part (9 cm); gray thrombolite in the lower part. 27 cm
6. Gray thick-bedded stromatolite, locally with gray thrombolite. 103 cm
5. Gray thick-bedded thrombolite, with ostracods *Bairdia* cf. *beedei* and *Paracypris jinyaensis.* 64 cm
4. Gray thick-bedded thrombolite. 115 cm
3. Gray medium-bedded algae limestone, locally with thrombolite and stromatolite. 20 cm
2. Gray medium-bedded bioclastic limestone, locally with stromatolite. Abundant microgastriopods were yielded. 15 cm

Hiatus
Changhsing Formation in the Upper Permian

1. Light gray massive bioclastic limestone with abundant fossils. Ostracods *Bairdiacypris fornicata*, *Rectobairdia* cf. *tantilla*, *Basslerella obesa*, *Basslerella* sp., *Silenites lenticularis*, *Liuzhinia subovata*, *Liuzhinia antalyaensis*, *Bairdia* cf. *beedei* and *Bairdia* sp. were recovered at the top part of the bed. 400 cm

3.20 Kangjiaping Section in Cili County

The Kangjiaping section is located in the Kangjiaping Village, Cili County, Hunan Province, South China (Fig. 3.47).

The sequence of the Kangjiaping section is composed of the Changhsing and Taye formations in an ascending order (Fig. 3.48). The Changhsing Formation is mainly composed of bioclastic limestones. The Taye Formation is characterized by argillaceous limestones, with microbialites at the basal part. Abundant foraminifers were discovered in the Changhsing Formation and a few conodonts were found in the Taye Formation (Fig. 3.48).

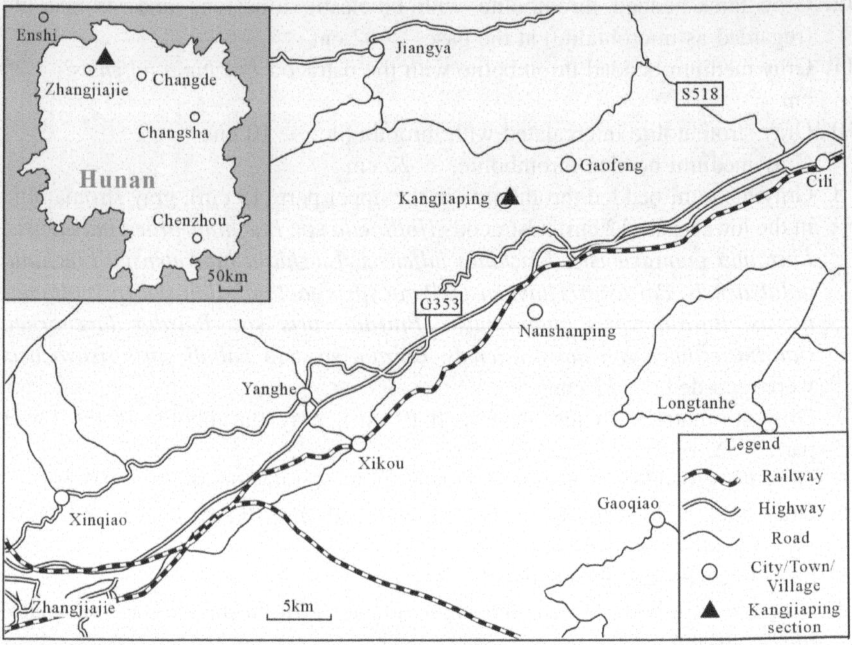

Fig. 3.47 Geographic location of the Kangjiaping section

Not measured upwards
Taye Formation in the Lower Triassic

6. Gray medium-bedded oolitic limestone, with a few foraminifers, ostracods, microgastropods, brachiopods and bivalves. 200 cm
5. Gray medium-bedded calcirudite, with conodonts *Hindeodus parvus*, *Hindeodus* sp., *Isarcicella staeschei* and *Isarcicella* sp., and a few foraminifers, ostracods, microgastropods and brachiopods. 320 cm
4. Gray thin-bedded vermicular limestone, with a few foraminifers, ostracods, microgastropods and brachiopods. 220 cm
3. Gray to dark gray oolitic limestone intercalated with vermicular limestone, with a few foraminifers and ostracods. 300 cm

Microbialite Unit (assigned to basal part of the Taye Formation)

2. Gray massive microbialite, with the conodont *Hindeodus parvus* and a few foraminifers, ostracods, microgastropods, brachiopods and bivalves. 460 cm

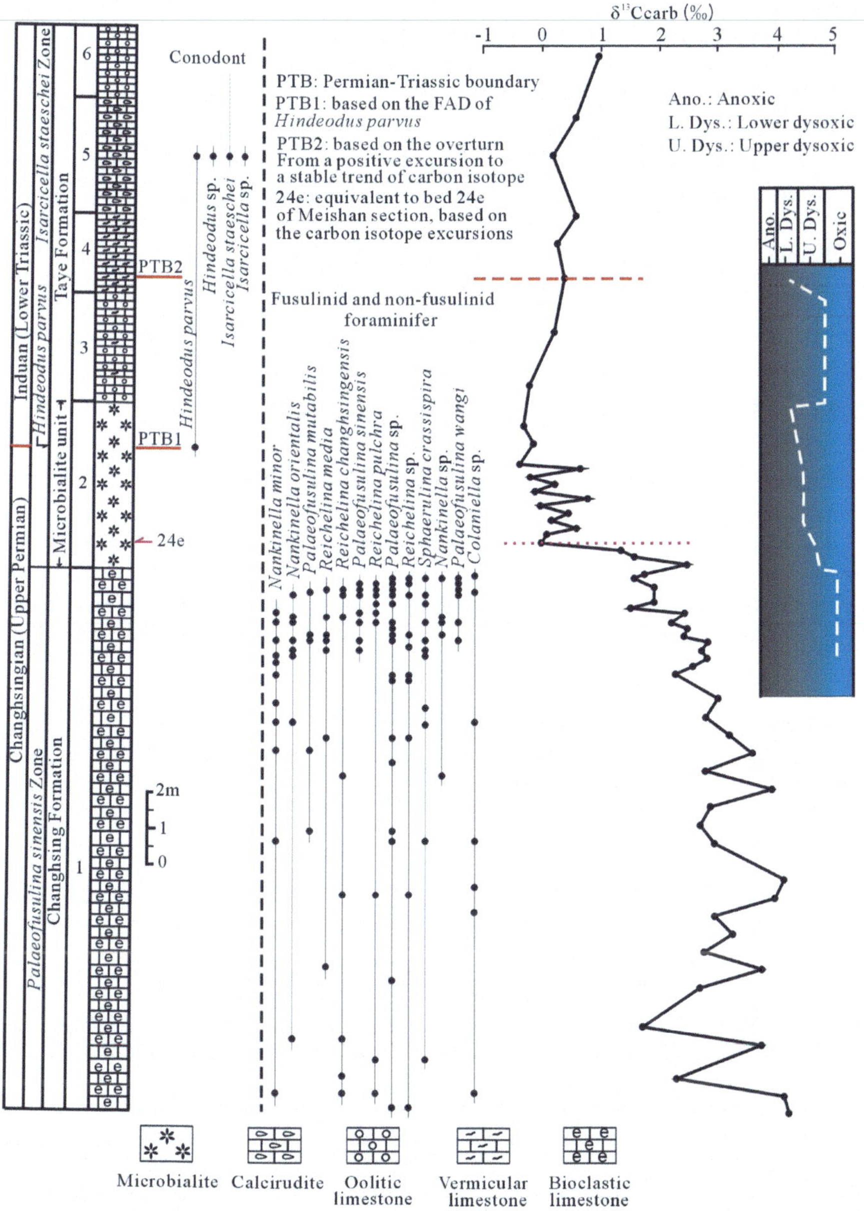

Fig. 3.48 Stratigraphic column and fossil occurrences at the Kangjiaping section. *Note* Data conodonts and carbon isotopes are after Wang et al. (2009), foraminifers after Yang et al. (2013), and redox conditions based on Liao (2020)

Conformity
Changhsing Formation in the Upper Permian

1. Gray to dark gray thick-bedded bioclastic limestone, with abundant ostracods, calcareous algaes (e.g., *Tubiphytes* sp.), echinoderms and bryozoans. Additionally, fusulinids *Nankinella minor*, *Nankinella orientalis*, *Nankinella* sp., *Palaeofusulina mutabilis*, *Palaeofusulina sinensis*, *Palaeofusulina* sp. and *Palaeofusulina wangi* were reported. Non-fusulinid foraminifers *Reichelina media*, *Reichelina changhsingensis*, *Reichelina pulchra*, *Reichelina* sp., *Sphaerulina crassispira* and *Colaniella* sp. were examined. 1500 cm

3.21 Laolongdong Section in Chongqing County

The Laolongdong section (29° 56′ 24″ N, 106° 33′ 36″ E) is located at the Laolongdong Village, Beibei District, Chongqing City, South China (Fig. 3.49).

The sequence of the Laolongdong section is composed of the Changhsing and Feixianguan formations in an ascending order (Fig. 3.50). The Changhsing Formation is mainly composed of bioclastic limestones. The Feixianguan Formation is characterized by silty mudstones, with microbialites at the basal part. Abundant

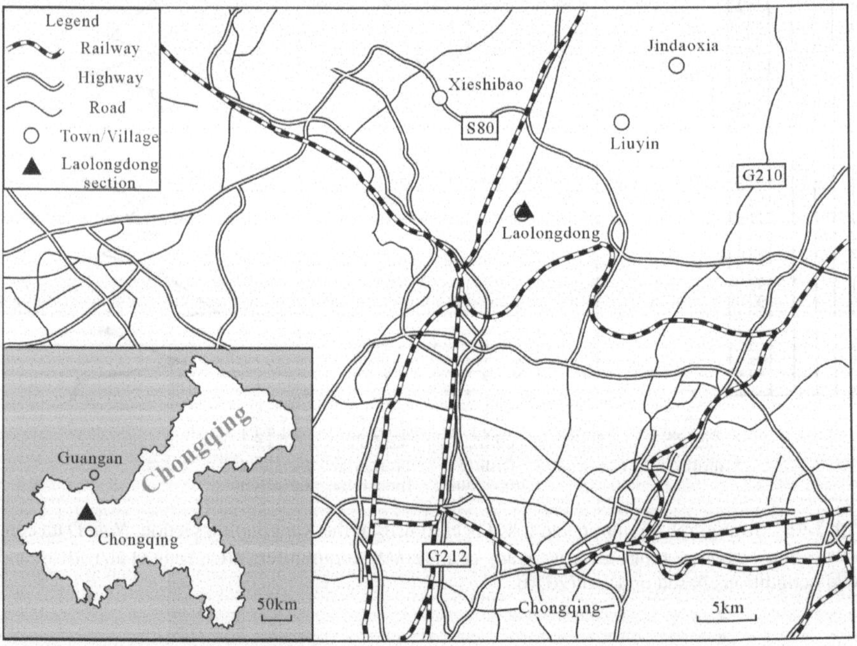

Fig. 3.49 Geographic location of the Laolongdong section

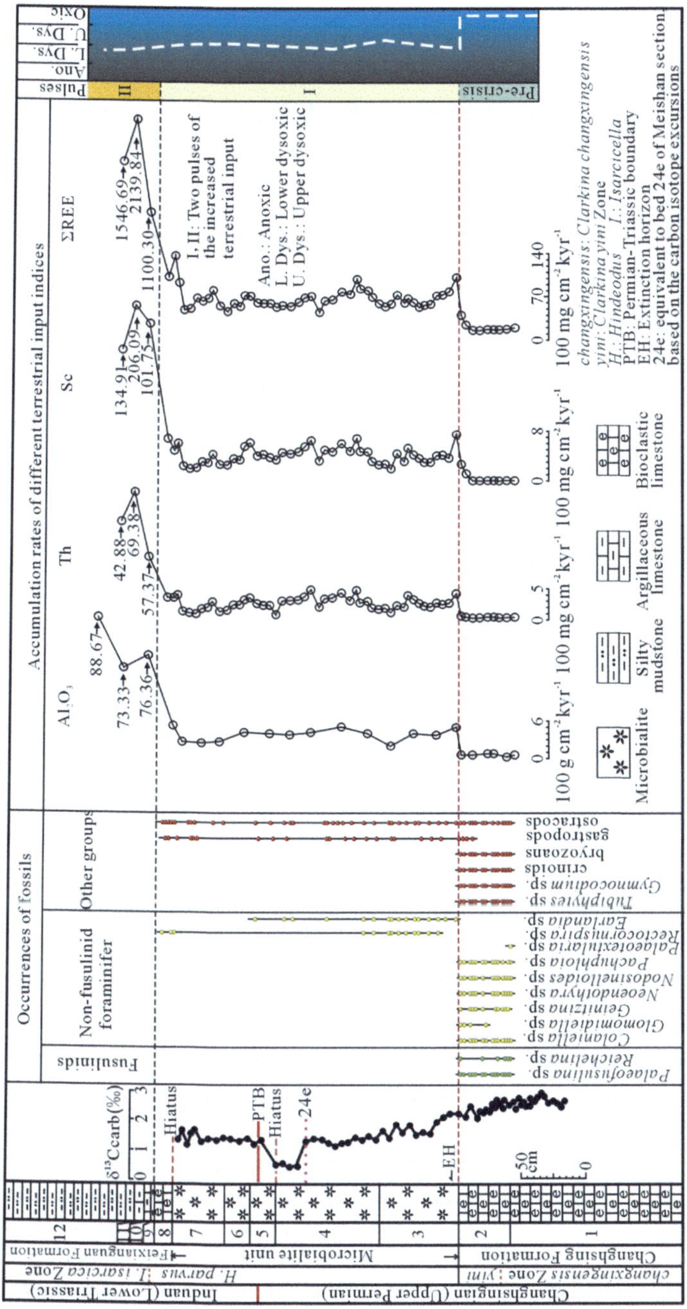

Fig. 3.50 Stratigraphic column and fossil occurrences at the Laolongdong section (Huang et al. 2022)

foraminifers were discovered in the Changhsing Formation and a few microgastriopods and ostracods were found in the microbialite of the Feixianguan Formation (Fig. 3.50).

Unexposed upwards
Feixianguan Formation in the Lower Triassic (Induan)

12. Grayish brown thin-bedded silty mudstone. 100 cm
11. Grayish green argillaceous limestone with a few microgastriopods. 7 cm
10. Grayish green argillaceous limestone with a few microgastriopods. 5 cm
9. Grayish yellow mudstone, abundantly with small bivalves and microgastriopods. 2 cm

Microbialite unit (assigned to basal part of the Feixianguan Formation)

8. Gray bioclasctic limestone, abundantly with microgastriopods, ostracods, brachiopod lingulids, bivalves, and non-fusulinid foraminifer *Rectocornuspira* sp.. 14 cm
7. Gray to dark gray microbialite (the thrombolite mosaiced by the dendrite), with an erosional surface at the top. Abundant coccoidal cyanobacteria fossils, microgastriopods, ostracods, brachiopod lingulids, and non-fusulinid foraminifer *Rectocornuspira* sp. were recorded. 40 cm
6. Gray to dark gray microbialite (the thrombolite mosaiced by the dendrite), with microgastriopods, ostracods, and non-fusulinid foraminifer *Rectocornuspira* sp.. 20 cm
5. Gray microbialite (the thrombolite mosaiced by the dendrite). 20 cm
4. Gray microbialite with an erosional surface at the top. Abundant ostracods, gastriopods, and non-fusulinid foraminifers *Rectocornuspira* sp. and *Earlandia* sp. were recorded. 80 cm
3. Gray to dark gray microbialite, with abundant coccoidal cyanobacteria fossils, gastriopods, ostracods, and non-fusulinid foraminifers *Rectocornuspira* sp. and *Earlandia* sp.. 60 cm

Conformity
Changhsing Formation in the Upper Permian (Changhsingian)

2. Light gray thick-bedded bioclastic limestone, abundantly with crinoids, bryozoans, calcareous sponges, brachiopods, gastriopods, ostracods, calcareous algaes *Gymnocodium* sp. and *Tubiphytes* sp. (reef-building organism), fusulinids *Palaeofusulina* sp. and *Reichelina* sp., non-fusulinid foraminifers *Colaniella* sp., *Glomomidiella* sp., *Geinitzina* sp., *Neoendothyra* sp., *Nodosinelloides* sp., *Pachuphloia* sp., *Palaeotextularia* sp., *Rectocornuspira* sp. and *Earlandia* sp.. 40 cm

1. Light gray massive bioclastic limestone, with foraminifers, crinoids, calcareous algaes (e.g., *Tubiphytes* sp. as reef-building organism), bryozoans, ostracods, brachiopods and non-fusulinid foraminifer *Colaniella* sp.. 135 cm

3.22 Dajiang Section in Luodian County

The Dajiang section is located at the Dajiang Village, Luodian County, Guizhou Province, southwestern China (Fig. 3.51).

The sequence of the Dajiang section is composed of the Wuchiaping and Taye formations in an ascending order (Fig. 3.52). The Wuchiaping Formation is mainly composed of thick-bedded bioclastic limestones. The Taye Formation is characterized by medium- to thin-bedded limestones, with microbialites at the basal part. Abundant foraminifers were found in the Wuchiaping Formation and conodonts in the Taye Formation (Fig. 3.52).

Not measured upwards
Taye Formation in the Lower Triassic

9. Grayish green to light gray medium- to thin-bedded packstone (argillaceous limestone). Conodonts *Hindeodus sosioensis*, *Parachirognathus* sp., *Clarkina*

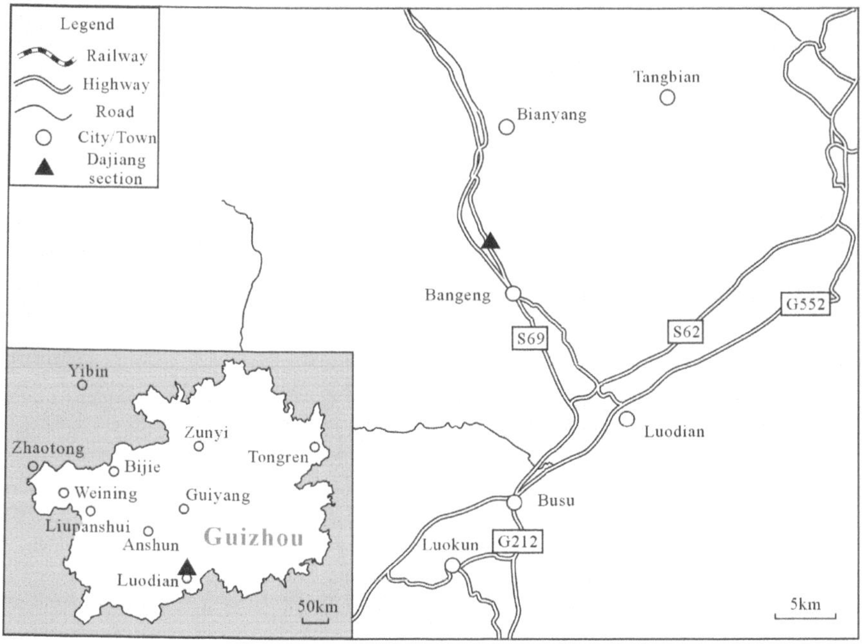

Fig. 3.51 Geographic location of the Dajiang section

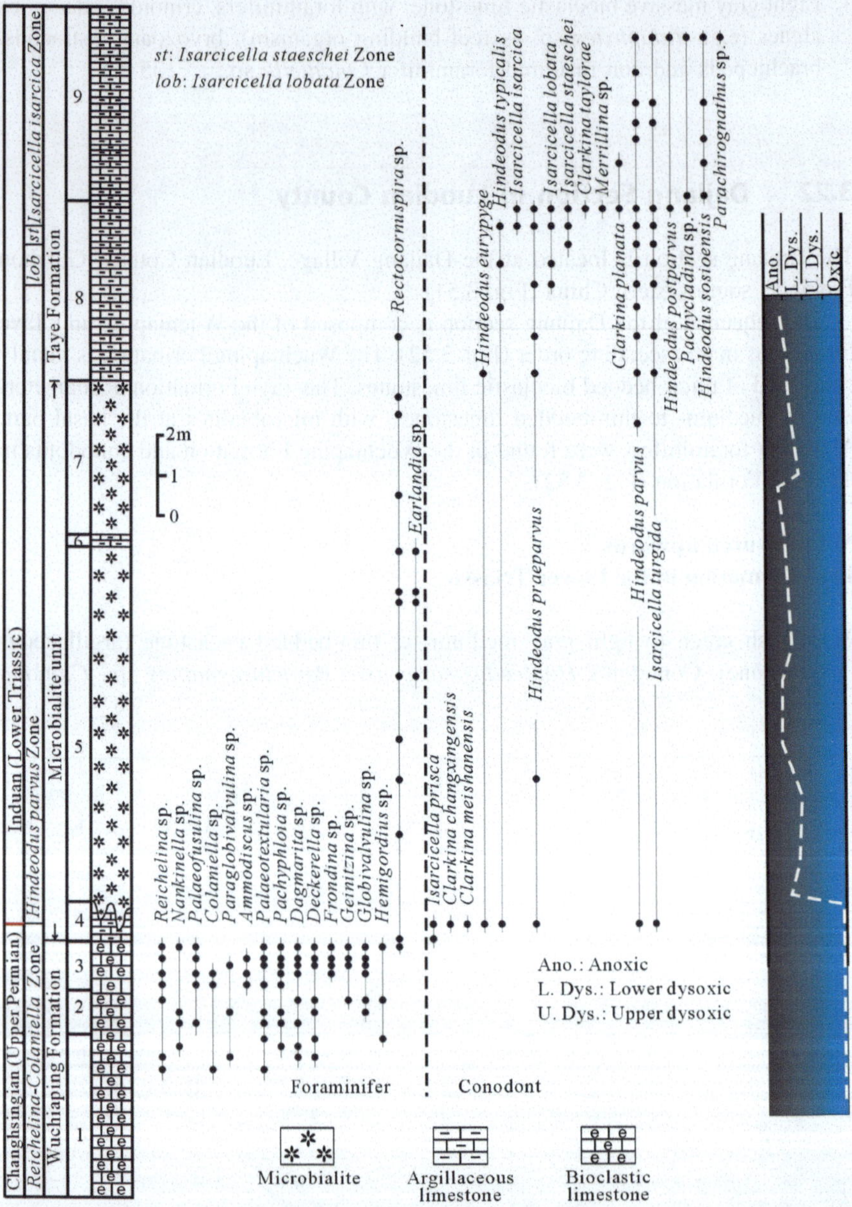

Fig. 3.52 Stratigraphic column and fossil occurrences at the Dajiang section. Data of foraminifers and redox are based on Liao (2020); conodonts are based on Jiang et al. (2014)

planata, *Hindeodus parvus*, *Hindeodus postparvus*, *Isarcicella turgida* and *Pachycladina* sp. were examined. > 480 cm

8. Grayish green to light gray medium-bedded packstone. Conodonts *Hindeodus eurypyge*, *Hindeodus typicalis*, *Hindeodus praeparvus*, *Hindeodus parvus*, *Hindeodus postparvus*, *Isarcicella isarcica*, *Isarcicella lobata*, *Isarcicella staeschei*, *Isarcicella turgida*, *Clarkina taylorae*, *Clarkina planata*, *Merrillina* sp. and *Pachycladina* sp. were examined. 450 cm

Microbialite Unit (assigned to basal part of the Taye Formation)

7. Gray to dark gray thick-bedded microbialite. The non-fusulinid foraminfer *Rectocornuspira* sp., conodont *Hindeodus parvus*, ostracods and microgastropods were yielded. 380 cm
6. Grayish green packstone. 30 cm
5. Grayish green to dark gray thick-bedded microbialite. Ostracods, microgastropods, conodont *Hindeodus praeparvus*, and non-fusulinid foraminifers *Rectocornuspira* sp. and *Earlandia* sp. were recovered. 910 cm
4. Grayish yellow packstone interfingered with microbialite. Ostracods and conodonts *Isarcicella prisca*, *Isarcicella turgida*, *Clarkina meishanensis*, *Clarkina changxingensis*, *Hindeodus eurypyge*, *Hindeodus typicalis*, *Hindeodus praeparvus* and *Hindeodus parvus* were examined in the packstone. 100 cm

Hiatus
Wuchiaping Formation in the Upper Permian

3. Light gray thick-bedded bioclastic limestone. Fusulinids *Reichelina* sp., *Nankinella* sp. and *Palaeofusulina* sp. and non-fusulinid foraminifers *Colaniella* sp., *Paraglobivalvulina* sp., *Ammodiscus* sp., *Palaeotextularia* sp., *Pachyphloia* sp., *Dagmarita* sp., *Deckerella* sp., *Frondina* sp., *Geinitzina* sp., *Globivalvulina* sp., *Hemigordius* sp. and *Rectocornuspira* sp. were discovered. In addition, brachiopods, gastropods, ostracods, calcareous sponges, bryozoans, calcareous algaes (e.g., *Tubiphytes* sp.) and crinoids were found in the bed. 120 cm
2. Light gray thick-bedded bioclastic limestone. Fusulinids *Reichelina* sp., *Nankinella* sp. and *Palaeofusulina* sp. and non-fusulinid foraminifers *Paraglobivalvulina* sp., *Palaeotextularia* sp., *Pachyphloia* sp., *Dagmarita* sp., *Deckerella* sp., *Geinitzina* sp., *Globivalvulina* sp. and *Hemigordius* sp. were examined. Additionally, abundant calcareous algaes (e.g., *Tubiphytes* sp.), gastropods and ostracods have been observed in the bed. 110 cm
1. Light gray thick-bedded bioclastic limestone. Fusulinids *Reichelina* sp. and *Nankinella* sp. and non-fusulinid foraminifers *Colaniella* sp., *Paraglobivalvulina* sp., *Palaeotextularia* sp., *Pachyphloia* sp., *Dagmarita* sp., *Deckerella*

sp., *Hemigordius* sp., *Rectocornuspira* sp. and *Earlandia* sp. were examined. Additionally, abundant calcareous sponges, bryozoans, calcareous algaes (e.g., *Tubiphytes* sp.), crinoids, gastropods and ostracods were discovered in the bed. 300 cm

3.23 Zuodeng Section in Baise City

The Zuodeng section (23° 27′ 00″ N, 106° 59′ 47″ E) is located at the Zuodeng Village, Tiandong County, Guangxi, South China (Fig. 3.53).

The sequence of the Zuodeng section is composed of the Heshan and Luolou formations in an ascending order (Fig. 3.54). The Heshan Formation is mainly composed of bioclastic limestones with abundant cherty nodules. The Luolou Formation is characterized by limestones, with interlayers of tuffs and mudstones in the middle and upper parts, and with microbialites at the basal part. Abundant ostracods were found in the Heshan and Luolou Formations and a few conodonts in the Luolou Formation (Fig. 3.54).

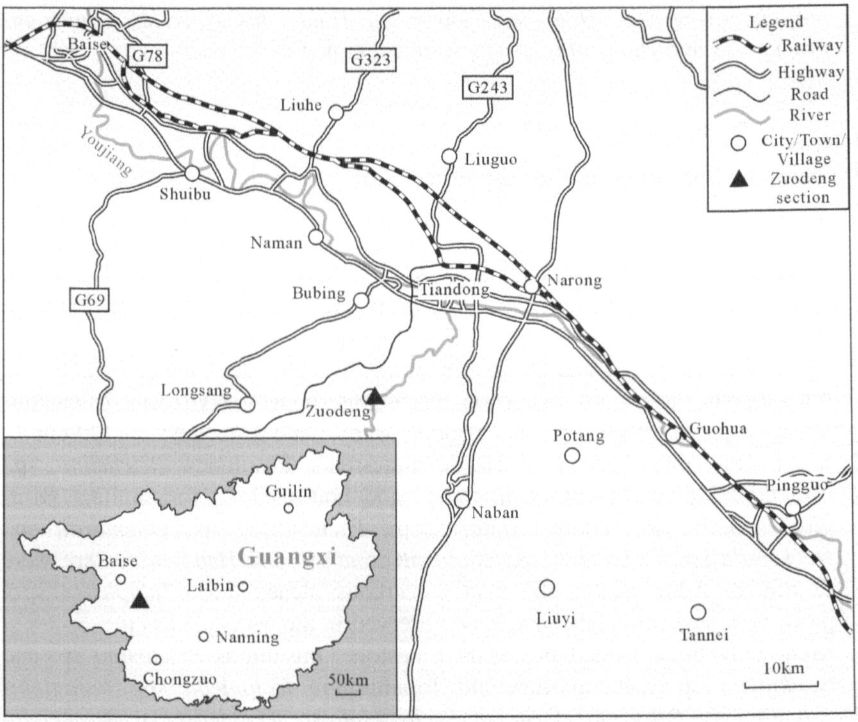

Fig. 3.53 Geographic location of the Zuodeng section

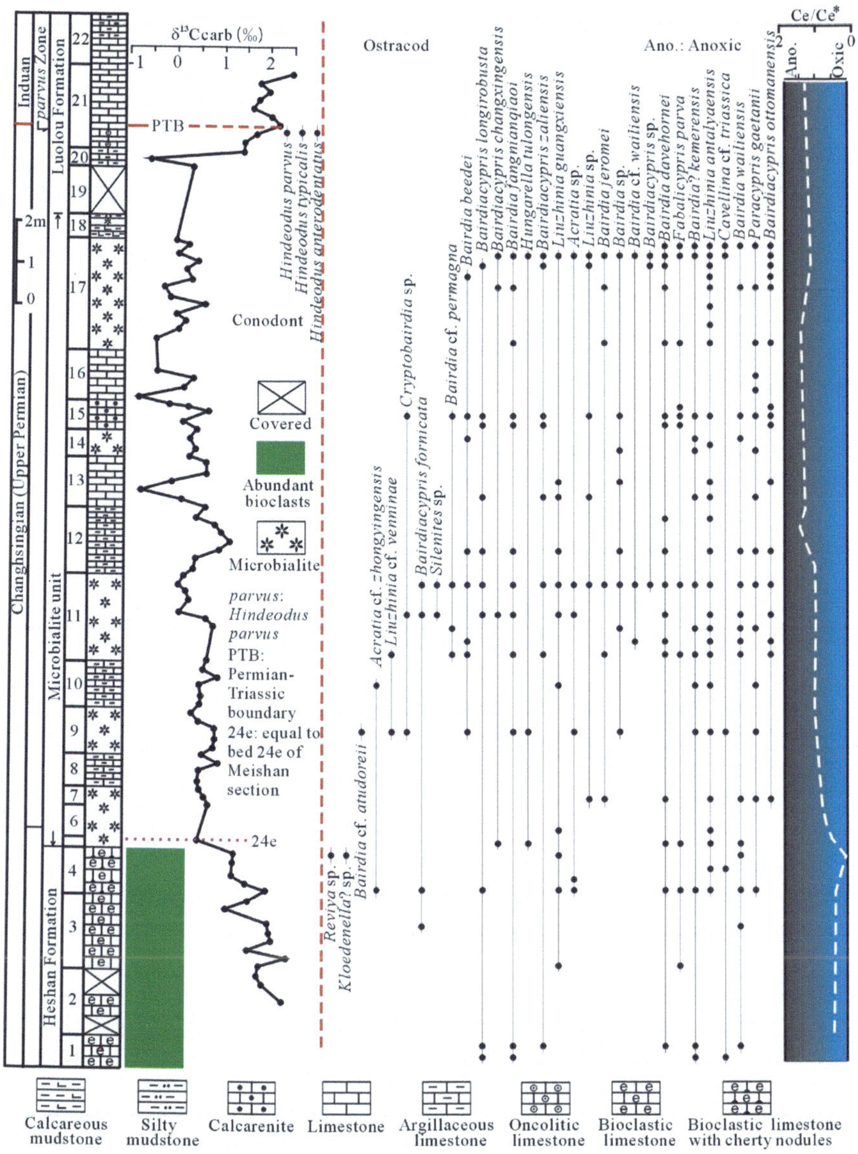

Fig. 3.54 Stratigraphic column and fossil occurrences at the Zuodeng section. *Note* Data of ostracods after Wan et al. (2019), carbon isotopes referred to Yang (2006) and redox conditions based on Liao (2020)

Not measured upwards
Luolou Formation in the Lower Triassic

22. Grayish green thin-bedded argillaceous limestone in the upper part (30 cm); grayish green thin-bedded limestone in the lower part. 120 cm
21. Grayish green medium- to thin-bedded limestone, with oncolitic limestone at the basal part. Conodonts *Hindeodus parvus, Hindeodus typicalis* and *Hindeodus anterodentatus* were examined at the basal part of the bed. 210 cm
20. Grayish green to grayish yellow argillaceous limestone. 40 cm
19. Covered by the Quaternary sediments. 110 cm

Microbialite Unit (assigned to basal part of the Luolou Formation)

18. Thin-bedded silty mudstone in the upper part (24 cm); grayish green microbialite in the middle part (12 cm); yellowish green thin-bedded calcareous mudstone in the lower part. 59 cm
17. Gray thick-bedded microbialite. Microgastropods, foraminfers and ostracods *Bairdia beedei, Bairdia fangnianqiaoi, Bairdia wailiensis, Bairdia jeromei, Bairdia davehornei, Bairdia? kemerensis, Bairdia* cf. *wailiensis, Bairdia* sp., *Bairdiacypris longirobusta, Bairdiacypris changxingensis, Bairdiacypris zaliensis, Bairdiacypris ottomanensis, Bairdiacypris* sp., *Hungarella tulongensis, Liuzhinia guangxiensis, Liuzhinia antalyaensis, Liuzhinia* sp., *Acratia* sp., *Fabalicypris parva, Cavellina* cf. *triassica* and *Paracypris gaetanii* were recorded. 270 cm
16. Light gray medium-bedded limestone. 120 cm
15. Light gray medium-bedded calcarenite. Ostracods *Bairdia wailiensis, Bairdia fangnianqiaoi, Bairdia* cf. *permagna, Bairdia? kemerensis, Bairdia beedei, Bairdia davehornei, Bairdia* sp., *Bairdiacypris longirobusta, Bairdiacypris zaliensis, Bairdiacypris ottomanensis, Liuzhinia antalyaensis, Liuzhinia* sp., *Fabalicypris parva, Paracypris gaetanii* and *Cryptobairdia* sp. were recovered. 70 cm
14. Gray thick-bedded microbialite. Ostracods *Bairdia beedei, Bairdia wailiensis, Bairdia? kemerensis, Bairdia* sp., *Liuzhinia antalyaensis, Paracypris gaetanii* were recorded. 64 cm
13. Light gray medium-bedded limestone. Ostracods *Bairdiacypris longirobusta, Bairdiacypris zaliensis, Bairdiacypris ottomanensis, Liuzhinia guangxiensis, Liuzhinia antalyaensis, Liuzhinia* sp., *Bairdia? kemerensis, Bairdia* sp. and *Paracypris gaetanii* were found. 120 cm
12. Grayish yellow thin-bedded argillaceous limestone. Ostracods *Bairdia beedei, Bairdia fangnianqiaoi, Bairdia davehornei, Bairdia wailiensis, Bairdia* sp., *Bairdiacypris longirobusta, Bairdiacypris ottomanensis, Liuzhinia guangxiensis, Liuzhinia antalyaensis* and *Paracypris gaetanii* were discovered. 160 cm
11. Dark gray thick-bedded microbialite. Ostracods *Liuzhinia guangxiensis, Liuzhinia antalyaensis, Liuzhinia* cf. *venninae, Liuzhinia* sp., *Cryptobairdia* sp., *Silenites* sp., *Bairdia* cf. *permagna, Bairdia? kemerensis, Bairdia*

cf. *wailiensis*, *Bairdia fangnianqiaoi*, *Bairdia davehornei*, *Bairdia beedei*, *Bairdia wailiensis*, *Bairdia jeromei*, *Bairdia* sp., *Bairdiacypris longirobusta*, *Bairdiacypris changxingensis*, *Bairdiacypris zaliensis*, *Bairdiacypris fornicate*, *Bairdiacypris ottomanensis*, *Bairdiacypris* sp., *Acratia* sp., *Fabalicypris parva* and *Paracypris gaetanii* were collected. 210 cm

10. Grayish yellow thin-bedded argillaceous limestone. Ostracods *Acratia* cf. *zhongyingensis*, *Liuzhinia guangxiensis*, *Liuzhinia antalyaensis*, *Bairdia*? *kemerensis* and *Paracypris gaetanii* were recorded. 110 cm
9. Dark gray thick-bedded microbialite. Ostracods *Bairdia beedei*, *Bairdia fangnianqiaoi*, *Bairdia* cf. *atudoreii*, *Bairdia* sp., *Bairdia*? *kemerensis*, *Liuzhinia antalyaensis*, *Liuzhinia* cf. *venninae*, *Cryptobairdia* sp., *Hungarella tulongensis*, *Acratia* sp., *Cavellina* cf. *triassica* and *Paracypris gaetanii* were recovered. 110 cm
8. Grayish yellow thin-bedded argillaceous limestone. 80 cm
7. Dark gray thick- to medium-bedded microbialite, with ostracods *Liuzhinia antalyaensis*, *Liuzhinia* sp., *Bairdia jeromei*, *Bairdia davehornei*, *Bairdia wailiensis*, *Paracypris gaetanii* and *Bairdiacypris ottomanensis*. 440 cm
6. Dark gray thick-bedded microbialite, with ostracods *Liuzhinia guangxiensis* and *Liuzhinia antalyaensis*. 76 cm
5. Dark gray microbialite, with ostracods *Bairdiacypris changxingensis*, *Hungarella tulongensis*, *Bairdia davehornei*, *Bairdia wailiensis*, *Fabalicypris parva* and *Liuzhinia antalyaensis*. 25 cm

Hiatus

Heshan Formation in the Upper Permian (Changhsingian)

4. Grayish black thick-bedded bioclastic limestone with abundant cherty nodules. Abundant fusulinids, gastropods and ostracods *Reviya* sp., *Kloedenella*? sp., *Acratia* cf. *zhongyingensis*, *Acratia* sp., *Bairdiacypris fornicata*, *Bairdiacypris longirobusta*, *Bairdiacypris zaliensis*, *Bairdia davehornei*, *Bairdia wailiensis*, *Bairdia*? *kemerensis*, *Liuzhinia antalyaensis*, *Cavellina* cf. *triassica*, *Paracypris gaetanii* and *Fabalicypris parva*, were recorded. 110 cm
3. Grayish black thick-bedded bioclastic limestone. Abundant algaes, fusulinids, gastropods, bivalves and ostracods *Bairdiacypris fornicata*, *Liuzhinia guangxiensis*, *Fabalicypris parva* and *Bairdia wailiensis* were discovered. 180 cm
2. Chert-nodule limestone (mostly covered by the Quaternary sediments). 160 cm
1. Light gray thick-bedded bioclastic limestone. Abundant fusulinids, gastropods, calcareous sponges and ostracods *Bairdiacypris longirobusta*, *Bairdiacypris zaliensis*, *Bairdia fangnianqiaoi*, *Bairdia davehornei*, *Bairdia wailiensis*, *Bairdia*? *kemerensis* and *Cavellina* cf. *triassica* were collected. 7 cm

Note: Listed ostracods are referred to Wan et al. (2019).

3.24 Majiashan Section in Chaohu City

The Majiashan section (117° 49′ 12″ E, 31° 37′ 33″ N) is located about 5 km northwest of Chaohu City, Anhui Province, South China (Fig. 3.55).

The sequence of the Majiashan section is composed of the upper Talung (=Dalong) Formation and the conformably overlying Yinkeng Formation (Fig. 3.56). The upper Talung Formation is dominated by the gray to dark gray thin-bedded siliceous mudstones intercalated with yellowish green calcareous mudstones, yellow mudstones and pale volcanic ash in the upper part and the grayish black thin-bedded carbonaceous mudstones intercalated with yellowish green calcareous mudstones, chert and pale volcanic ash in the lower part. Horizontal bedding is commonly recorded in the mudstone. The basal part of the Yinkeng Formation is outcropped and dominated by grayish green thin- to medium-bedded argillaceous limestones interbedded with thin-bedded calcareous mudstones. Abundant brachiopods, ammonoids and radiolarians were discovered in the Talung Formtion (Fig. 3.56).

Yinkeng Formation in the Lower Triassic
Not described upwards

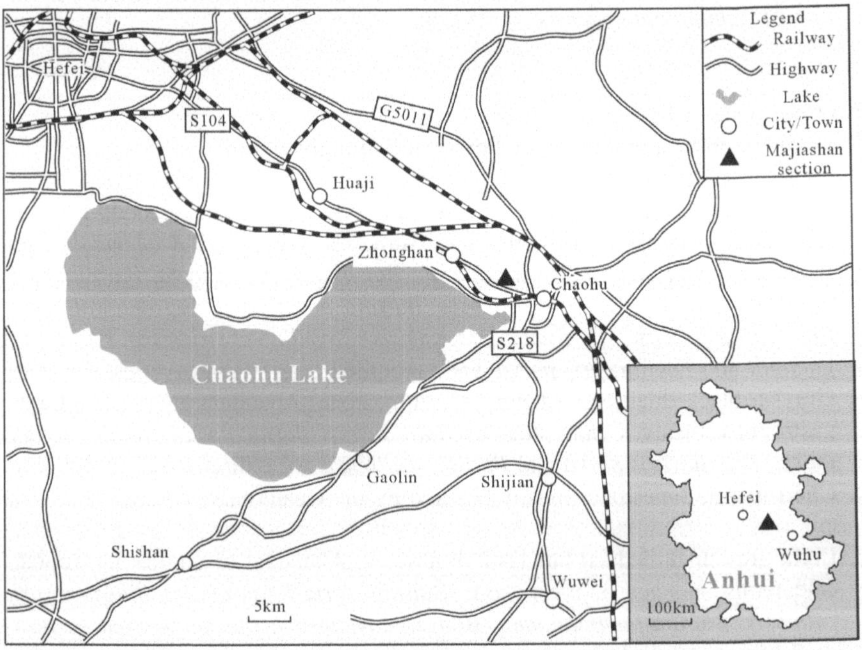

Fig. 3.55 Geographic location of the Majiashan section

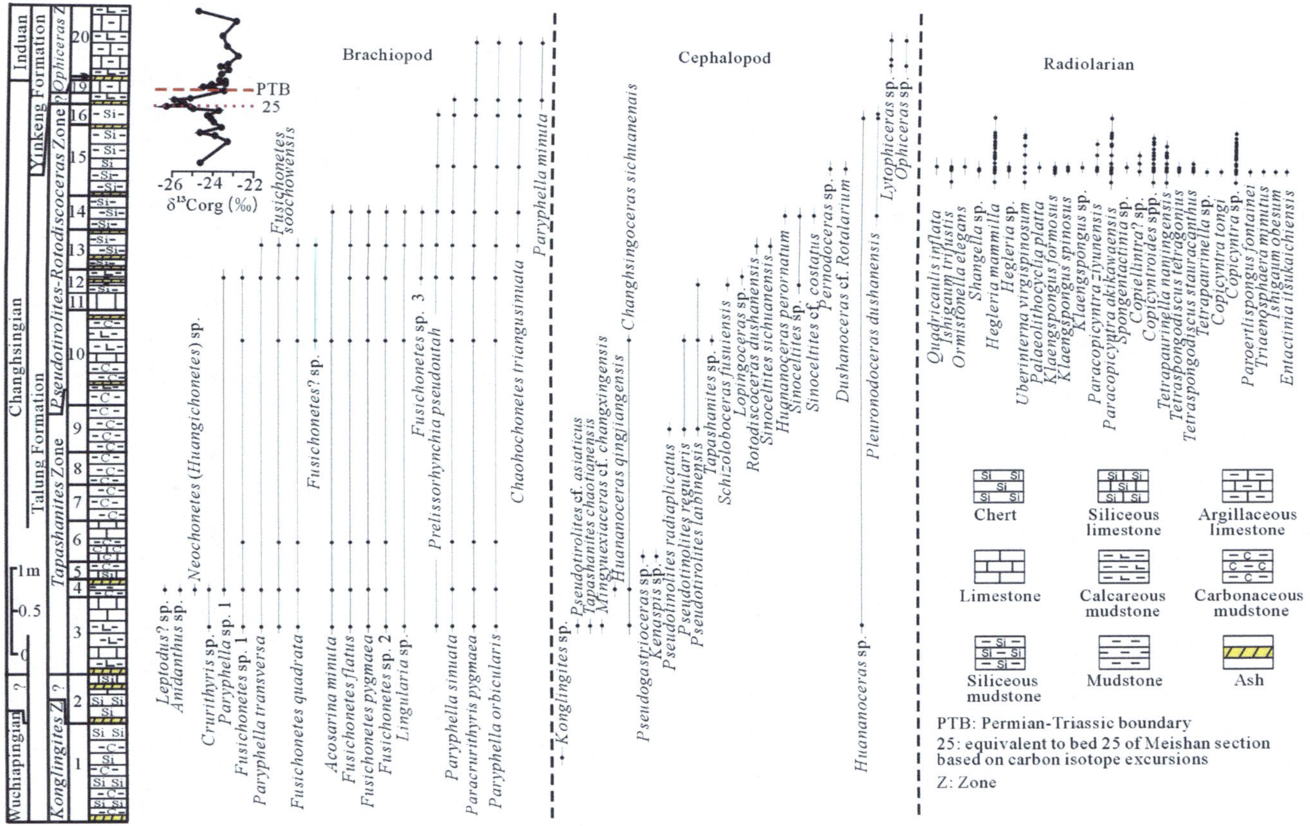

Fig. 3.56 Stratigraphic column and fossil occurrences at the Majiashan section (data of radiolarians after He et al. 2008 and Gui et al. 2009; carbon isotopes based on Fang 2021)

20-4. Yellowish green medium-bedded argillaceous limestone with the brachiopod *Fusichonetes* sp. and ammonoids *Ophiceras* sp. and *Lytophiceras* sp.. 45 cm
20-3. Yellowish green calcareous mudstone with the ammonoid *Lytophiceras* sp.. 4 cm
20-2. Yellowish green medium-bedded argillaceous limestone with ammonoids *Ophiceras* sp. and *Lytophiceras* sp.. 30 cm
20-1. Brown thin-bedded mudstone (2 cm thick) in the upper part; pale volcanic ash (1 cm thick) in the lower part. 3 cm
19. Grayish green thin-bedded argillaceous limestone with a few brachiopods *Fusichonetes* sp. and *Paryphella orbicularis.* 20 cm
18. Grayish green thin-bedded calcareous mudstone. 10 cm
17-2. Yellow thin-bedded mudstone. 7 cm
17-1. Pale volcanic ash. 2 cm

Conformity
Talung Formation in the Upper Permian

16-3. Grayish black thin-bedded siliceous mudstone, with the ammonoid *Pleuronodoceras dushanensis.* 10 cm
16-2. Grayish black thin-bedded siliceous mudstone, with ammonoids *Huananoceras* sp. and *Pleuronodoceras dushanensis.* 6 cm
16-1. Grayish black mudstone. 3.5 cm
15-10. Pale volcanic ash intercalated with yellow mudstone. 4.5–10 cm
15-9. Grayish black calcareous and siliceous mudstone. 11 cm
15-8. Grayish black thin-bedded siliceous mudstone. 4 cm
15-7. Light gray calcareous and siliceous mudstone. 9 cm
15-6. Light gray calcareous and siliceous mudstone. 3 cm
15-5. Redish brown chert. 8 cm
15-4. Gray siliceous mudstone. 6 cm
15-3. Light gray mudstone with laminations. 3 cm
15-2. Grayish black siliceous and calcareous mudstone with laminations. 9 cm
15-1. Gray thin-bedded mudstone intercalated with yellow mudstone, with pale volcanic ash at the basal part of the bed. 3–13 cm
A few ammonoids *Dushanoceras* cf. *rotalarium* and *Perndoceras* sp. were collected in bed 15 (bed 15 not subdivided when collecting macrofossils).
14-5. Gray black thin-bedded calcareous and siliceous mudstone with laminations. 3 cm
14-4. Light gray siliceous mudstone. 7 cm
14-3. Gray siliceous mudstone, with pale volcanic ash (< 1 cm) at the base. 7 cm
14-2. Light gray mudstone. 8 cm

14-1. Light gray siliceous mudstone. 13 cm
Ammonoids *Sinoceltites* cf. *costatus, Sinoceltites* sp., *Huananoceras perornatum* and *Pleuronodoceras dushanensis* were collected in bed 14.
13-7. Pale volcanic ash intercalated with yellow mudstone. 2 cm
13-6. Gray calcareous and siliceous mudstone with laminations. 5 cm
13-5. Light gray thin-bedded calcareous and siliceous mudstone. 12 cm
13-4. Light gray thin-bedded siliceous mudstone with laminations. 10 cm
13-3. Pale volcanic ash intercalated with yellow mudstone. 3 cm
13-2. Yellowish green calcareous and siliceous mudstone. 8 cm
13-1. Brown thin-bedded calcareous mudstone intercalated with pale volcanic ash. 4 cm
A few ammonoids *Sinoceltites sichuanensis* and *Rotodiscoceras dushanensis* were collected in bed 13.
12-5. Yellowish green calcareous mudstone with the cephalopod *Lopingoceras* sp.. 9 cm
12-4. Grayish green calcareous mudstone. 8 cm
12-3. Grayish green thin-bedded calcareous mudstone intercalated with pale volcanic ash. 5 cm
12-2. Yellow mudstone intercalated with pale volcanic ash. 3 cm
12-1. Grayish green siliceous mudstone. 4 cm
Ammonoids *Pseudotirolites* cf. *laibinensis, Schizoloboceras fusuiensis* and *Sinoceltites* sp. were collected in bed 12.
11. Gray medium-bedded micrite in the upper part; argillaceous limestone in the lower part. 20 cm
10-15. Gray black carbonaceous mudstone intercalated with pale to yellow volcanic ash, with pale volcanic ash at the base. 5 cm
10-14. Yellowish green thin-bedded carbonaceous mudstone. 8 cm
10-13. Gray micrite. 10 cm
10-12. Yellowish green thin-bedded carbonaceous mudstone. 7 cm
10-11. Yellowish brown carbonaceous mudstone with the ammonoid *Mingyuexiaceras* sp.. 8 cm
10-10. Dark gray argillaceous limestone. 10 cm
10-9. Pale volcanic ash intercalated with yellow mudstone. 3 cm
10-8. Grayish black carbonaceous mudstone. 7 cm
10-7. Grayish green shale. 6 cm
10-6. Grayish black carbonaceous mudstone. 8 cm
10-5. Grayish black thin-bedded carbonaceous mudstone. 9 cm
10-4. Grayish black thin-bedded carbonaceous mudstone. 8 cm
10-3. Grayish black thin-bedded carbonaceous mudstone. 6 cm
10-2. Grayish black thin-bedded carbonaceous mudstone. 8 cm
10-1. Gray thin-bedded limestone intercalated with mudstone, with laminations in the mudstone. 9 cm
Abundant ammonoids *Tapashanites* sp., *Pseudotirolites laibinensis, Pseudotinolites* cf. *regularis* and *Changsingoceras* cf. *sichuangensis* were recovered in bed 10.

9-5. Black calcareous and carbonaceous mudstone with laminations. 7 cm
9-4. Black carbonaceous mudstone. 6 cm
9-3. Grayish black thin-bedded carbonaceous mudstone. 19 cm
9-2. Dark gray carbonaceous mudstone. 5 cm
9-1. Grayish black medium-bedded carbonaceous limestone. 19 cm
Abundant ammonoids *Pseudotirolites laibinensis*, *Pseudotinolites regularis* and *Pseudotinolites radiaplicatus* were collected in bed 9.
8-4. Dark gray carbonaceous mudstone. 5 cm
8-3. Grayish black medium-bedded carbonaceous limestone. 19 cm
8-2. Dark gray carbonaceous mudstone with laminations. 9 cm
8-1. Dark gray carbonaceous mudstone. 5 cm
7-3. Dark gray medium-bedded carbonaceous limestone with laminations. 21 cm
7-2. Black carbonaceous mudstone. 8 cm
7-1. Black medium-bedded carbonaceous limestone with laminations. 13 cm
6-4. Gray micrite. 2 cm
6-3. Brown carbonaceous mudstone. 9–25 cm
6-2. Black manganous and carbonaceous mudstone. 9 cm
6-1. Black carbonaceous limestone. 10 cm
A few ammonoids *Kenaspis* sp. and *Pseudogastrioceras* sp. were discovered in bed 6.
5-2. Black thin-bedded carbonaceous mudstone. 10 cm
5-1. Dark gray thin-bedded siliceous limestone. 10 cm
4-3. Pale volcanic ash intercalated with yellow and black mudstone. 3 cm
4-2. Black thin-bedded carbonaceous mudstone. 14 cm
4-1. Yellow volcanic ash. 3 cm
Cephalopods *Changhsingoceras sichuanenais* and *Huananoceras qiingjiangensis* were collected in beds 4 and 5.
3-3. Gray medium-bedded limestone intercalated with siliceous mudstone (4 cm thick). 41 cm
3-2. Grayish green thin-bedded calcareous mudstone. 41 cm
3-1. Pale volcanic ash intercalated with yellow mudstone in the upper part (totally about 6 cm thick); gray medium-bedded limestone in the middle part (14 cm thick); pale volcanic ash in the lower part (about 6 cm thick); thin-bedded limestone at the base (4 cm thick). 10–30 cm
Abundant cephalopods *Mingyuexiaceras* cf. *changxingensis*, *Huananoceras* sp., *Tapashanites chaotianmensis*, *Pseudotirolites* cf. *asiaticus* and *Changhsingoceras sichuanenais* were recorded in bed 3.
2-2. Medium-bedded siliceous limestone in the upper part (13 cm thick); pale volcanic ash in the middle part (1 cm thick); thin-bedded siliceous limestone in the lower part (12 cm thick); pale volcanic ash at the base (6 cm thick). 32 cm
2-1. Thin-bedded chert in the upper part (14 cm thick); yellow and pale volcanic ash in the middle part (6 cm thick); grayish black volcanic ash in the lower part (4 cm). 24 cm

1-9. Black thin-bedded siliceous and carbonaceous mudstone with the ammonoid *Konglingites* sp.. 70 cm
1-8. Black thin-bedded siliceous and carbonaceous mudstone. 96 cm
1-7. Gray thin- to medium-bedded siliceous and carbonaceous mudstone. 150 cm
1-6. Black thin- to medium-bedded siliceous and carbonaceous mudstone, with medium-bedded chert at the base (20 cm thick). 132 cm.
1-5. Black thin-bedded siliceous and carbonaceous mudstone, intercalated with thin- to medium-bedded carbonaceous chert in the lower part. 122 cm
1-4. Black thin- to medium-bedded carbonaceous chert. 93 cm
1-3. Black medium-bedded siliceous and carbonaceous mudstone in the upper part; black carbonaceous and calcareous mudstone in the lower part; yellow volcanic ash at the base (3 cm thick). 32 cm
1-2. Thin-bedded carbonaceous and calcareous mudstone, with pale volcanic ash at the base (4 cm thick). 16 cm
1-1. Medium-bedded carbonaceous and siliceous mudstone in the upper part; thin-bedded carbonaceous mudstone in the lower part; pale volcanic ash at the base. 70 cm

A few ammonoids *Konglingites* sp., *Planodiscoceras* sp. and *Planodiscoceras involutum* were collected in bed 1.

Covered by the Quaternary downwards

3.25 Jingshan Section in Jingmen City

The Jingshan section is located in the Jingshan County (part of Jingmen City), along the 240 National Highway, 140 km northwest to the Wuhan City, Hubei Province, South China (Fig. 3.57).

The sequence of the Jingshan section is composed of the Talung and Taye formations in an ascending order (Fig. 3.58). The Talung Formation is mainly composed of mudstones. The Taye Formation is characterized by thin-bedded argillaceous limestones and mudstones. A few brachiopods, ammonoids and bivalves were discovered in the section (Fig. 3.58).

Taye Formation in the Lower Triassic (Induan)

36. Grayish yellow thin-bedded mudstone with calcareous nodules in the upper part; grayish green thin-bedded calcareous mudstone intercalated with mudstone in the lower part. The ammonoid *Ophiceras* sp. was examined. 410 cm
35. Grayish green medium- to thin-bedded argillaceous limestone interbedded with grayish black mudstone. The bivalve *Claraia* sp. and ammonoid *Ophiceras* sp. were identified. 485 cm

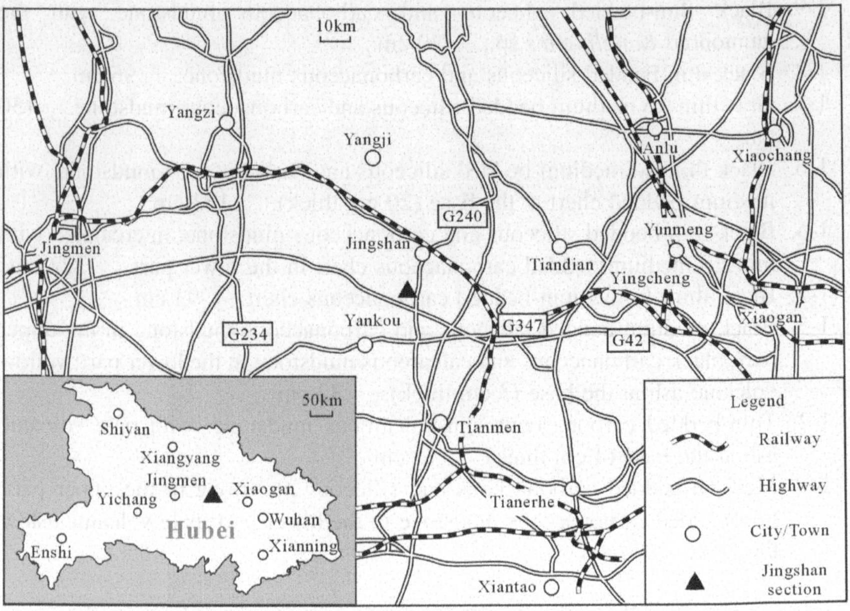

Fig. 3.57 Geographic location of the Jingshan section

Conformity
Talung Formation in the Upper Permian (Changhsingian)

34. Black thin-bedded chert intercalated with mudstone in the upper part; black mudstone in the lower part (8 cm); yellow volcanic ash at the basal part (4 cm). 39 cm
33. Seven cycles were deposited and each cycle includes: black mudstone or calcareous mudstone in the upper part; yellow volcanic ash in the lower part. The trilobite *Pseudophillipsia subcircularis* was found. 100 cm
32. Five cycles were deposited in a descending order: (1) Gray medium-bedded limestone in the upper part (13 cm); black mudstone in the lower part (3 cm); yellow volcanic ash at the basal part (1 cm). (2) Black mudstone in the upper part (31 cm); yellow volcanic ash in the lower part (4 cm). (3) Gray lens-shaped limestone in the upper part (13 cm); black mudstone in the middle part (15 cm); yellow volcanic ash in the lower part (2 cm). (4) Gray limestone in the upper part (18 cm); black mudstone in the lower part (6 cm); yellow volcanic ash at the basal part (1 cm). (5) Black mudstone in the upper part (17 cm); yellow volcanic ash in the lower part (3 cm). 126 cm
31. Grayish black siliceous mudstone with the brachiopod *Paryphella* sp. in the upper part (6 cm); black mudstone in the middle part (10 cm); yellow volcanic ash in the lower part (5 cm). 21 cm
30. Grayish yellow to black mudstone with ammonoids. 105 cm

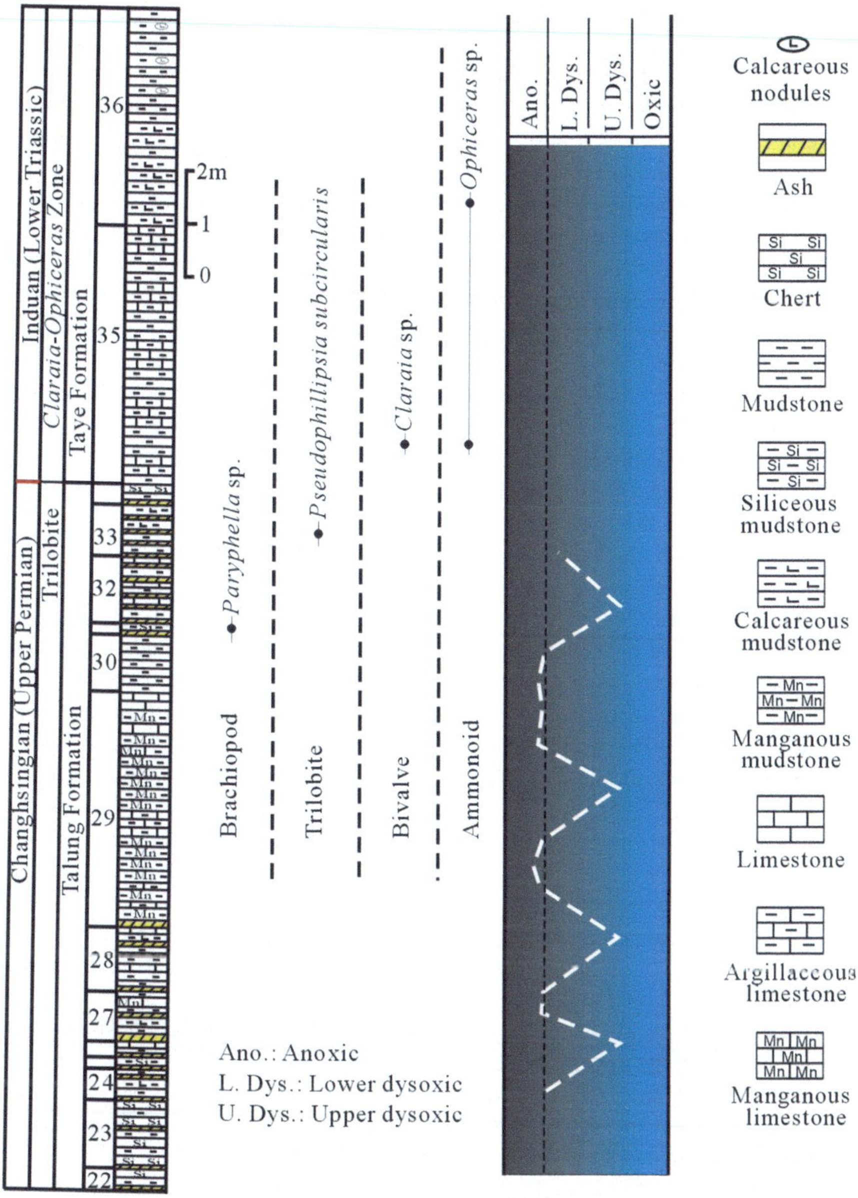

Fig. 3.58 Stratigraphic column and fossil occurrences at the Jingshan section (Data of fossils and redox conditions based on framboid pyrites are after Li 2016)

29. Six cycles were deposited in a descending order: (1) Grayish black limestone in the upper part (35 cm); brown manganous mudstone in the lower part (21 cm). (2) Grayish black limestone in the upper part (25 cm); brown manganous mudstone in the lower part (20 cm). (3) Brownish yellow lens-shaped argillaceous and manganous limestone in the upper part (10 cm); brown manganous mudstone in the lower part (110 cm). (4) Grayish yellow argillaceous limestone in the upper part (45 cm); brown manganous mudstone in the lower part (90 cm). (5) Grayish yellow argillaceous limestone in the upper part (16 cm); brown manganous mudstone in the lower part (25 cm). (6) Grayish yellow argillaceous limestone in the upper part (20 cm); brown manganous mudstone in the lower part (25 cm); yellow volcanic ash at the basal part (about 1 cm). 443 cm
28. Three cycles were deposited in a descending order: (1) Gray limestone in the upper part (30 cm); yellowish green calcareous mudstone in the lower part (4 cm); yellow volcanic ash at the basal part (1 cm). (2) Black mudstone in the upper part (7 cm); black mudstone in the middle part (5 cm); yellow volcanic ash in the lower part (2 cm). (3) Black mudstone in the upper part (40 cm); grayish black lens-shaped limestone in the middle part (14 cm); black mudstone in the lower part (18 cm); yellow volcanic ash at the basal part (4 cm). 121 cm
27. Two cycles were deposited in a descending order: (1) Black mudstone in the upper part (7 cm); brown manganous limestone in the middle part (16 cm); black mudstone in the lower part (25 cm); yellow volcanic ash at the basal part (2 cm). (2) Black calcareous mudstone in the upper part (24 cm); black mudstone in the middle part (12 cm); yellow volcanic ash in the lower part (9 cm). 95 cm
26. Grayish black argillaceous limestone in the upper part (20 cm); black mudstone in the lower part (8 cm); yellow volcanic ash at the basal part (2 cm). 30 cm
25. Grayish black siliceous mudstone with small brachiopods in the upper part (6 cm); black mudstone in the lower part (3 cm); yellow volcanic ash at the basal part (0.5 cm). 9.5 cm
24. Two cycles were deposited in a descending order: (1) Black mudstone with laminations in the upper part (18 cm); yellow volcanic ash in the lower part (4 cm). (2) Black calcareous mudstone in the upper part (32 cm); grayish black mudstone in the lower part (12 cm); yellow volcanic ash at the basal part (2 cm). 68 cm
23. Four cycles were deposited in a descending order: (1) Grayish black chert in the upper part (12 cm); black mudstone with laminations in the lower part (20 cm). (2) Grayish black chert in the upper part (8 cm); black mudstone with laminations in the middle part (2 cm); yellow volcanic ash in the lower part (1 cm). (3) Grayish black mudstone intercalated with eight layers of cherts (68 cm). (4) Grayish black chert in the upper part (9 cm); black mudstone with laminations in the middle part (4 cm); yellow volcanic ash in the lower part (3 cm). 127 cm

22. Two cycles were deposited in a descending order: (1) Black chert with abundant radiolarians in the upper part (10 cm); black mudstone in the middle part (9 cm); yellow volcanic ash in the lower part (4 cm). (2) Black chert with abundant radiolarians in the upper part (14 cm); black mudstone in the middle part (2 cm); yellow volcanic ash in the lower part (1 cm). 40 cm

Not described downwards

3.26 Xiejiaping Section in Wufeng County

The Xiejiaping section (30° 20′ 12″ N, 110° 35′ 53″ E) is located at Huoshan, 3 km north to the Xiejiaping Village, Wufeng County, Hubei Province, South China (Fig. 3.59).

The sequence of the Xiejiaping section is composed of the Wuchiaping, Talung, Changhsing and Taye formations in an ascending order (Fig. 3.60). The Wuchaiping Formation is dominated by gray thick-bedded dolomitized limestones. The Talung Formation is characterized by thin-bedded carbonaceous mudstones and cherts intercalated with siliceous limestones. The Changhsing Formation is mainly

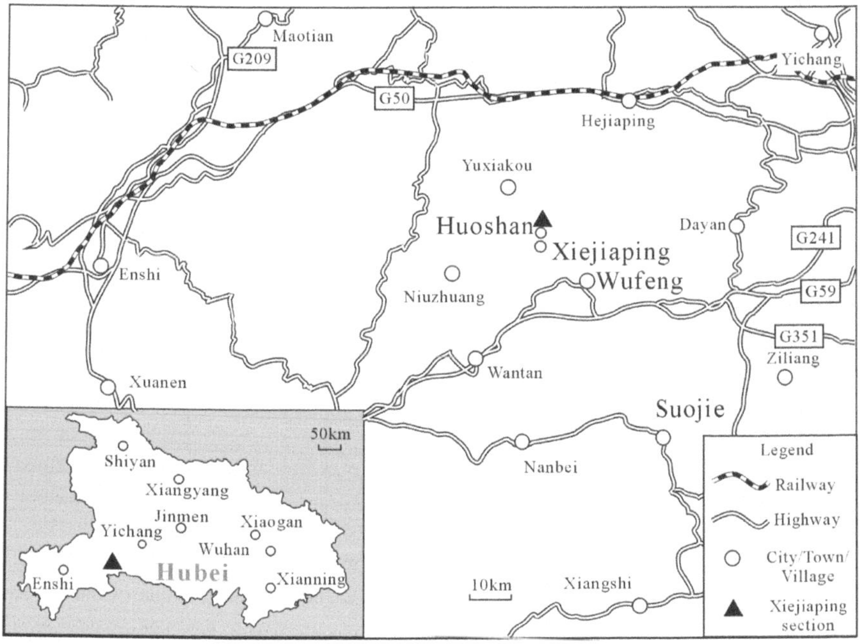

Fig. 3.59 Geographic location of the Xiejiaping section

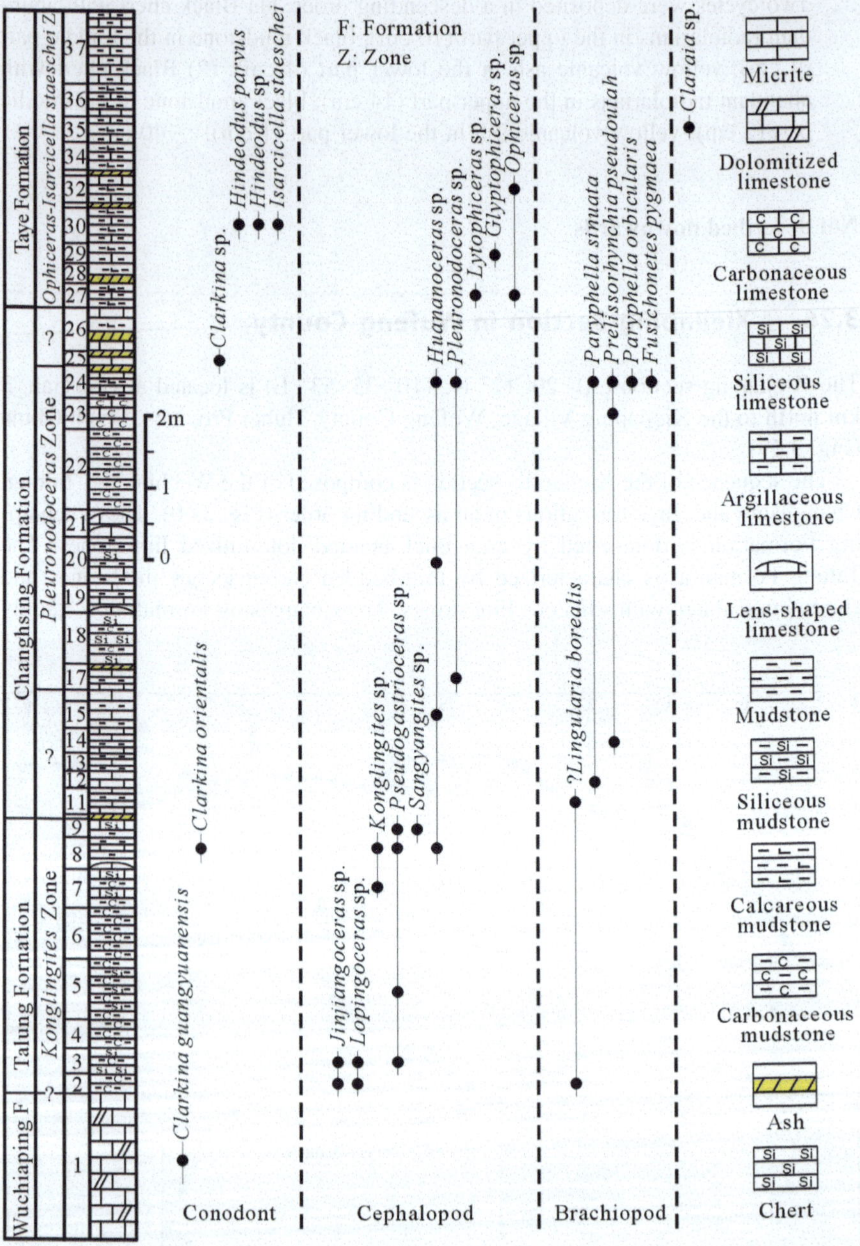

Fig. 3.60 Stratigraphic column and fossil occurrences at the Xiejiaping section (brachiopods referred to He et al. 2014)

composed of medium- to thin-bedded micrites. The Taye Formation is characterized by medium- to thin-bedded argillaceous limestones intercalated with calcareous mudstones.

Taye Formation in the Lower Triassic

37. Gray medium-bedded argillaceous limestone. > 100 cm
36. Gray thin-bedded argillaceous limestone intercalated with four layers of grayish green calcareous mudstone. 44 cm
35. Gray thin-bedded argillaceous limestone intercalated with four layers of calcareous mudstone. Bivalves *Claraia* spp. were found in the basal part. 35 cm
34. Gray thin-bedded argillaceous limestone interbedded with calcareous mudstone. 38 cm
33. Yellow volcanic ash. 8 cm
32. Gray thin-bedded argillaceous limestone interbedded with calcareous mudstone, with the ammonoid *Ophiceras* sp.. 38 cm
31. Yellow volcanic ash. 7 cm
30. Gray medium- to thin-bedded argillaceous limestone intercalated with calcareous mudstone. Conodonts *Hindeodus parvus*, *Hindeodus* sp. and *Isarcicella staeschei* were discovered in the bed. 50 cm
29. Gray medium-bedded argillaceous limestone, with calcareous mudstone at the top part. The ammonoid *Glyptophiceras* sp. was found in the basal part. 28 cm
28. Grayish green thin-bedded calcareous mudstone, with grayish green volcanic ash at the basal part (3 cm). 28 cm
27. Dark gray medium-bedded argillaceous limestone interbedded with calcareous mudstone with laminations. Abundant ammonoids *Lytophiceras* spp. and *Ophiceras* spp. were discovered. 33 cm

Conformity

Changhsing Formation in the Upper Permian

26. Dark gray medium-bedded micrite intercalated with calcareous mudstone with laminations, with a layer of volcanic ash in the lower part. 60 cm
25. Gray medium-bedded micrite with pale volcanic ash at the top. Conodonts *Clarkina* spp. was examined. 20 cm
24. Dark gray thin-bedded micrite interbedded with calcareous mudstone (with laminations), with pale volcanic ash at the top. Brachiopods *Paryphella orbicularis*, *Fusichonetes pygmaea*, *Paryphella sinuata* and *Prelissorhynchia pseudoutah*, and ammonoids *Huananoceras* sp. and *Pleuronodoceras* sp. were identified. 46 cm
23. Dark gray medium-bedded carbonaceous limestone with the brachiopod *Prelissorhynchia pseudoutah*. 42 cm

22. Black thin-bedded carbonaceous mudstone with laminations. 112 cm
21. Dark gray lens-shaped micrite. 38 cm
20. Grayish black thin-bedded mudstone intercalated with chert and micrite, with the ammonoid *Huananoceras* sp.. 64 cm
19. Grayish black thin-bedded micrite intercalated with mudstone, with lens-shaped limestone at the basal part. 40 cm
18. Black thin-bedded chert interbedded with carbonaceous mudstone, with laminations. 70 cm
17. Dark gray thin-bedded micrite interbedded with mudstone, with pale volcanic ash at the top. The ammonoid *Pleuronodoceras* sp. was examined. 36 cm
16. Dark gray medium-bedded micrite. 16 cm
15. Grayish black thin-bedded mudstone intercalated with micrite, with the ammonoid *Huananoceras* sp.. 38 cm
14. Dark gray thin-bedded micrite, intercalated with mudstone in the middle part. The brachiopod *Prelissorhynchia pseudoutah* was examined. 33 cm
13. Dark thin-bedded mudstone intercalated with micrite. 26 cm
12. Grayish black thin-bedded micrite with fragments of brachiopods. 24 cm
11. Grayish black thin-bedded micrite intercalated with mudstone. The brachiopod ?*Lingularia borealis* was found. 36 cm
10. Pale volcanic ash. 7 cm

Conformity
Talung Formation in the Upper Permian

9. Black thin-bedded mudstone interbedded with thin-bedded siliceous limestone, with laminations. The ammonoids *Pseudogastrioceras* sp. and *Sangyangites* sp. were examined. 26 cm
8. Grayish black thin-bedded mudstone intercalated with thin-bedded micrite (with laminations) in the upper part; dark gray lens-shaped micrite in the lower part (10 cm). The ammonoids *Huananoceras* sp., *Konglingites* sp. and *Pseudogastrioceras* sp. and the conodont *Clarkina orientalis* were discovered. 43 cm
7. Grayish black thin-bedded siliceous limestone interbedded with carbonaceous mudstone with laminations. The ammonoid *Konglingites* sp. was collected. 53 cm
6. Black thin-bedded carbonaceous mudstone with laminations. 70 cm
5. Grayish black siliceous mudstone interbedded with carbonaceous mudstone with laminations. The ammonoid *Pseudogastrioceras* sp. was found. 83 cm
4. Black thin-bedded carbonaceous mudstone with laminations. 40 cm
3. Black thin-bedded chert intercalated with carbonaceous mudstone, with laminations in the mudstone. Ammonoids *Jinjiangoceras* sp., *Pseudogastrioceras* sp. and *Lopingoceras* sp. were recovered. 33 cm

2. Black thin-bedded carbonaceous mudstone with laminations. The brachiopod ?*Lingularia borealis* and ammonoids *Jinjiangoceras* sp. and *Lopingoceras* sp. were examined. 28 cm

Conformity
Wuchiaping Formation in the Upper Permian

1. Light gray thick-bedded dolomitized limestone. The conodont *Clarkina guangyuanensis* was identified. > 200 cm

3.27 Shangsi Section in Guangyuan City

The Shangsi section is located about 1 km north to the Shangsi Village, 38 km southwest to the Guangyuan City, Sichuan Province, South China (Fig. 3.61).

The sequence of the Shangsi section is composed of the Feixianguan and Talung formations in a descending order (Fig. 3.62). The Talung Formation is dominated by grayish black to dark gray medium- to thin-bedded carbonaceous cherts and

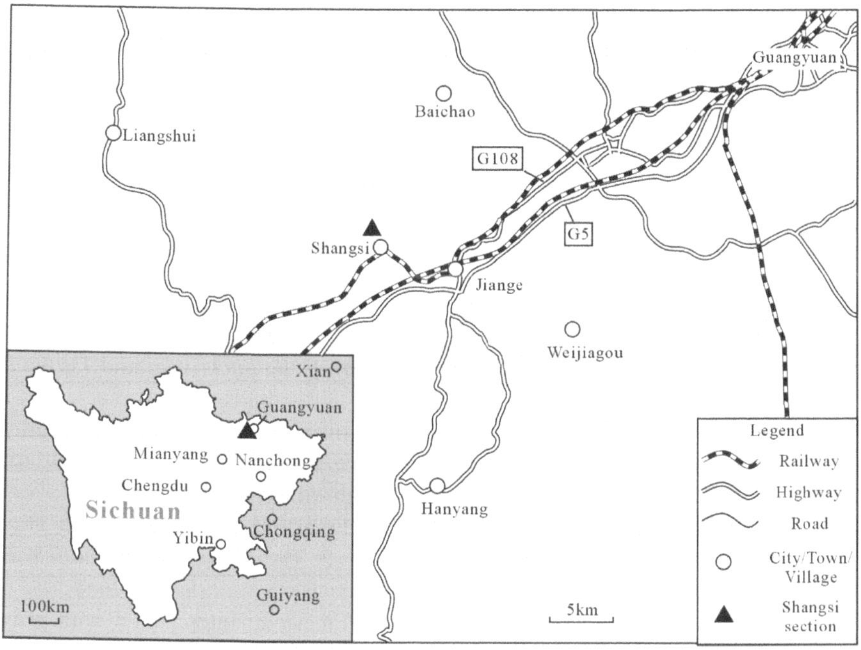

Fig. 3.61 Geographic location of the Shangsi section

siliceous limestones, intercalated with carbonaceous shales and siliceous mudstones. Feixianguan Formation is mainly composed of yellowish green thin-bedded argillaceous limestones and calcareous mudstones, intercalated with pale volcanic ash. Abundant conodonts, a few bivalves and spores and pollens were discovered in the Feixianguan Formation, abundant cephalopods and a few brachiopods and foraminifers were found in the Talung Formation (Figs. 3.62, 3.63, 3.64). In addition, abundant radiolarians were reported in the Talung Formation.

Feixianguan Formation in the Lower Triassic
Not described upwards

34. Gray thin-bedded micritic limestone intercalated with shale, with laminations. 120 cm
33. Gray medium-bedded calcirudite with **conodonts** *Hindeodus parvus, Hindeodus eurypyge, Hindeodus anterodentatus, Hindeodus inflatus, Isarcicella turgida, I. lobata, I. staeschei, I. inflata, I. isarcica* and *I. pecularity.* 268 cm
32. Gray medium-bedded algal limestone with **conodonts** *Hindeodus praeparvus, Hindeodus parvus, Hindeodus eurypyge, Hindeodus anterodentatus, Hindeodus inflatus, Isarcicella turgida, I. lobata, I. staeschei, I. inflata* and *I. isarcica.* 96 cm

31-2. Yellowish green thin-bedded argillaceous limestone interbedded with gray algal limestone, with birdeye structure at the top. **Conodonts** *Hindeodus praeparvus, Hindeodus parvus, Hindeodus eurypyge, Hindeodus anterodentatus, Hindeodus inflatus, Isarcicella turgida, I. lobata, I. staeschei, I. inflata, I. isarcica* and *I. pecularity* were identified. 135 cm

31-1. Yellowish green thin-bedded argillaceous limestone interbedded with gray algal limestone, with birdeye structure at the top. **Conodonts** *Hindeodus parvus, Hindeodus inflatus, Hindeodus eurypyge, Hindeodus anterodentatus, H. praeparvus, Isarcicella turgida, I. prisca, I. lobata* and *I. inflata* were found. 135 cm

30-4. Yellowish green thin-bedded argillaceous limestone intercalated with gray argillaceous limestone with birdeye structure. **Conodonts** *Hindeodus parvus, H. eurypyge, H. inflatus, H. praeparvus, Isarcicella huckriedei* and *I. lobata* were examined. 45 cm

30-3. Yellowish green thin-bedded argillaceous limestone intercalated with gray argillaceous limestone with birdeye structure. **Conodonts** *Hindeodus parvus, H. anterodentatus* and *H. huckriedei* were discovered. 45 cm

30-2. Yellowish green thin-bedded argillaceous limestone intercalated with gray argillaceous limestone with birdeye structure. The **conodont** *I. lobata* was identified. 45 cm

30-1. Yellowish green thin-bedded argillaceous limestone intercalated with gray argillaceous limestone with birdeye structure. The **conodont** *Clarkina* sp.,

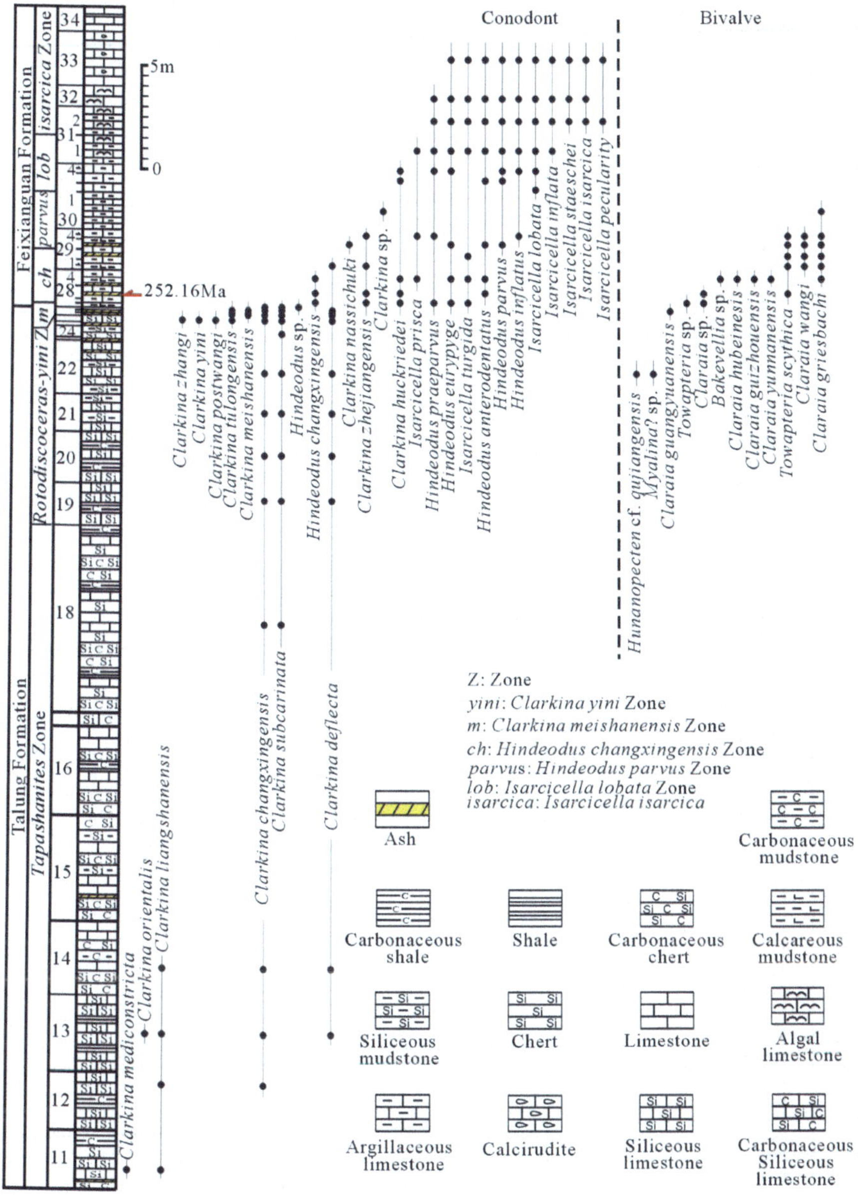

Fig. 3.62 Stratigraphic column and fossil occurrences (conodont, bivalve) at the Shangsi section. *Note* fossils in beds 11–21 referred to Li et al. (1989); 252.16 Ma of U–Pb date after Shen et al. (2011)

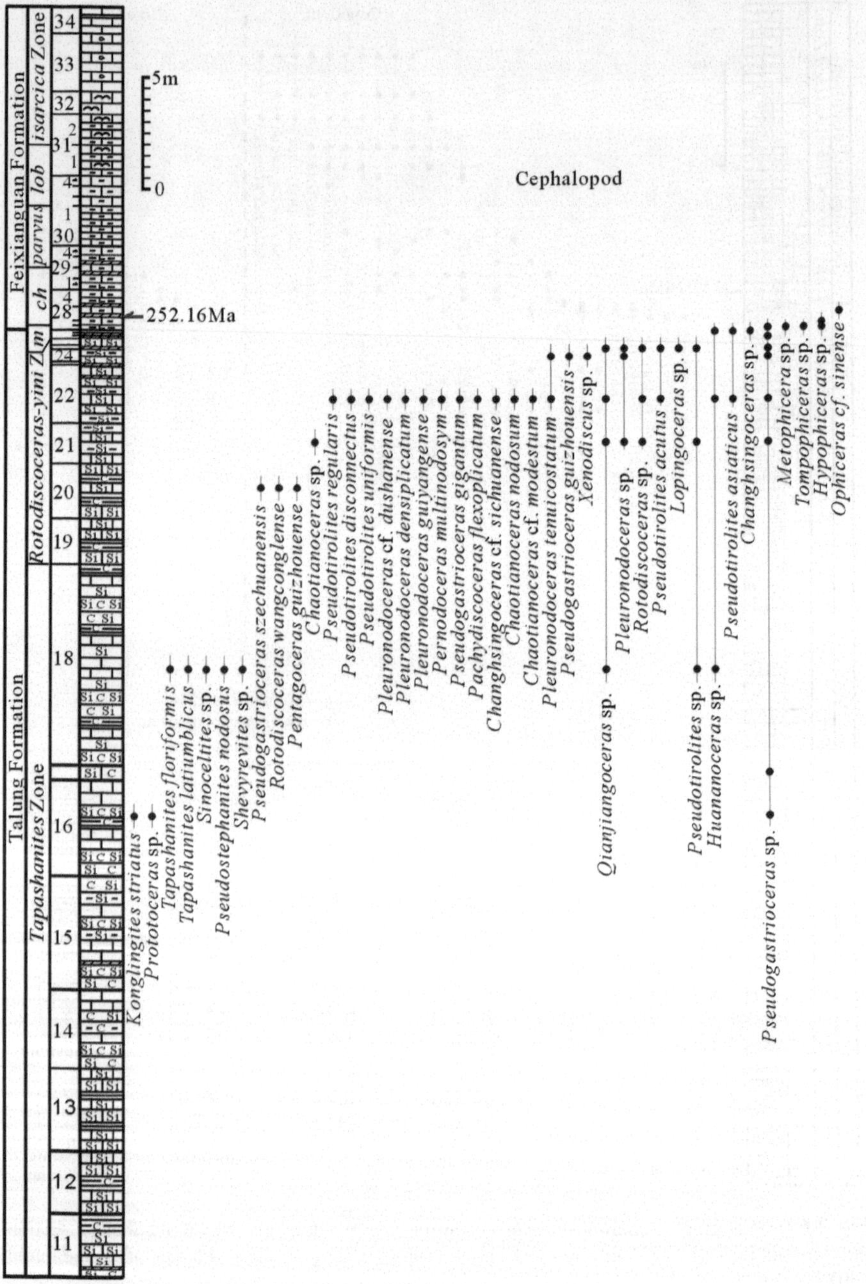

Fig. 3.63 Stratigraphic column and fossil occurrences (cephalopod) at the Shangsi section. *Note* cephalopods in beds 11–21 referred to Li et al. (1989); legends and abbreviations same to Fig. 3.62

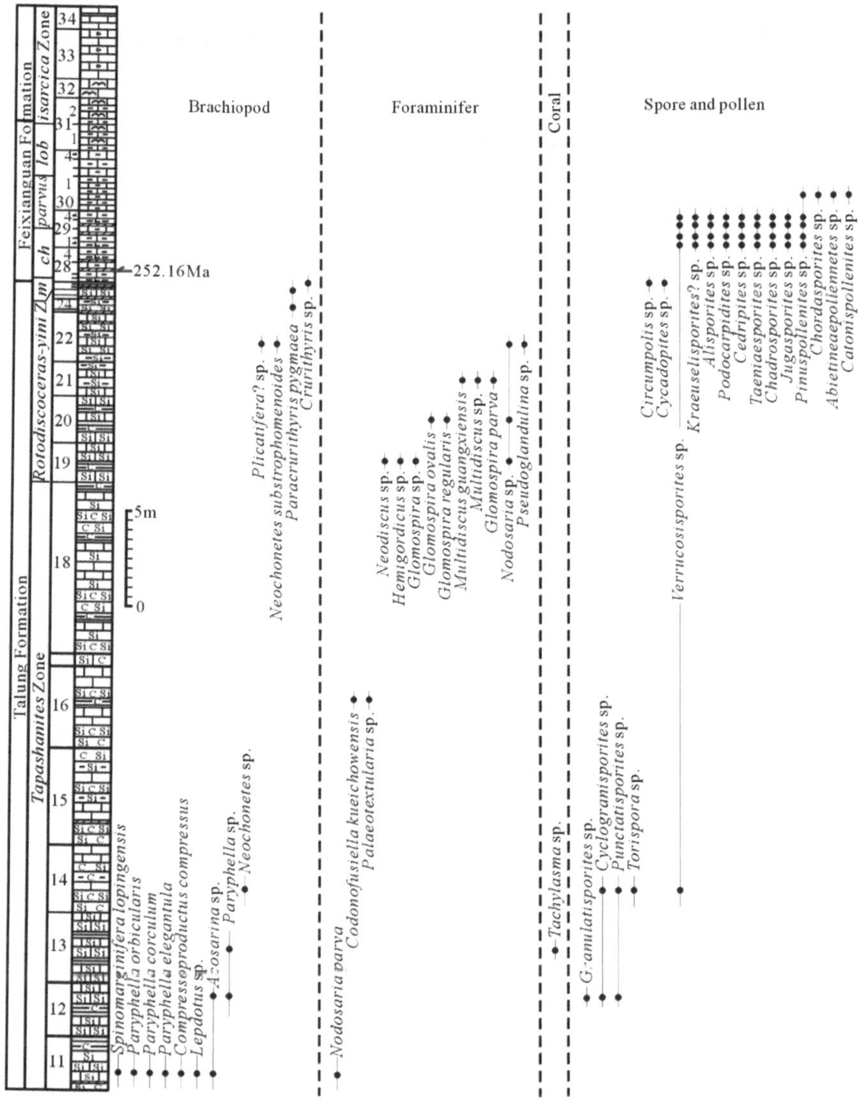

Fig. 3.64 Stratigraphic column and fossil occurrences (brachiopod, foraminifer, coral, spore and pollen) at the Shangsi section. *Note* fossils in beds 11–21 referred to Li et al. (1989); legends and abbreviations same to Fig. 3.62

bivalve *Claraia griesbachi*, **spores and pollens** *Chordasporites* sp., *Abietineaepollennetes* sp., *Catonispollenites* sp. and *Pinuspollenites* sp. were examined. 180 cm

29-4. Yellowish green thin-bedded calcareous mudstone and argillaceous limestone intercalated with volcanic ash, with laminations in the mudstone. **Conodonts** *Hindeodus inflatus*, *H. praeparvus*, *Isarcicella prisca* and *Clarkina zhejiangensis*, **bivalves** *Claraia griesbachi*, *Claraia wangi* and *Towapteria scythica*, and **spores and pollens** *Verrucosisporites* sp., *Kraeuselisporites*? sp., *Alisporites* sp., *Pinuspollenites* sp., *Podocarpidites* sp., *Cedripites* sp., *Taeniaesporites* sp., *Chadrosporites* sp. and *Jugasporites* sp. were recovered. 48 cm

29-3. Yellowish green thin-bedded calcareous mudstone and argillaceous limestone intercalated with volcanic ash, with laminations in the mudstone. **Conodonts** *Hindeodus parvus*, *H. eurypyge*, *H. anterodentatus* and *C. nassichuki*, **bivalves** *Claraia griesbachi*, *Claraia wangi* and *Towapteria scythica*, and **spores and pollens** *Verrucosisporites* sp., *Kraeuselisporites*? sp., *Alisporites* sp., *Pinuspollenites* sp., *Podocarpidites* sp., *Cedripites* sp., *Taeniaesporites* sp., *Chadrosporites* and *Jugasporites* sp. were found. 48 cm

29-2. Yellowish green thin-bedded calcareous mudstone and argillaceous limestone intercalated with volcanic ash, with laminations in the mudstone. The **conodont** *Isarcicella turgida*, **bivalves** *Claraia griesbachi*, *Claraia wangi* and *Towapteria scythica*, and **spores and pollens** *Verrucosisporites* sp., *Kraeuselisporites*? sp., *Alisporites* sp., *Pinuspollenites* sp., *Podocarpidites* sp., *Cedripites* sp., *Taeniaesporites* sp., *Chadrosporites* sp. and *Jugasporites* sp. were collected. 48 cm

29-1. Yellowish green thin-bedded calcareous mudstone and argillaceous limestone intercalated with volcanic ash, with laminations in the mudstone. **Conodonts** *Clarkina zhejiangensis* and *C. deflecta*, **bivalves** *Claraia griesbachi*, *Claraia wangi* and *Towapteria scythica*, and **spores and pollens** *Verrucosisporites* sp., *Kraeuselisporites*? sp., *Alisporites* sp., *Pinuspollenites* sp., *Podocarpidites* sp., *Cedripites* sp., *Taeniaesporites* sp., *Chadrosporites* sp. and *Jugasporites* sp. were examined. 48 cm

28-4. Yellowish green thin-bedded calcareous mudstone and argillaceous limestone intercalated with volcanic ash, with laminations in the mudstone. **Conodonts** *Hindeodus huckriedei*, *H. changxingensis*, *H. eurypyge*, *H. anterodentatus*, *Isarcicella turgida*, *I. prisca*, and **bivalves** *Bakevellia* sp., *Claraia griesbachi*, *C. hubeinesis*, *C. guizhouensis* and *C. yunnanensis* were obtained. 74 cm

28-3. Yellowish green thin-bedded calcareous mudstone and argillaceous limestone intercalated with volcanic ash, with laminations in the mudstone. The **conodonts** *Hindeodus changxingensis*, *Hindeodus huckriedei*, *H. anterodentatus* and *Clarkina zhejiangensis*, **bivalves** *Claraia* sp. and *Towapteria scythica*, and the **ammonoid** *Ophiceras* cf. *sinense* were discovered. 64 cm

28-2. Yellowish green thin-bedded calcareous mudstone and argillaceous limestone intercalated with volcanic ash, with laminations in the mudstone. **Conodonts** *Hindeodus huckriedei*, *H. changxingensis*, *H. eurypyge*, *H.*

praeparvus, *Clarkina zhejiangensis* and *Isarcicella turgida*, **bivalves** *Claraia* sp. and *Towapteria* sp., and the **ammonoid** *Hypophiceras* sp. were found. 24 cm

28-1. Yellowish green thin-bedded calcareous mudstone and argillaceous limestone intercalated with volcanic ash, with laminations in the mudstone. **Conodonts** *Hindeodus* sp., *Clarkina changxingensis* and *C. subcarinata*, **ammonoids** *Hypophiceras* sp., *Metophicera* sp., *Tompophiceras* sp. and *Pseudogastrioceras* sp., the **brachiopod** *Crurithyris* sp., and **spores and pollens** *Circumpolis* sp. and *Cycadopites* sp. were examined. 16 cm

Conformity
Talung Formation in the Upper Permian

27-3. Grayish black calcareous mudstone. **Conodonts** *Clarkina meishanensis*, *C. tulongensis*, *C. changxingensis*, *C. deflecta* and *C. subcarinata*, the **bivalve** *Claraia guangyuanensis*, and **ammonoids** *Huananoceras* sp., *Pseudotirolites asiaticus* and *Changhsingoceras* sp. were identified. 4 cm

27-2. Pale volcanic ash, with **conodonts** *Clarkina meishanensis*, *C. tulongensis* and *C. deflecta*. 6 cm

27-1. Pale volcanic ash, with **conodonts** *Clarkina tulongensis*, *C. changxingensis*, *C. deflecta*, *C. subcarinata* and *C. meishanensis*, the **brachiopod** *Paracrurithyris pygmaea*, and the **ammonoid** *Pseudogastrioceras* sp.. 5 cm

26. Gray thin-bedded siliceous limestone, with **conodonts** *Clarkina tulongensis*, *C. changxingensis*, *C. deflecta*, *C. subcarinata*, *C. zhangi*, *C. yini* and *C. postwangi*. 25 cm

25. Pale volcanic ash, with **cephalopods** *Pleuronodoceras* sp., *Rotodiscoceras* sp., *Pseudotirolites acutus*, *P.* sp., *Pseudogastrioceras* sp., *Qianjiangoceras* sp. and *Lopingoceras* sp.. 15 cm

24. Dark gray thin-bededed chert intercalated with siliceous mudstone, with laminations. **Ammonoids** *Pleuronodoceras* sp., *P. tenuicostatum*, *Pseudogastrioceras* sp., *P.* aff. *guizhouensis* and *Xenodiscus* sp., the **brachiopod** *Paracrurithyris pygmaea*, and the **conodont** *Clarkina subcarinata* were examined. 59 cm

23. Pale volcanic ash. 6 cm

22. Dark gray thin-bedded chert, siliceous limestone and siliceous mudstone, intercalated with pale volcanic ash in the upper part (ash is 65 cm below the top surface of the bed). **Ammonoids** *Pseudotirolites regularis*, *P. disconnectus*, *P. asiaticus*, *P. acutus*, *P. uniformis*, *Pleuronodoceras* cf. *dushanense*, *P. densiplicatum*, *P. tenuicostatum*, *P. guiyangense*, *Pernodoceras multinodosym*, *Pseudogastrioceras* sp., *P.* cf. *gigantum*, *Pachydiscoceras flexoplicatum*, *Changhsingoceras* cf. *sichuanense*, *Chaotianoceras nodosum*, *C.* cf. *modestum*, *Huananoceras* sp., *Rotodiscoceras* sp. and *Qianjiangoceras*

sp., **brachiopods** *Plicatifera*? sp. and *Neochonetes substrophomenoides*, **bivalves** *Hunanopecten* cf. *qujiangensis* and *Myalina*? sp., **foraminifers** *Nodosaria* sp. and *Pseudoglandulina* sp., and **conodonts** *Clarkina changxingensis, C. deflecta* and *C. subcarinata* were recovered. 261 cm
Additionally, **radiolarians** *Albaillella* sp., *Neoalbaillella antaixiangi, Entactinia* cf. *itsukaichiensis, Entactinia* sp., *Copicyntra* spp., *Copicyntra*? sp., *Nazarovella*? sp. and *Trilonche* spp. were recorded in the chert and siliceous limestone of upper 100 cm of beds 22–26 (Yao and Kuwahara 1999).

21. Gray to dark gray thin-bedded siliceous limestone intercalated with grayish green siliceous mudstone. **Ammonoids** *Pseudogastrioceras* sp., *Pseudotirolites* sp., *Pleuronodoceras* sp., *Qianjiangoceras* sp., *Rotodiscoceras* sp. and *Chaotianoceras* sp., **foraminifers** *Multidiscus guangxiensis, Multidiscus* sp. and *Glomospira parva*, and **conodonts** *Clarkina subcarinata, Clarkina deflecta* and *Clarkina changxingensis* were identified. 178 cm
20. Gray to dark gray thin-bedded siliceous limestone intercalated with carbonaceous shale with laminations. **Ammonoids** *Pseudogastrioceras szechuanensis, Rotodiscoceras wangconglense* and *Pentagoceras guizhouense*, **foraminifers** *Glomospira ovalis, Glomospira regularis* and *Nodosaria* sp., and **conodonts** *Clarkina deflecta, C. changxingensis* and *C. subcarinata* were examined. 248 cm
19. Gray to dark gray medium-bedded siliceous limestone intercalated with carbonaceous shale. **Foraminifers** *Nodosaria* sp., *Neodiscus* sp., *Hemigordicus* sp. and *Glomospira* sp., and **conodonts** *Clarkina deflecta, C. changxingensis* and *C. subcarinata* were identified. 205 cm
18. Grayish black thin-bedded carbonaceous chert, micritic limestone and chert intercalated with carbonaceous shale, with laminations. **Ammonoids** *Tapashanites floriformis, Tapashanites latiumblicus, Sinoceltites* sp., *Pseudostephanites nodosus, Shevyrevites* sp., *Qianjiangoceras* sp., *Huananoceras* sp. and *Pseudotirolites* sp., and **conodonts** *Clarkina changxingensis* and *Clarkina subcarinata* were found. 898 cm
Additionally, **radiolarians** *Entactinia* sp., *Copicyntra*? sp., *Latentifistula* spp., *Nazarovella gracilis, Nazarovella*? sp., *Ishigaum trifustis* and *Ishigaum obesum* were recorded in the chert, silicecous limestone and siliceous mudstone of beds 18–21 (Yao and Kuwahara 1999).
17. Dark gray thin-bedded carbonaceous siliceous limestone. The **ammonoid** *Pseudogastrioceras* sp. was identified. 66 cm
16. Dark gray thin-bedded carbonaceous chert and micritic limestone intercalated with carbonaceous shale, with pyrites and laminations. The **ammonoids** *Pseudogastrioceras* sp., *Konglingites striatus* and *Prototoceras* sp., and **foraminifers** *Codonofusiella kueichowensis* and *Palaeotextularia* sp. were discovered. 434 cm
15. Dark gray medium- to thin-bedded carbonaceous chert and micritic limestone intercalated with siliceous mudstone with laminations. Additionally, a layer of pale volcanic ash was intercalated in the lower part of the bed. 500 cm

14. Dark gray thin-bedded carbonaceous chert and micritic limestone intercalated with carbonaceous mudstone with pyrite nodules. The **brachiopod** *Neochonetes* sp., and **conodonts** *Clarkina changxingensis*, *C. liangshanensis*, *C. deflecta*, and **spores and pollens** *Torispora* sp., *Cyclogranisporites* sp., *Verrucosisporites* sp. and *Punctatisporites* sp. were examined. 356 cm
13. Dark gray lens-shaped siliceous limestone intercalated with shale. The **coral** *Tachylasma* sp., the **brachiopod** *Paryphella* sp., and **conodonts** *Clarkina changxingensis*, *Clarkina orientalis*, *Clarkina liangshanensis* and *Clarkina deflecta* were found. 373 cm
12. Gray to dark gray medium-bedded siliceous limestone intercalated with carbonaceous shale. **Conodonts** *Clarkina changxingensis* and *C. liangshanensis*, **brachiopods** *Acosarina* sp. and *Paryphella* sp., and **spores and pollens** *Cyclogranisporites* sp., *Punctatisporites* sp. and *Granulatisporites* sp. were recovered. 287 cm
11. Dark gray medium- to thin-bedded carbonaceous chert and siliceous limestone intercalated with chert and carbonaceous shale, with pyrite nodules and laminations. In addition, a layer of pale volcanic ash was intercalated in the lower part. **Brachiopods** *Acosarina* sp., *Spinomarginifera lopingensis*, *Paryphella orbicularis*, *Paryphella corculum*, *Paryphella elegantula*, *Compressoproductus compressus* and *Lepdotus* sp., **foraminifer** *Nodosaria parva*, and **conodonts** *Clarkina liangshanensis* and *Clarkina mediconstricta* were examined. 287 cm

Conformity

Wuchiaping Formation in the Upper Permian
Note: Listed radiolarians are based on Yao and Kuwahara (1999); information of volcanic ash in the interval from beds 11–12 are after Yang et al. (1987); lithological and other palaeontological informaton of beds 11–21 are based on Li et al. (1989), but revised and updated in taxonomy.

3.28 Rencunping Section at Rencunping Village, Liangshuikou Town

The Rencunping section (110° 06′ 02″ E, 29° 34′ 49″ N) is located at the Rencunping Village, about 4 km north of Liangshuikou Town, Sangzhi County, northwest Hunan Province, South China (Fig. 3.65).

The sequence of the Rencunping section is composed of the Talung Formation and the conformably overlying Taye Formation (Fig. 3.66). The Talung Formation is dominated by dark gray siliceous mudstones, siliceous limestones and carbonaceous mudstones, intercalated with calcareous mudstones and pale volcanic ash. Laminations are commonly recorded in mudstones and Bouma sequences in siliceous limestones. The basal part of the Taye Formation is outcropped and

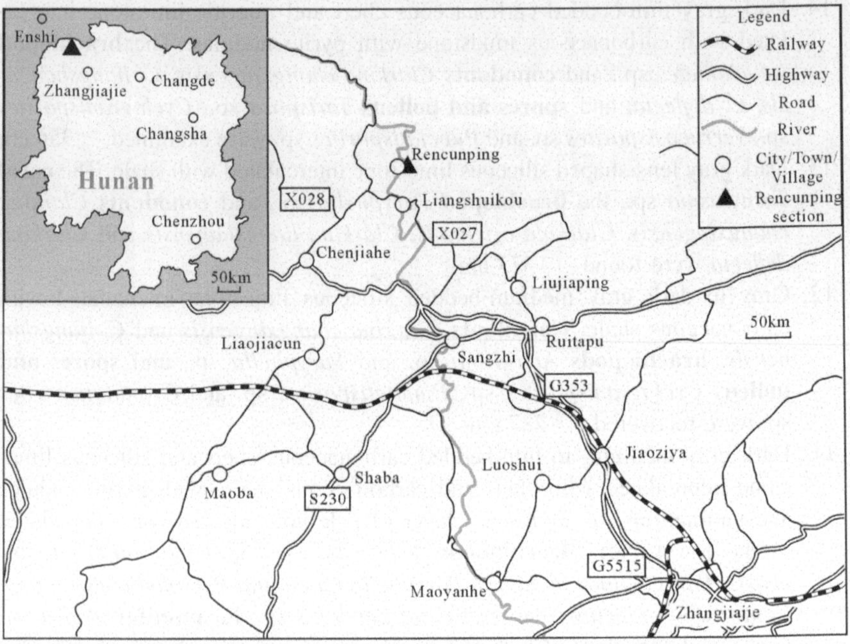

Fig. 3.65 Geographic location of the Rencunping section

dominated by grayish green calcareous mudstones interbedded with argillaceous limestones. Abundant brachiopods, ammonoids, conodonts and radiolarians were yielded in the Talung Formation (Fig. 3.66).

Taye Formation in the Lower Triassic
Not described upwards

34. Grayish green medium-bedded micritic limestone altered with dark gray argillaceous limestone. Thickness not measured.
33. Grayish green to grayish brown argillaceous limestone intercalated with light gray medium-bedded micritic limestone. 343 cm
32. Grayish green to grayish brown argillaceous limestone intercalated with light gray thin-bedded micritic limestone. 125 cm
31. Grayish green to grayish brown argillaceous limestone intercalated with grayish green micritic limestone. **Ammonoids** *Ophiceras* sp. and *Lytophiceras* sp. were collected in the mudstone and **conodonts** *Hindeodus parvus*, *H. parvus anterodentatus* and *H. typicalis* were collected in the micritic limestone. 202 cm
30. Grayish green to grayish yellow argillaceous limestone intercalated with micritic limestone; a layer of pale volcanic ash (3 cm thick) at the basal part, with the **conodont** *Clarkina* sp.. 95 cm

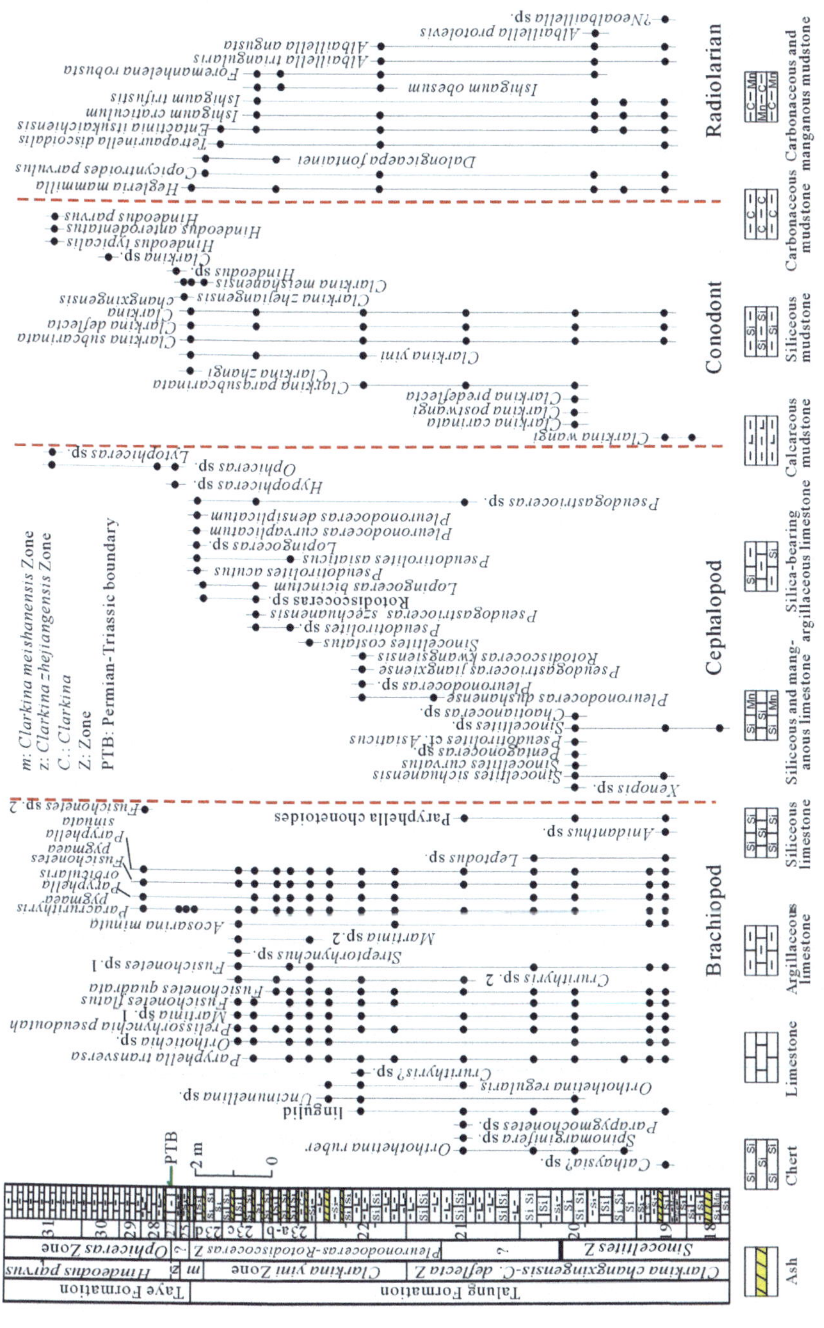

Fig. 3.66 Stratigraphic column and fossil occurrences at the Rencunping section

29. Grayish brown to grayish yellow argillaceous limestone intercalated with micritic limestone, with a few **brachiopods** *Paracrurithyris pygmaea, Paryphella orbicularis, Fusichonetes pygmaea, Paryphella sinuata* and *Fusichonetes* sp. in the basal part of the bed. 42 cm
28. Grayish green to grayish yellow calcareous mudstone altered with micritic limestone, with a pale volcanish ash at the base (but laterally discontinuous). The **ammonoid** *Ophiceras* sp. was examined in the mudstone. 65 cm
27b. Grayish green thin-bedded argillaceous limestone with **ammonoids** *Ophiceras* sp. and *Hypophiceras* sp. and the **conodont** *Hindeodus* sp.. 8 cm
27a. Grayish green thin-bedded argillaceous limestone. 8 cm
26. Grayish brown thin-bedded calcareous mudstone with a few **conodonts** *Clarkina meishanensis* and *C. zhejiangensis* and the **brachiopod** *Paracrurithyris pygmaea.* 8 cm
25-3. Pale volcanic ash. 9 cm
25-2. Grayish black thin-bedded siliceous mudstone with **conodonts** *Clarkina changxingensis, C. deflecta, C. meishanensis, C. subcarinata, C. yini* and *C. zhangi*, the **brachiopod** *Paracrurithyris pygmaea*, and the **radiolarian** *Hegleria mammilla.* 6 cm
25-1. Pale volcanic ash. 5 cm

Conformity
Talung Formation in the Upper Permian

24. Grayish black medium-bedded limestone with **cephalopods** *Pseudogastrioceras* sp., *Pleuronodoceras densiplicatum, Pl. curvaplicatum, Lopingoceras* sp., *Pseudotirolites asiaticus* and *Ps. acutus*, and the **brachiopod** *Paracrurithyris pygmaea.* 24 cm
23d-3. Grayish black medium-bedded siliceous limestone, with a pale volcanic ash (laterally continuous) at the top. The **conodont** *Clarkina meishanensis* and **radiolarians** *Copicyntroides parvulus* and *Dalongicaepa fontainei* were discovered. 7 cm
23d-2. Black volcanic ash altered with pale volcanic ash. 19 cm
23d-1. Black thin-bedded chert with **cephalopods** *Lopingoceras bicinctum, Rotodiscoceras* sp. and **radiolarians** *Entactinia itsukaichiensis* and *Tetrapaurinella discoidalis.* 8 cm
23c-7. Black volcanic ash altered with pale volcanic ash. 4 cm
23c-6. Grayish brown medium-bedded siliceous limestone. 16 cm
23c-5. Yellowish and pale volcanic ashes. 13 cm
23c-4. Grayish brown medium-bedded siliceous limestone. 11 cm
23c-3. Pale volcanic ash. 4 cm
23c-2. Black volcanic ash. 3 cm

23c-1. Grayish brown thick-bedded siliceous limestone with **cephalopods** *Lopingoceras bicinctum, Rotodiscoceras* sp., *Pseudogastrioceras* sp., *Pseudogastrioceras szechuanensis* and *Pseudotirolites* sp., and **conodonts** *Clarkinia changxingensis, C. deflecta* and *C. yini.* 55 cm
Additionally, a few **brachiopods** *Prelissorhynchia pseudoutah, Orthotichia* sp., *Martinia* sp., *Fusichonetes flatus, F. pygmaea, F.* sp., *Crurithyris* sp., *Streptorhynchus* sp., *Acosarina minuta, Paracrurithyris pygmaea* and *Paryphella sinuata* were found in the upper part of bed 23c; brachiopods *Paryphella transversa, P. orbicularis, P. sinuata Prelissorhynchia pseudoutah, Orthotichia* sp., *Martinia* sp., *Paracrurithyris pygmaea, Fusichonetes flatus* and *F. pygmaea* were discovered in the lower part of bed 23c; and **radiolarians** *Entactinia itsukaichiensis, Ishigaum craticulum, I. trifustis, I. obesum* and *Foremanhelena robusta* were collected in bed 23c.
23b-9. Grayish brown medium-bedded chert. 18 cm
23b-8. Pale volcanic ash. 2 cm
23b-7. Grayish black medium-bedded siliceous limestone with **radiolarians** *Dalongicaepa fontainei* and *Hegleria mammilla.* 31 cm
23b-6. Pale volcanic ash. 3 cm
23b-5. Grayish black medium-bedded siliceous limestone with **radiolarians** *Ishigaum obesum* and *Foremanhelena robusta.* 18 cm
23b-4. Dark brown ash. 5 cm
23b-3. Grayish black thin-bedded chert. 4 cm
23b-2. Yellow ash. 3 cm
23b-1. Grayish black thin- to medium-bedded siliceous limestone with **ammonoids** *Pseudotirolites asiaticus* and *Pseudotirolites* sp.. 31 cm
Additionally, a few **brachiopods** *Prelissorhynchia pseudoutah, Fusichonetes quadrata, F. pygmaea, Crurithyris* sp., *Paracrurithyris pygmaea, Paryphella orbicularis* and *P. sinuata* were collected in the upper part of bed 23b; brachiopods *Paryphella transversa, P. orbicularis, P. sinuata Prelissorhynchia pseudoutah, Orthotichia* sp., *Martinia* sp., *Fusichonetes flatus, F. quadrata, F. pygmaea, F.* sp. and *Paracrurithyris pygmaea* were found in the middle part of bed 23b.
23a-4. Pale volcanic ash. 7 cm
23a-3. Black thin-bedded carbonaceous mudstone. 6 cm
23a-2. Pale volcanic ash. 8 cm
23a-1. Grayish black thin-bedded siliceous mudstone, with the **ammonoid** *Sinoceltites costatus.* 8 cm
In addition, a few **brachiopods** *Paryphella transversa, P. orbicularis, P. sinuata, Prelissorhynchia pseudoutah, Orthotichia* sp., *Martinia* sp., *Fusichonetes flatus, F. quadrata, F. pygmaea, F.* sp., *Crurithyris* sp. and *Paracrurithyris pygmaea* were collected in bed 23a.
22. Black medium- to thick-bedded siliceous and calcareous mudstone intercalated with light gray lens-shaped siliceous limestone. Abundant **radiolarians** *Copicyntroides parvulus, Hegleria mammilla, Tetrapaurinella discoidalis, Entactinia itsukaichiensis, I. craticulum, I. obesum, Foremanhelena*

robusta, *Albaillella triangularis* and *A. angusta*, **ammonoids** *Rotodiscoceras kwangsiensis*, *Pseudogastrioceras jiangxiense*, *Pleuronodoceras* sp. and *Pl. dushanense*, and **conodonts** *Clarkina changxingensis*, *C. deflecta*, *C. parasubcarinata*, *C. subcarinata* and *C. yini* were recovered. 230 cm
In addition, **brachiopods** *Uncinunellina* sp., *Orthothetina ruber*, *Paryphella transversa*, *Prelissorhynchia pseudoutah*, *Orthotichia* sp., *Martinia* sp., *Fusichonetes flatus*, *F. quadrata*, *F. pygmaea*, *Paracrurithyris pygmaea*, *Paryphella orbicularis* and *P. sinuata* were found in the upper part of bed 22; brachiopods lingulids, *Uncinunellina* sp., *Orthothetina ruber*, *Crurithyris*? sp., *Paryphella transversa*, *P. orbicularis*, *P. sinuata*, *Prelissorhynchia pseudoutah*, *Martinia* sp., *Fusichonetes flatus*, *F. quadrata*, *F. pygmaea*, *Crurithyris* sp. and *Paracrurithyris pygmaea* were obtained in the middle part of bed 22; brachiopods *Paryphella transversa*, *P. orbicularis*, *P. sinuata*, *Prelissorhynchia pseudoutah*, *Fusichonetes flatus*, *F. quadrata*, *F. pygmaea*, *Acosarina minuta* and *Paracrurithyris pygmaea* were discovered in the lower part of bed 22.

21. Black medium- to thick-bedded siliceous and calcareous mudstone intercalated with light gray lens-shaped siliceous limestone. **Conodonts** *Clarkina changxingensis*, *C. deflecta*, *C. parasubcarinata* and *C. subcarinata*, **ammonoids** *Pseudogastrioceras* sp. and *Pleuronodoceras dushanense*, and a few **brachiopods** lingulids, *Uncinunellina* sp., *Orthothetina ruber*, *F. flatus*, *F. quadrata*, *F. pygmaea*, *Crurithyris*? sp., *Paryphella transversa*, *P. orbicularis*, *P. sinuata*, *Prelissorhynchia pseudoutah*, *Martinia* sp., *Crurithyris* sp. and *Paracrurithyris pygmaea* were collected. 310 cm
20. Grayish black thin-bedded chert and manganese-bearing chert intercalated with brown manganous mudstone and lens-shaped siliceous limestone in the upper part; grayish black thin-bedded chert intercalated with siliceous mudstone, carbonaceous and manganous mudstone and gray lens-shaped siliceous limestone in the lower part. Abundant **radiolarians** *Copicyntroides parvulus*, *Hegleria mammilla*, *Entactinia itsukaichiensis*, *I. craticulum*, *Ishigaum trifustis*, *Foremanhelena robusta*, *Albaillella triangularis*, *A. angusta* and A. *protolevis*, **ammonoids** *Chantianoceras* sp. *Sinoceltites* sp., *Pseudotirolite* cf. *asiaticus*, *Pentagonoceras* sp., *Sinoceltites curvatus*, *Sinoceltites sichuanensis* and *Xenapis* sp., and **conodonts** *C. changxingensis*, *C. deflecta*, *C. parasubcarinata*, *C. predeflecta*, *C. postwangi*, *C. subcarinata* and *C. carinata* were obtained. 305 cm
Additionally, a few **brachiopods** *Orthothetina ruber*, lingulid, *Paryphella transversa*, *P. orbicularis*, *P. sinuata*, *Prelissorhynchia pseudoutah*, *Martinia* sp., *Fusichonetes flatus*, *F. quadrata*, *F. pygmaea*, *F.* sp., *Paracrurithyris pygmaea* and *Leptodus* sp. were collected in the upper part of bed 20; brachiopods *Orthothetina ruber*, lingulids, *Uncinunellina* sp., *Paryphella transversa*, *Paryphella orbicularis*, *P. sinuata*, *P. chonetoides*, *Orthotichia* sp., *Prelissorhynchia pseudoutah*, *Martinia* sp., *Fusichonetes*

flatus, *F. quadrata*, *F. pygmaea* and *Paracrurithyris pygmaea* were recovered in the middle part of bed 20; brachiopods *Orthothetina ruber* and *Paryphella transversa* were found in the lower part of bed 20.

19. Grayish black thin-bedded chert intercalated with brown manganous mudstone, black carbonaceous mudstone and lens-shaped limestone in the upper part; pale volcanic ash in the middle (16 cm thick) part; dark gray siliceous limestone intercalated with carbonaceous and manganous mudstone in the lower part. A large number of **radiolarians** *Copicyntroides parvulus*, *Hegleria mammilla*, *Tetrapaurinella discoidalis*, *Entactinia itsukaichiensis*, *I. craticulum*, *Ishigaum trifustis*, *Albaillella triangularis*, *A. angusta* and *Neoalbaillella* sp., **ammonoids** *Sinoceltites sichuanensis* and *Sinoceltites* sp., and **conodonts** *C. changxingensis*, *C. deflecta*, *C. wangi* and *C. subcarinata* were collected. 128 cm

    In addition, a few **brachiopods** *Paryphella transversa*, *P. orbicularis*, *P. sinuata*, *Prelissorhynchia pseudoutah*, *Orthotichia* sp., *Martinia* sp., *Fusichonetes flatus*, *F. quadrata*, *F. pygmaea*, *F.* sp., *Acosarina minuta*, and *Paracrurithyris pygmaea* were found at the top part of bed 19; brachiopods *Cathaysia*? sp., lingulid, *Paryphella transversa*, *P. orbicularis*, *P. sinuata*, *P. chonetoides*, *Prelissorhynchia pseudoutah*, *Orthotichia* sp., *Martinia* sp., *Fusichonetes flatus*, *F. quadrata*, *F. pygmaea*, *F.* sp., *Acosarina minuta*, *Paracrurithyris pygmaea*, *Leptodus* sp. and *Anidanthus* sp. were discovered in the lower middle part of bed 19.

18-4. Grayish black medium-bedded silica-bearing argillaceous limestone, with the **conodont** *Clarkina wangi*. 43 cm

18-3. Grayish black thin-bedded silica-bearing argillaceous limestone. 7 cm

18-2. Pale and yellow ashes intercalated with a layer of black carbonaceous ash. 26 cm

18-1. Grayish black medium-bedded siliceous and manganous limestone with the **ammonoid** *Sinoceltites* sp.. 35 cm

17. Grayish black thin-bedded chert and siliceous mudstone altered with yellow, pale and black ashes in the upper part (66 cm); brownish red ashes altered with yellow and pale ashes in the middle part (27 cm); black carbonaceous and manganous mudstone altered with the yellow ash in the lower part (43 cm). 136 cm

16. Black thin-bedded carbonaceous mudstone, with interlayers of thin-bedded siliceous mudstone in the lower part, with the **ammonoid** *Sanyangites* sp. in the carbonaceous mudstone. 183 cm

15. Black thin-bedded carbonaceous mudstone intercalated with thin-bedded carbonaceous limestone and siltstone, with carbonaceous limestone (8 cm thick) at the top. **Ammonoids** *Pseudogastriocera* sp., *Planodiscoceras* sp., *Anderssonoceras* sp. and *Prototoceras* sp. were collected. 235 cm

14. Black thin-bedded carbonaceous mudstone intercalated with thin-bedded siliceous limestone and siltstone, with carbonaceous limestone (12 cm thick) at the top. **Cephalopods** *Lopingoceras* sp., *Retiogastrioceras* sp., *Planodiscoceras* sp. and *Prototoceras* sp. were found. 190 cm

13. Black thin-bedded carbonaceous mudstone intercalated with thin-bedded siliceous limestone, with **ammonoids** *Prototoceras* sp., *Kiangsiceras* sp., *Anderssonoceras* sp., *Leptogyroceras* sp., *Pseudogastriocera* sp., *Planodiscoceras* sp. and *Konglingites striatus.* 130 cm
12. Black thin-bedded carbonaceous mudstone, with thin-bedded siliceous limestone at the top. 185 cm
11. Black thin-bedded carbonaceous mudstone, with grayish black medium-bedded calcarenite (14 cm thick) at the top. 157 cm
10. Black thin-bedded carbonaceous mudstone, with dark brown carbonaceous siltstone (10 cm thick) at the top. 79 cm
9. Black carbonaceous shale altered with grayish black thin- to medium-bedded siliceous mudstone in the upper part (four cycles formed); black thin-bedded carbonaceous mudstone (50 cm thick) altered with grayish black medium-bedded carbonaceous micritic limestone (34 cm thick) in the lower part. 237 cm
8. Grayish black thin-bedded siliceous mudstone altered with black thin-bedded carbonaceous mudstone. 171 cm
7. Black thin-bedded carbonaceous mudstone intercalated with light gray thin-bedded micritic limestone and siliceous mudstone. Abundant **brachiopod** *Martinia*? sp. formed a shelly bed (4 cm thick) in the siliceous mudstone (the shelly bed being 243 cm above the base of the bed). 443 cm
6. Light gray thick-bedded micritic limestone. 50 cm
5. Black thin-bedded carbonaceous mudstone in the upper part (50 cm thick); grayish black thin-bedded chert altered with black thin-bedded carbonaceous mudstone in the lower part. Abundant radiolarians were yielded in the chert. 210 cm

Conformity
Wuchiaping Formation in the Upper Permian

4. Black thin-bedded carbonaceous mudstone altered with grayish black thin-bedded micritic limestone, with abundant brachiopods (formed shelly bed). 104 cm
3. Grayish black medium-bedded micritic limestone intercalated with black thin-bedded carbonaceous mudstone (239 cm thick); carbonaceous mudstone at the basal part (6 cm thick), with four cycles formed by the thickness variation (thickening upwards for each cycle). 245 cm
2. Gray medium-bedded micritic limestone. 220 cm
1. Gray medium-bedded limestone with lens-shaped cherts.

Not described downwards
Note: Listed brachiopods are based on He et al. (2015), but revised and updated in taxonomy in this book; cephalopods are after Zhang et al. (2009).

3.29 Xinmin Section at Xinmin Village, Puding County

The Xinmin section (26° 22′ 35″ N, 105° 55′ 05″ E) is located at the Xinmin Village, Puding County, Anshun City, Guizhou Province, South China (Fig. 3.67).

The sequence of the Xinmin section is composed of the Changhsing Formation, Talung Formation and the Taye Formation (Fig. 3.68). The Changhsing Formation is dominated by gray bioclastic limestones. The Talung Formation is dominated by siliceous mudstones and siliceous limestones intercalated with calcareous mudstones and volcanic ashes. The Taye Formation is dominated by grayish green argillaceous limestones intercalated with siltstones and silty limestones. Thin-bedded calcareous mudstones was more abundantly intercalated in the lower part of the Taye Formation. Abundant ammonoids and brachiopods, and a few conodonts were recovered in the Talung Formation (Figs. 3.68, 3.69).

Taye Formation in the Lower Triassic
Covered by the Quaternary at the top

73. Grayish green thin- to medium-bedded argillaceous limestone. 36 cm
72. Grayish green medium-bedded argillaceous limestone. 187 cm
71. Grayish green thin- to medium-bedded argillaceous limestone. 188 cm
70. Grayish green thin- to medium-bedded argillaceous limestone. 152 cm
69. Light gray shale. 8 cm

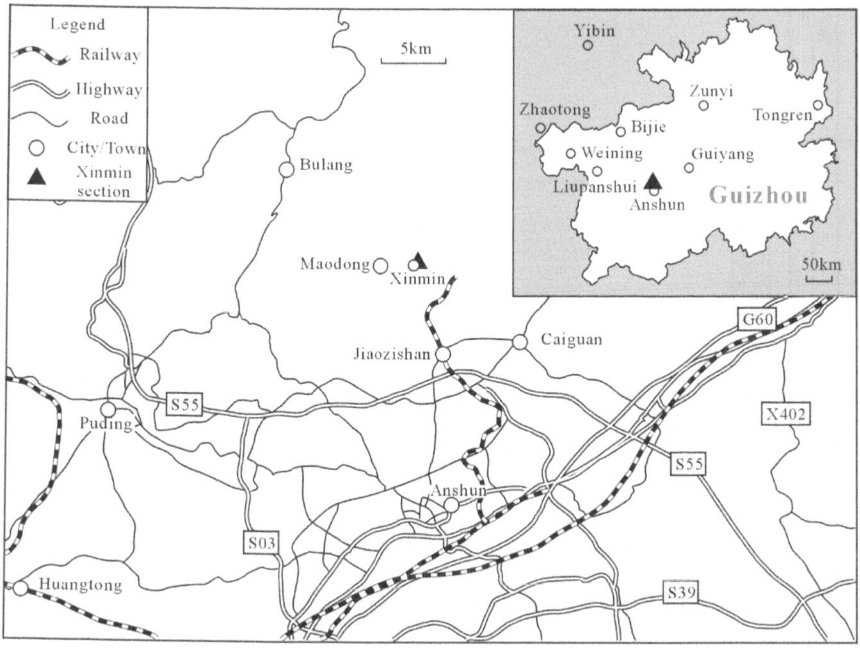

Fig. 3.67 Geographic location of the Xinmin section

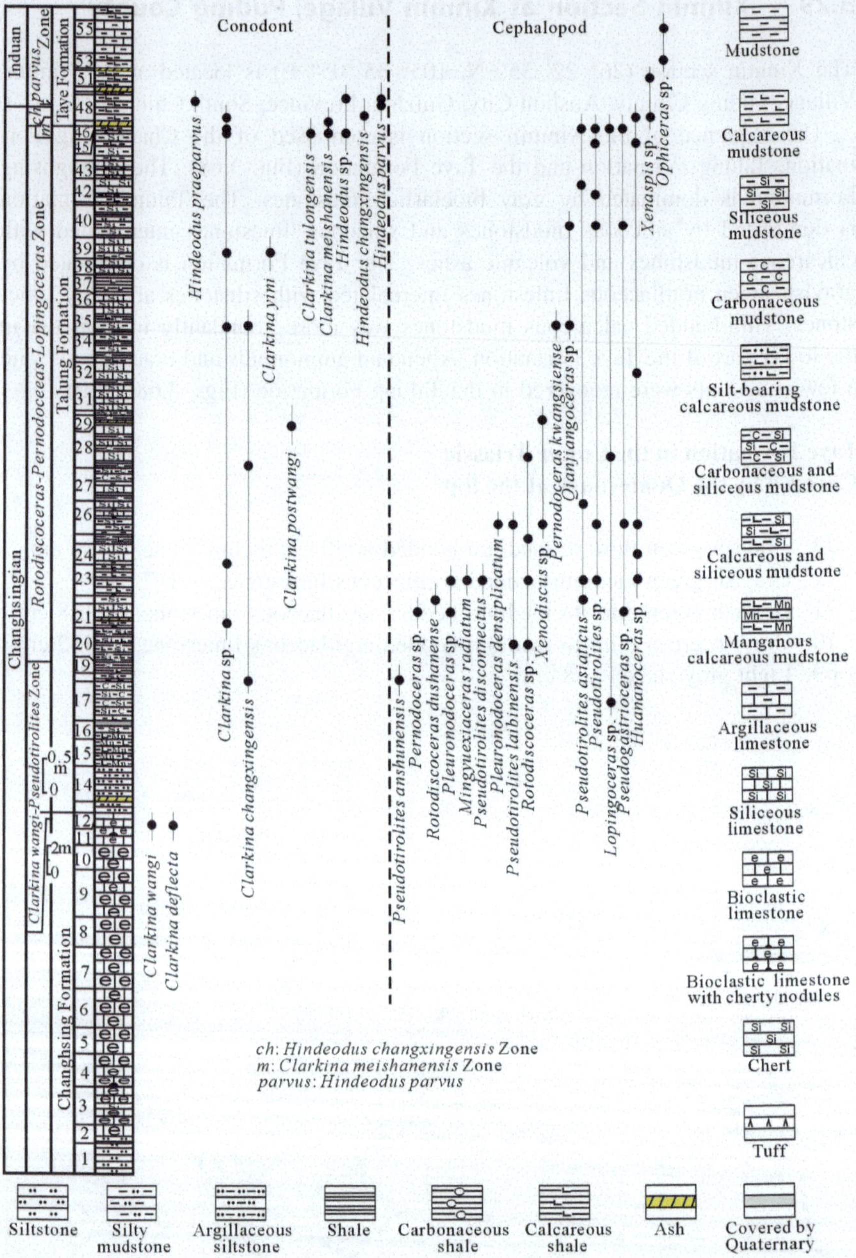

Fig. 3.68 Stratigraphic column and fossil occurrences (conodont, cephalopod) at the Xinmin section. *Note* conodonts after Zhang et al. (2014c)

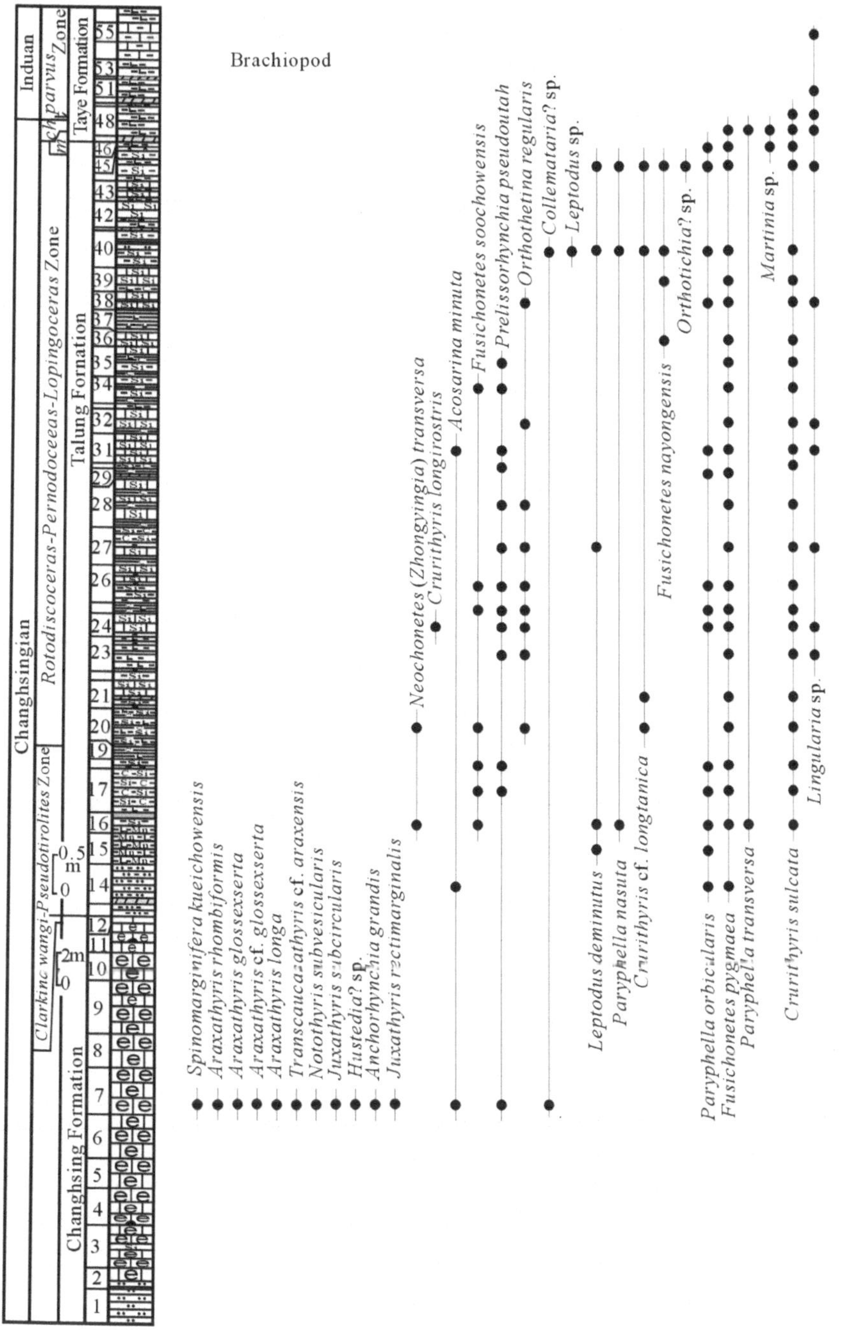

Fig. 3.69 Stratigraphic column and fossil occurrences (brachiopod) at the Xinmin section. *Note* brachiopods after Wu et al. (2018); legends and abbreviations same to Fig. 3.68

68. Grayish green medium- to thick-bedded argillaceous limestone, with thin-bedded argillaceous limestone at the basal part. 50 cm
67. Dark brown medium- to thick-bedded argillaceous limestone. 286 cm
66. Dark brown medium-bedded argillaceous limestone. 92 cm
65. Light gray silty limestone altered with grayish green thin-bedded silty and calcareous mudstone. 123 cm
64. Grayish green thin-bedded silty and calcareous mudstone in the middle part, light gray thin-bedded silty and calcareous mudstone at the top and light gray medium-bedded silty limestone at the base. 43 cm
63. Argillaceous limestone altered with manganous sandstone, layers thinning upwards. 93 cm
62. Argillaceous limestone altered with manganous siltstone, with grayish green calcareous shale at the base. 83 cm
61. Dark brown medium-bedded mudstone intercalated with argillaceous limestone. 40 cm
60. Thin-bedded argillaceous limestone altered with grayish green calcareous shale. 79 cm
59. Dark gray shale and black carbonaceous shale in the upper part; grayish green thin-bedded calcareous mudstone in the lower part. 69 cm
58. Dark gray medium-bedded silty and calcareous mudstone. 27 cm
57. Grayish yellow medium-bedded silty and calcareous mudstone. 115 cm
56. Yellowish green medium-bedded calcareous mudstone. 18 cm
55. Grayish green medium-bedded argillaceous limestone, with the brachiopod *Lingularia* sp. and ammonoid *Ophiceras* sp.. 24 cm
54. Grayish green medium-bedded argillaceous limestone. 19 cm
53. Yellowish brown medium-bedded calcareous mudstone, with the ammonoid *Ophiceras* sp.. 21 cm
52. Grayish green volcanic ash. 2 cm

51-3. Grayish green thin-bedded calcareous mudstone, with the ammonoid *Xenaspis* sp.. 10 cm

51-2. Grayish green thin-bedded calcareous mudstone. 5 cm

51-1. Grayish green thin-bedded calcareous mudstone, with the brachiopod *Lingularia* sp. and ammonoid *Xenaspis* sp.. 9 cm

50. Pale volcanic ash. 5 cm
49. Grayish green thin-bedded calcareous mudstone. 2.5 cm
48. Grayish green calcareous mudstone. Ammonoids *Xenaspis* sp. and *Huananoceras* sp. were found in the bed; brachiopods *Crurithyris sulcata* and *Lingularia* sp. and conodonts *Hindeodus praeparvus*, *H.* sp. and *H. parvus* were discovered in the upper part of bed 48; brachiopods *Fusichonetes pygmaea*, *Paryphella transversa*, *Martinia* sp., *Crurithyris sulcata* and *Lingularia* sp. and conodonts *Clarkina* sp. and *Hindeodus changxingensis* were identified in the lower part of bed 48. 13 cm
47. Grayish green to reddish brown ash with the conodont *Clarkina meishanensis*. 7 cm

Conformity
Talung Formation in the Upper Permian

46. Black thin-bedded carbonaceous mudstone. Brachiopods *Paryphella orbicularis*, *Fusichonetes pygmaea*, *Martinia* sp. and *Crurithyris sulcata*, ammonoids *Pseudogastrioceras* sp., *Lopingoceras* sp., *Pseudotirolites* sp. and *Pseudotirolites asiaticus*, and conodonts *Clarkina* sp., *C. tulongensis*, *C. meishanensis* and *Hindeodus* sp. were examined. 8 cm
45. Yellowish green tuff. 1 cm
44-2. Grayish black medium-bedded siliceous mudstone, with yellowish brown calcareous mudstone at the base. Brachiopods *Leptodus deminutus*, *Paryphella nasuta*, *Crurithyris* cf. *longtanica*, *Fusichonetes nayongensis*, *Orthotichia*? sp., *Paryphella orbicularis*, *Fusichonetes pygmaea*, *Crurithyris sulcata* and *Lingularia* sp. and ammonoids *Pseudotirolites* sp. and *Huananoceras* sp. were collected. 31 cm
44-1. Grayish black thin-bedded siliceous mudstone, with grayish brown calcareous mudstone at the base. 10 cm
43. Yellowish brown thin-bedded siliceous limestone, with yellowish brown thin-bedded calcareous mudstone at the base. 26 cm
42-2. Grayish black medium-bedded chert, with yellow calcareous mudstone at the base. The ammonoid *Pseudotirolites asiaticus* was examined. 18 cm
42-1. Yellowish brown medium-bedded calcareous mudstone with the ammonoid *Pseudotirolites* sp.. 19 cm
41. Yellow volcanic ash. 2 cm
40-2. Grayish brown thin-bedded siliceous mudstone, with yellowish brown shale at the base. 5 cm
40-1. Grayish brown medium-bedded siliceous mudstone, with grayish black shale at the base. Brachiopods *Collemataria*? sp., *Leptodus* sp., *Leptodus deminutus*, *Paryphella nasuta*, *Crurithyris* cf. *longtanica*, *Fusichonetes nayongensis*, *Paryphella orbicularis*, *Fusichonetes pygmaea* and *Lingularia* sp., and ammonoids *Pseudotirolites* sp., *Pseudogastrioceras* sp. and *Qianjiangoceras* sp. were discovered. 42.5 cm
39. Grayish black medium-bedded siliceous limestone, with carbonaceous and calcareous mudstone at the base. Brachiopods *Fusichonetes nayongensis*, *F. pygmaea* and *Crurithyris sulcata* and conodonts *Clarkina* sp. and *C. yini* were examined. 31 cm
38-2. Dark gray thin-bedded siliceous limestone, with grayish yellow calcareous shale and grayish black carbonaceous shale at the base. Brachiopods *Orthothetina regularis*, *Paryphella orbicularis*, *Fusichonetes pygmaea*, *Crurithyris sulcata* and *Lingularia* sp. were discovered. 10 cm
38-1. Grayish black medium-bedded siliceous mudstone, with grayish brown carbonaceous shale at the base. 13.5 cm
37-5. Yellowish brown calcareous shale. 3.5 cm
37-4. Yellowish brown calcareous shale. 6 cm
37-3. Yellowish brown calcareous shale. 3.5 cm

37-2. Yellowish brown calcareous shale. 3.5 cm
37-1. Yellowish brown calcareous shale. 12.5 cm
36. Black medium-bedded siliceous limestone with the Bouma Sequence (normal graded bedding and parallel bedding). Brachiopods *Fusichonetes nayongensis, F. pygmaea* and *Crurithyris sulcata* were examined. 18 cm
35-5. Grayish black medium-bedded siliceous limestone, with grayish yellow calcareous shale and grayish black carbonaceous shale at the basal part. 14.5 cm
35-4. Grayish black thin-bedded siliceous mudstone, with dark gray carbonaceous shale at the basal part. 7.5 cm
35-3. Grayish black siliceous mudstone; grayish black mudstone intercalated with yellow shale at the base. Brachiopods *Prelissorhynchia pseudoutah, Fusichonetes pygmaea* and *Crurithyris sulcata* and ammonoids *Pernodoceras kwangsiense* and *Qianjiangoceras* sp. were identified. 9.5 cm
35-2. Dark gray thin-bedded carbonaceous mudstone, with carbonaceous shale at the base. 2.5 cm
35-1. Gray thin-bedded siliceous mudstone, with dark gray carbonaceous shale at the base. Conodonts *Clarkina changxingensis* and *C.* sp. were examined. 5.5 cm
34-2. Grayish black medium-bedded siliceous mudstone, with black carbonaceous shale at the base. 16.5 cm
34-1. Grayish black medium-bedded siliceous mudstone, with carbonaceous shale at the base. Brachiopods *Fusichonetes soochowensis, Prelissorhynchia pseudoutah, Fusichonetes pygmaea* and *Crurithyris sulcata* were found. 20 cm
33-2. Gray thin-bedded calcareous mudstone, with grayish black carbonaceous shale at the base. 2.5 cm
33-1. Gray thin-bedded calcareous mudstone, with grayish black carbonaceous shale at the base. 2.5 cm
32. Dark gray medium-bedded siliceous limestone, with grayish black carbonaceous shale at the base. Brachiopods *Orthothetina regularis, Fusichonetes pygmaea, Crurithyris sulcata* and *Lingularia* sp., and the ammonoid *Huananoceras* sp. were recorded. 32 cm
31. Grayish black thin-bedded siliceous limestone, with brachiopods *Acosarina minuta, Prelissorhynchia pseudoutah, Paryphella orbicularis, Fusichonetes pygmaea, Crurithyris sulcata* and *Lingularia* sp.. 38 cm
30. Grayish black thin-bedded carbonaceous and siliceous mudstone, with brachiopods *Prelissorhynchia pseudoutah, Paryphella orbicularis* and *Fusichonetes pygmaea.* 7 cm
29. Grayish black carbonaceous shale in the upper part, yellow ash in the middle part, and grayish yellow calcareous shale in the lower part. The ammonoid *Xenodiscus* sp. was found in the bed. 14 cm

28-3. Grayish black medium-bedded siliceous limestone, with carbonaceous shale at the base. The conodont *Clarkina postwangi* was identified in the bed. 26.5 cm
28-2. Grayish black thin-bedded siliceous limestone, with carbonaceous shale at the base. 8.5 cm
28-1. Grayish black medium-bedded siliceous limestone, with carbonaceous shale at the base. Brachiopods *Prelissorhynchia pseudoutah*, *Orthothetina regularis*, *Fusichonetes pygmaea* and *Crurithyris sulcata*, and cephalopods *Lopingoceras* sp. and *Rotodiscoceras* sp., and conodont *Clarkina changxingensis* were recorded. 27.5 cm
27-2. Grayish black medium-bedded carbonaceous and siliceous mudstone, with grayish black mudstone at the base. 20.5 cm
27-1. Grayish black medium-bedded siliceous limestone, with grayish black carbonaceous shale at the base. Brachiopods *Prelissorhynchia pseudoutah*, *Orthothetina regularis*, *Leptodus deminutus*, *F. pygmaea*, *Crurithyris sulcata* and *Lingularia* sp. were examined. 28 cm
26-3. Grayish black medium-bedded siliceous limestone, with black mudstone at the base. 18.2 cm
26-2. Grayish black thin-bedded siliceous limestone, with grayish black mudstone at the base. Brachiopods *Fusichonetes soochowensis*, *Prelissorhynchia pseudoutah*, *Orthothetina regularis*, *Paryphella orbicularis*, *Fusichonetes pygmaea* and *Crurithyris sulcata* and the ammonoid *Pseudotirolites asiaticus* were found. 3.2 cm
26-1. Grayish black medium-bedded siliceous mudstone, with mudstone at the base. Ammonoids *Pseudotirolites* sp., *Pseudotirolites laibinensis*, *Huananoceras* sp., *Pseudogastrioceras* sp., *Xenodiscus* sp. and *Pleuronodoceras densiplicatum* were discovered. 27.2 cm
25. Grayish black carbonaceous shale in the upper part, yellow green calcareous shale in the middle part, and grayish black carbonaceous shale in the lower part. Brachiopods *Fusichonetes soochowensis*, *Prelissorhynchia pseudoutah*, *Orthothetina regularis*, *Paryphella orbicularis*, *Fusichonetes pygmaea* and *Crurithyris sulcata* and the ammonoid *Rotodiscoceras* sp. were collected. 14 cm
24. Dark gray thick-bedded siliceous limestone, with yellow mudstone at the base. Brachiopods *Crurithyris longirostris*, *Prelissorhynchia pseudoutah*, *Orthothetina regularis*, *Paryphella orbicularis*, *Fusichonetes pygmaea*, *Crurithyris sulcata* and *Lingularia* sp., ammonoid *Huananoceras* sp. and conodont *Clarkina* sp. were recovered. 30 cm
23-3. Grayish black calcareous mudstone, with grayish yellow mudstone at the base. 5.5 cm
23-2. Dark gray thin-bedded calcareous mudstone, with grayish green shale at the base. 10 cm

23-1. Grayish black medium-bedded calcareous mudstone, with yellowish green mudstone at the base. Brachiopods *Prelissorhynchia pseudoutah*, *Orthothetina regularis*, *Fusichonetes pygmaea*, *Crurithyris sulcata* and *Lingularia* sp. and ammonoids *Pseudotirolites* sp., *Pseudotirolites disconnectus*, *Pseudotirolites acutus*, *Rotodiscoceras* sp., *Mingyuexiaceras radiatum*, *Hunanoceras* sp. and *Pleuronodoceras* sp. were identified. 30 cm

22. Grayish black medium-bedded siliceous mudstone with ammonoids *Pseudotirolites asiaticus* and *Rotodiscoceras dushanense*. 12 cm

21-2. Gray medium-bedded siliceous limestone, with grayish green volcanish ash at the base. Brachiopods *Crurithyris* cf. *longtanica*, *Fusichonetes pygmaea* and *Crurithyris sulcata*, ammonoids *Pernodoceras* sp., *Pseudogastrioceras* sp. and *Pleuronodoceras* sp., and the conodont *Clarkina* sp. were found. 21 cm

21-1. Grayish black medium-bedded siliceous mudstone, with grayish yellow mudstone at the base. 14.5 cm

20. Eight cycles were formed and each cycle consists: black thin-bedded calcareous and siliceous mudstone in the upper; black thin-bedded carbonaceous mudstone in the lower. Brachiopods *Neochonetes (Zhongyingia) transversa*, *Fusichonetes soochowensis*, *Orthothetina regularis*, *Crurithyris* cf. *longtanica*, *Fusichonetes pygmaea* and *Crurithyris sulcata* and ammonoids *Pseudotirolites laibinensis*, *Rotodiscoceras* sp. were examined. 49.5 cm

19. Grayish black carbonaceous shale at the top, grayish yellow calcareous shale in the upper, grayish black carbonaceous shale in the lower and grayish yellow calcareous shale at the base. 16 cm

18-2. Dark gray thin-bedded siliceous mudstone, with grayish black mudstone at the base. 6.2 cm

18-1. Dark gray thin-bedded siliceous mudstone, with grayish yellow mudstone at the base. Brachiopods *Fusichonetes soochowensis*, *Prelissorhynchia pseudoutah*, *Paryphella orbicularis*, *Fusichonetes pygmaea* and *Crurithyris sulcata*, the ammonoid *Pseudotirolites anshunensis* and conodont *Clarkina changxingensis* were discovered. 6.2 cm

17. Dark gray medium-bedded carbonaceous and siliceous mudstone, with grayish yellow calcareous mudstone at the base. Brachiopods *Fusichonetes soochowensis*, *Prelissorhynchia pseudoutah*, *Paryphella orbicularis*, *Fusichonetes pygmaea* and *Crurithyris sulcata* and the cephalopod *Lopingoceras* sp. were identified. 57.2 cm

16. Grayish black thin-bedded siliceous mudstone in the upper part; brown thin-bedded manganese and calcareous mudstone in the lower part. Brachiopods *Neochonetes (Zhongyingia) transversa*, *Fusichonetes soochowensis*, *Leptodus deminutus*, *Paryphella nasuta*, *P. orbicularis*, *Fusichonetes pygmaea*, *Paryphella transversa* and *Crurithyris sulcata* were collected. 60 cm
The Quaternary covered the interval (with a thickness of about 800 cm) between beds 15 and 16.

15. Brown thin-bedded manganese and calcareous mudstone, with brachiopods *Leptodus deminutus* and *Paryphella orbicularis*. 40 cm
14. Yellowish brown argillaceous siltstone, with pale volcanic ash at the base. Brachiopods *Acosarina minuta*, *Paryphella orbicularis* and *Fusichonetes pygmaea* were examined. 50 cm
13. Yellowish green thin-bedded silty mudstone with the brachiopod *Spinomarginifera* sp.. 15 cm

Conformity
Changhsing Formation in the Upper Permian

12. Gray thin-bedded micritic limestone with conodonts *Clarkina wangi* and *C. deflecta*. 35 cm
11. Gray medium-bedded bioclastic limestone with cherty nodules. 180 cm
10. Gray thick-bedded bioclastic limestone with the brachiopod *Juxathyris rectimarginalis*. 160 cm
9. Light gray thick-bedded bioclastic micritic limestone with the brachiopod *Juxathyris rectimarginalis*. 320 cm
8. Light gray thick-bedded bioclastic micritic limestone, with corals, gastropods, crinoids and abundant brachiopods *Spinomarginifera kueichowensis*, *Araxathyris rhombiformis*, *A. glossexserta*, *A.* cf. *glossexserta*, *A. longa*, *Transcaucasathyris* cf. *araxensis*, *Notothyris subvesicularis*, *Juxathyris subcircularis*, *Hustedia*? sp., *Anchorhynchia grandis*, *Juxathyris rectimarginalis*, *Acosarina minuta*, *Prelissorhynchia pseudoutah* and *Collemataria*? sp.. 200 cm
7. Light gray medium- to thick-bedded bioclastic micritic limestone with abundant sponges. 280 cm
6. Light gray thin- to medium-bedded bioclastic micritic limestone. 270 cm
5. Dark gray thin- to medium-bedded micritic limestone. 180 cm
4. Dark gray thin- to medium-bedded micritic limestone with cherty nodules. 210 cm
3. Grayish black thin- to medium-bedded micritic limestone with cherty nodules and bandings. 250 cm
2. Grayish brown thick-bedded micritic limestone, with medium-bedded micritic limestone at the base. 110 cm
1. Yellowish brown carbonaceous, calcareous and argillaceous siltstone.

Unexposed at the base
Note: Listed brachiopods are based on Wu et al. (2018), but updated in taxonomy; data of conodonts are after Zhang et al. 2014c.

3.30 Kejiao Section in Huishui County

The Kejiao section (25° 49′ 37″ N, 106° 32′ 26″ E) is nearby the Kejiao Village, Duanshan Town, Huishui County, Guizhou Province, South China (Fig. 3.70).

The sequence of the Kejiao section is composed of the Talung Formation and the conformably overlying Luolou Formation (Fig. 3.71). The Talung Formation is dominated by grayish black to yellowish green thin-bedded siliceous mudstones, carbonaceous mudstones and calcareous mudstones, commonly intercalated with pale volcanic ashes. The Luolou Formation is characterized by grayish green argillaceous limestones, yellowish green calcareous mudstones, and purplish red tuff-bearing mudstones, intercalated with pale volcanic ashes. Abundant ammonoids and a few radiolarians, brachiopods and bivalves were yielded in the Talung Formation (Fig. 3.71).

Luolou Formation in the Lower Triassic (Induan)

93. Yellow thin-bedded argillaceous limestone intercalated with calcareous mudstone. 100 cm
92. Grayish green calcareous shale, with grayish green volcanic ash at the base. 150 cm
91. Yellowish brown medium- to thin-bedded argillaceous limestone. 220 cm

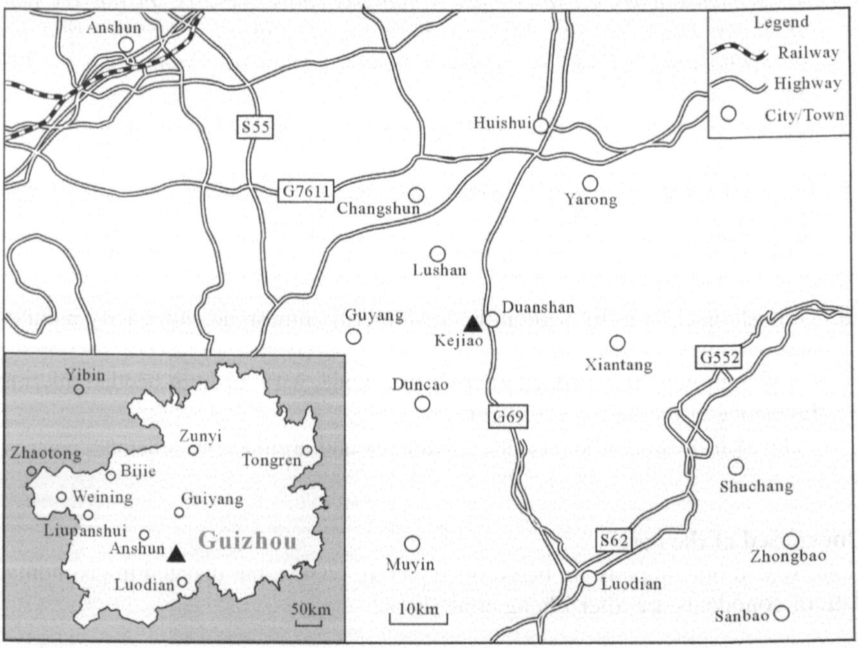

Fig. 3.70 Geographic location of the Kejiao section

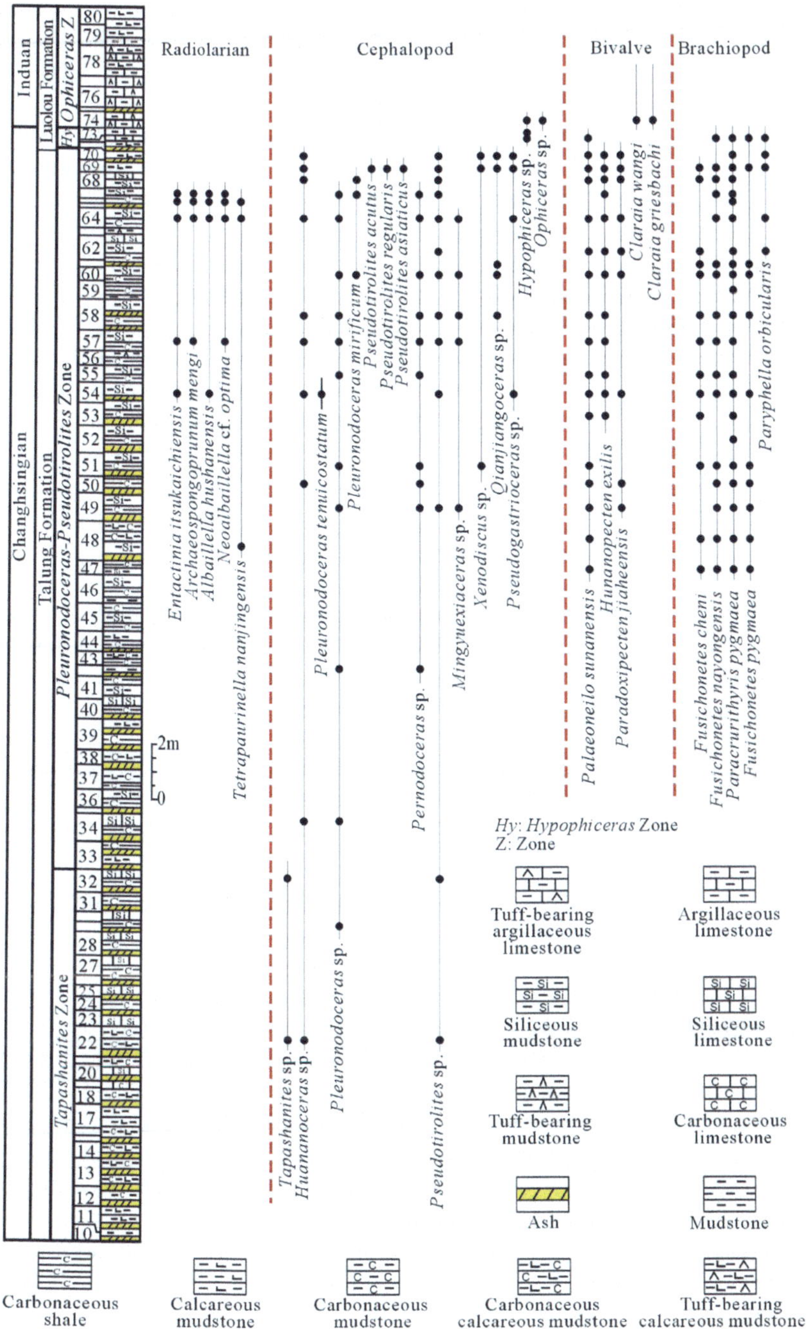

Fig. 3.71 Stratigraphic column and fossil occurrences at the Kejiao section (fossils referred to Wang et al. 2023)

90. Yellowish green thin-bedded argillaceous limestone intercalated with calcareous mudstone, with grayish green volcanic ash at the base. The bivalves *Posidonia* sp. and *Pteria* sp. and the ammonoid *Ophiceras* sp. were collected. 275 cm
89. Yellow to black thin-bedded manganous argillaceous limestone. The bivalves *Posidonia* sp. and *Pteria* sp. and the ammonoid *Ophiceras* sp. were found in the basal and uppermost parts of the bed. 145 cm
88. Yellow thin-bedded calcareous mudstone intercalated with argillaceous limestone. The bivalves *Posidonia* sp. and *Pteria* sp. and the ammonoid *Ophiceras* sp. were recovered in the lower part of the bed. 180 cm
87. Yellow thin-bedded calcareous mudstone intercalated with argillaceous limestone, with grayish green tuff-bearing argillaceous limestone in the lower part and pale volcanic ash (40 cm thick) at the base. Bivalves *Posidonia* sp., *Pteria* sp. and *Claraia griesbachi* and the ammonoid *Ophiceras* sp. were collected. 310 cm
86. Yellow thin-bedded calcareous mudstone. 90 cm
85. Yellow thin-bedded calcareous mudstone, with grayish green volcanic ash (25 cm thick) at the base. Bivalves *Claraia griesbachi* and *C. wangi* were identified. 55 cm
84. Grayish green thin-bedded tuff-bearing argillaceous limestone in the upper part; yellow thin-bedded argillaceous limestone intercalated with calcareous mudstone in the middle part; pale tuff-bearing thin-bedded calcareous mudstone and tuffaceous mudstone (30 cm thick) in the lower part. 180 cm
83. Yellow tuff-bearing argillaceous limestone intercalated with calcareous mudstone and pale volcanic ash. 330 cm
82. Yellow tuff-bearing calcareous mudstone intercalated with yellowish green calcareous mudstone and argillaceous limestone. 115 cm
81. Grayish green thin-bedded tuff-bearing argillaceous limestone intercalated with tuff-bearing calcareous mudstone. 45 cm
80. Yellow thin-bedded calcareous mudstone. 30 cm
79. Yellow thin-bedded calcareous mudstone in the upper part (10 cm thick); yellowish gray thin-bedded argillaceous limestone in the lower part (40 cm thick). 50 cm
78. Grayish green thin-bedded tuff-bearing argillaceous limestone and tuff-bearing calcareous mudstone in the upper part (40 cm); pale thin-bedded argillaceous limestone in the lower part (40 cm). 80 cm
77. Yellowish to grayish green thin-bedded tuff-bearing argillaceous limestone intercalated with tuff-bearing mudstone. 34 cm
76. Grayish to yellowish green medium- to thin-bedded tuff-bearing argillaceous limestone. 76 cm
75. Pale to light green volcanic ash. 6 cm
74. Yellowish green thin-bedded argillaceous limestone, with grayish green tuff-bearing argillaceous limestone at the uppermost (2 cm thick). Bivalves *Claraia*

griesbachi and *Claraia wangi* and ammonoids *Ophiceras* sp. and *Hypopiceras* sp. were collected. 60 cm

73. Yellowish green thin-bedded argillaceous limestone with the ammonoid *Hypopiceras* sp.. 31 cm
72. Grayish green thin-bedded argillaceous limestone. Brachiopods *Fusichonetes nayongensis*, *Fusichonetes pygmaea*, *Paryphella orbicularis* and *Paracrurithyris pygmaea*, the bivalve *Palaeoneilo sunanensis*, and the ammonoid *Hypopiceras* sp. were discovered. 29 cm
71. Grayish black calcareous mudstone in the upper part (1–3 cm thick, laterally varied in the thickness); pale volcanic ash at the base (2–20 cm thick, laterally varied in the thickness). 23 cm

Conformity
Talung Formation in the Upper Permian (Changhsingian)

70. Yellowish green thin-bedded calcareous mudstone, with pale volcanic ash at the base. The ammonoids *Huananoceras* sp., *Xenodiscus* sp., *Pseudotirolites* sp., *Qianjiangoceras* sp. and *Pseudogastrioceras* sp., bivalves *Palaeoneilo sunanensis*, *Paradoxipecten jiaheensis*, and *Hunanopecten exilis*, and brachiopods *Paryphella orbicularis* and *Paracrurithyris pygmaea* were collected. 45 cm
69. Yellowish green thin-bedded calcareous mudstone with abundant pyrite nodules. Ammonoids *Huananoceras* sp., *Pseudotirolites acutus*, *Pseudotirolites regularis*, *Pseudotirolites asiaticus*, *Pseudotirolites* sp., *Xenodiscus* sp., *Qianjiangoceras* sp., and *Pseudogastrioceras* sp., bivalves *Palaeoneilo sunanensis*, *Paradoxipecten jiaheensis*, and *Hunanopecten exilis*, and brachiopods *Fusichonetes cheni*, *F. nayongensis*, *F. pygmaea*, *Paracrurithyris pygmaea* and *Paryphella orbicularis* were found. 100 cm
68. Dark gray thin-bedded siliceous mudstone and siliceous limestone intercalated with grayish black carbonaceous shale, with pale volcanic ash in the upper part. Ammonoids *Pseudotirolites* sp., *Huananoceras* sp. and *Pleuronodoceras mirificum*, bivalves *Palaeoneilo sunanensis*, *Hunanopecten exilis* and *Paradoxipecten jiaheensis*, and brachiopods *Fusichonetes cheni*, *F. nayongensis* and *Paracrurithyris pygmaea* were obtained. 52 cm
67. Dark gray thin-bedded siliceous mudstone intercalated with grayish black carbonaceous shale and pale volcanic ash, with abundant pyrite nodules in the shale. Radiolarians *Entactinia itsukaichiensis*, *Archaeospongoprunum mengi*, *Albaillella hushanensis* and *Neoalbaillella* cf. *optima*, bivalve *Hunanopecten exilis*, ammonoids *Pleuronodoceras mirificum*, *Pl.* sp., *Pernodoceras* sp. and *Pseudotirolites* sp., and brachiopods *Fusichonetes nayongensis* and *Paracrurithyris pygmaea* were examined. 54 cm
66. Dark gray thin-bedded siliceous mudstone, with grayish black carbonaceous shale in the uppermost (4 cm thick). Fragments of plant fossils were found. Radiolarians *Entactinia itsukaichiensis*, *Archaeospongoprunum*

mengi, Albaillella hushanensis, Neoalbaillella cf. *optima* and *Tetrapaurinella nanjingensis*, and the brachiopod *Paracrurithyris pygmaea* were recorded. 31 cm

65. Yellowish green thin-bedded tuff-bearing mudstone intercalated with pale volcanic ash. 15 cm
64. Grayish green thin-bedded siliceous mudstone intercalated with grayish black carbonaceous shale and yellow calcaceous shale. Radiolarians *Entactinia itsukaichiensis, Archaeospongoprunum mengi, Albaillella hushanensis, Neoalbaillella* cf. *optima* and *Tetrapaurinella nanjingensis*, ammonoids *Pleuronodoceras* sp., *Pernodoceras* sp., *Huananoceras* sp., *Mingyuexiaceras* sp., *Pseudogastrioceras* sp. and *Pseudotirolites* sp., bivalves *Palaeoneilo sunanensis, Hunanopecten exilis* and *Paradoxipecten jiaheensis*, and brachiopods *Fusichonetes nayongensis, Paryphella orbicularis* and *Paracrurithyris pygmaea* were collected. 72 cm.
63. Yellow thin-bedded silt-bearing calcareous mudstone in the upper part (22 cm); tuff-bearing mudstone in the lower part (8 cm). 30 cm
62. Grayish green thin-bedded siliceous mudstone and siliceous limestone intercalated with pale volcanic ash and grayish black carbonaceous shale. Bivalves *Hunanopecten exilis, Palaeoneilo sunanensis* and *Paradoxipecten jiaheensis*, ammonoid *Pseudotirolites* sp. and brachiopods *Fusichonetes cheni, Paracrurithyris pygmaea* and *Paryphella orbicularis* were recovered. 87 cm
61. Grayish green medium-bedded siliceous limestone in the upper part (40 cm); grayish black carbonaceous shale in the middle part (6 cm); yellowish brown volcanic ash in the lower part (2 cm). The ammonoid *Qianjiangoceras* sp. and brachiopods *Fusichonetes cheni, F. nayongensis, F. pygmaea* and *Paracrurithyris pygmaea* were examined. 48 cm
60. Grayish green thin-bedded silicecous mudstone and calcareous shale intercalated with silty mudstone and carbonaceous shale. Ammonoids *Pleuronodoceras mirificum, Pleuronodoceras* sp., *Pernodoceras* sp., *Pseudotirolites* sp., *Mingyuexiaceras* sp. and *Qianjiangoceras* sp., bivalves *Palaeoneilo sunanensis, Paradoxipecten jiaheensis* and *Hunanopecten exilis*, and brachiopods *Fusichonetes cheni, F. pygmaea, F. nayongensis* and *Paracrurithyris pygmaea* were collected. 45 cm
59. Yellowish green thin-bedded calcaceous mudstone intercalated with lens-shaped manganous and siliecous limestone in the upper part (60 cm); grayish black carbonaceous shale in the lower part (6 cm); yellow mudstone at the base (2 cm). The brachiopod *Paracrurithyris pygmaea* was found. 68 cm
58. Grayish green thin-bedded silicecous mudstone intercalated with calcareous mudstone and grayish black carbonaceous shale, with interlayers of pale volcanic ashes in the lower part. Ammonoids *Pleuronodoceras* sp., *Pernodoceras* sp., *Huananoceras* sp., *Pseudotirolites* sp., *Mingyuexiaceras* sp. and *Qianjiangoceras* sp., bivalves *Palaeoneilo sunanensis* and *Hunanopecten exilis*, and brachiopods *Fusichonetes cheni, F. nayongensis, F. pygmaea* and *Paracrurithyris pygmaea* were discovered. 115 cm

57. Grayish green medium- to thin-bedded siliceous mudstone, intercalated with grayish black carbonaceous shale (12 cm thick) and pale volcanic ash in the middle part. Radiolarians *Entactinia itsukaichiensis*, *Archaeospongoprunum mengi* and *Neoalbaillella* cf. *optima*, ammonoids *Pleuronodoceras* sp., *Pernodoceras* sp., *Pseudotirolites* sp., *Huananoceras* sp. and *Mingyuexiaceras* sp., bivalves *Hunanopecten exilis* and *Palaeoneilo sunanensis*, and brachiopods *Fusichonetes cheni*, *F. nayongensis* and *Paracrurithyris pygmaea* were examined. 90 cm
56. Black carbonaceous shale in the upper part (5 cm); yellow tuff-bearing mudstone in the middle part (45 cm); black carbonaceous shale in the lower part (3 cm). 53 cm
55. Grayish green medium- to thin-bedded siliceous mudstone intercalated with grayish black carbonaceous shale. Bivalves *Palaeoneilo sunanensis* and *Hunanopecten exilis*, ammonoids *Pernodoceras* sp. and *Pleuronodoceras* sp., and brachiopods *Fusichonetes cheni*, *F. nayongensis* and *Paracrurithyris pygmaea* were collected. 63 cm
54. Grayish green medium- to thin-bedded siliceous mudstone intercalated with grayish black carbonaceous shale in the upper part; pale volcanic ash intercalated with black carbonaceous shale in the lower part. Radiolarians *Entactinia itsukaichiensis* and *Albaillella hushanensis*, ammonoids *Huananoceras* sp., *Pleuronodoceras tenuicostatum*, *Pseudogastrioceras* sp. and *Pseudotirolites* sp., bivalves *Hunanopecten exilis*, *Palaeoneilo sunanensis* and *Paradoxipecten jiaheensis*, and brachiopods *Fusichonetes cheni*, *F. nayongensis*, *F. pygmaea* and *Paracrurithyris pygmaea* were recovered. 65 cm
53. Grayish green medium- to thin-bedded siliceous mudstone intercalated with pale volcanic ash and grayish black carbonaceous shale in the upper part; pale volcanic ash interbedded with black carbonaceous shale in the lower part. Bivalves *Palaeoneilo sunanensis* and *Hunanopecten exilis*, and brachiopods *Fusichonetes cheni* and *Paracrurithyris pygmaea* were found. 84 cm
52. Grayish green medium- to thin-bedded siliceous mudstone intercalated with pale volcanic ash and grayish black carbonaceous shale in the upper part; pale volcanic ash interbedded with black carbonaceous shale in the lower part. The brachiopod *Paracrurithyris pygmaea* was collected. 104 cm
51. Grayish green medium- to thin-bedded siliceous mudstone intercalated with black carbonaceous shale, with pale volcanic ash at the base. The bivalve *Palaeoneilo sunanensis*, ammonoids *Xenodiscus* sp., *Pleuronodoceras* sp., *Pernodoceras* sp. and *Pseudogastrioceras* sp., and brachiopods *Fusichonetes cheni*, *F. nayongensis*, *F. pygmaea* and *Paracrurithyris pygmaea* were discovered. 83 cm
50. Grayish green medium- to thin-bedded siliceous mudstone intercalated with grayish black carbonaceous shale, with interlayers of pale volcanic ashes in the lower part. Ammonoids *Huananoceras* sp. and *Pernodoceras* sp., bivalves *Palaeoneilo sunanensis* and *Paradoxipecten jiaheensis*, and

brachiopods *Fusichonetes nayongensis*, *Fusichonetes pygmaea* and *Paracrurithyris pygmaea* were obtained. 61 cm

49. Grayish green medium- to thin-bedded siliceous mudstone intercalated with grayish black carbonaceous shale, with interlayers of pale volcanic ashes in the lower part. Ammonoids *Pseudotirolites* sp., *Pernodoceras* sp., *Pleuronodoceras* sp. and *Mingyuexiaceras* sp., bivalves *Palaeoneilo sunanensis* and *Paradoxipecten jiaheensis*, and brachiopods *Fusichonetes nayongensis*, *Fusichonetes pygmaea* and *Paracrurithyris pygmaea* were found. 104 cm
48. Dark gray carbonaceous calcareous mudstone in the middle and upper parts (100 cm); grayish green siliceous mudstone in the lower part (40 cm); pale volcanic ash in the basal part (3 ~ 5 cm). The radiolarian *Tetrapaurinella nanjingensis*, bivalve *Palaeoneilo sunanensis*, and brachiopods *Fusichonetes cheni*, *F. nayongensis*, *F. pygmaea* and *Paracrurithyris pygmaea* were collected. 138 cm
47. Grayish black carbonaceous shale, intercalated with carbonaceous siliceous mudstone in the upper part and with pale volcanic ash in the middle part. The bivalve *Palaeoneilo sunanensis*, and brachiopods *Fusichonetes cheni*, *F. nayongensis*, *F. pygmaea* and *Paracrurithyris pygmaea* were recovered. 67 cm
46. Grayish green thin-bedded siliceous mudstone interbedded with black carbonaceous shale, with yellow mudstone (1 cm thick) at the base. 104 cm
45. Grayish green medium- to thin-bedded siliceous mudstone interbedded with black carbonaceous shale. 102 cm
44. Dark gray thin-bedded carbonaceous calcareous mudstone in the middle and upper parts (70 cm); black carbonaceous shale in the lower part (7 cm); pale volcanic ash at the base (2 cm). 79 cm
43. Dark gray thin-bedded carbonaceous calcareous mudstone in the middle and upper parts (30 cm); black carbonaceous shale in the lower part (20 cm); yellow mudstone at the base (2 cm). 52 cm
42. Yellowish green medium-bedded siliceous mudstone in the upper part; black carbonaceous shale interbedded with pale volcanic ash in the lower part. The ammonoids *Pleuronodoceras* sp. and *Pernodoceras* sp. were collected. 40 cm
41. Dark gray medium- to thin-bedded siliceous mudstone interbedded with black carbonaceous shale. 80 cm
40. Dark gray medium-bedded siliceous limestone interbedded with black carbonaceous shale, with pale volcanic ash at the base. 70 cm
39. Dark gray siliceous limestone in the uppermost (7 cm); yellowish green thin-bedded calcareous mudstone in the middle and upper parts (90 cm); yellowish green calcarous mudstone and black carboceous shale intercalated with yellow volcanic ash in the lower part. 113.5 cm
38. Grayish green calcareous siliceous mudstone, with yellow volcanic ash at the base (1 cm). 56 cm
37. Grayish black thin-bedded carbonaceous calcareous mudstone in the middle and upper parts (67 cm); yellowish brown thin-bedded mudstone in

the lower part (17 cm); grayish black carbonaceous shale at the base (6 cm). 90 cm

36. Gray medium-bedded siliceous mudstone intercalated with yellow mudstone in the upper part; black carbonaceous shale in the middle part; yellow volcanic ash intercalated with gypsum in the lower part. 68 cm
35. Black carbonaceous shale in the upper part (13 cm); yellow volcanic ash in the lower part (7 cm). 20 cm
34. Dark gray medium-bedded siliceous limestone intercalated with carbonaceous calcareous mudstone, with pale to yellow volcanic ash at the base. The ammonoids *Huananoceras* sp. and *Pleuronodoceras* sp. were examined. 105 cm
33. Dark gray medium- to thin-bedded calcareous mudstone intercalated with pale volcanic ash, with carbonaceous shale and siliceous limestone in the uppermost (15 cm). 100.5 cm
32. Three cycles were deposited and each cycle consists: dark gray silicecous limestone in the upper part; black carbonaceous shale in the middle part; pale volcanic ash in the lower part. The ammonoids *Tapashanites* sp., *Pseudotirolites dushanensis*, *Pseudotirolites acutus* and *Pseudotirolites* sp. were recovered. 82.5 cm
31. Four cycles were deposited and each cycle was divided into dark gray medium- to thin-bedded siliceous limestone and black carbonaceous shale in descending order, pale volcanic ash intercalated within the middle and basal parts. The ammonoid *Pseudotirolites asiaticus* was collected. 69.5 cm
30. Five cycles were deposited and each cycle was divided into dark gray thin-bedded siliceous limestone and black carbonaceous shale in descending order, with pale volcanic ash intercalated within the lower part. 40 cm
29. Three cycles were formed and each cycle was divided into dark gray thin-bedded siliceous limestone and yellow volcanic ash in descending order, with black carbonaceous shale in the middle and upper parts. The ammonoid *Pleuronodoceras* sp. was examined. 39 cm
28. Black carbonaceous shale interbedded with pale volcanic ash to form five cycles, with gray thin-bedded limestone in the middle part. 83.5 cm
27. Dark gray thin-bedded siliceous limestone interbedded with black carbonaceous shale to form six cycles, with a layer of gypsum at the base. 72 cm
26. Dark gray medium-bedded siliceous limestone and thin-bedded carbonaceous calcareous mudstone intercalated with pale volcanic ash and black carbonaceous shale. 35.5 cm
25. Dark gray thin-bedded siliceous limestone interbedded with black carbonaceous shale to form five cycles, with pale volcanic ash in the lower part. 41 cm
24. Dark gray thin-bedded siliceous limestone interbedded with black carbonaceous shale to form three cycles, with pale volcanic ash in the lower part. 48 cm

23. Gray to grayish black medium- to thin-bedded siliceous limestone (six layers), intercalated with a layer of carbonaceous shale in the middle part. 55 cm
22. Grayish black thin-bedded carbonaceous calcareous mudstone, with a layer of pale volcanic ash (1 cm thick) at the base. The ammonoids *Tapashanites* sp., *Huananoceras* sp. and *Pseudotirolites* sp. were collected. 114 cm
21. Grayish black thin-bedded carbonaceous calcareous mudstone. 27 cm
20. Dark gray medium-bedded carbonaceous calcareous mudstone in the upper (20 cm); gray medium-bedded siliceous limestone in the middle (36 cm); grayish black carbonaceous shale in the lower (3 cm); pale volcanic ash at the base (1 cm). 60 cm
19. Gray medium-bedded carbonaceous limestone. 25 cm
18. Three cycles were deposited and each cycle was divided into grayish black medium- to thin-bedded carbonaceous, calcareous mudstone and pale volcanic ash in descending order. 62 cm
17. Two cycles were deposited and each cycle composed of yellowish green calcareous mudstone in the upper and grayish black carbonaceous calcareous mudstone in the lower. 90 cm
16. Brown medium-bedded manganous mudstone. 25 cm
15. Yellowish green medium-bedded calcareous mudstone (17 cm thick), grayish black carbonaceous calcareous mudstone (2 cm thick) and pale volcanic ash (3 cm thick) in descending order. 22 cm
14. Yellowish green medium-bedded calcareous mudstone (47 cm), grayish black carbonaceous calcareous mudstone (5 cm), tuffaceous calcareous mudstone (4 cm) and pale volcanic ash (2 cm) in descending order. 58 cm

13-2. Yellowish green thin-bedded calcareous mudstone, intercalated with black carbonaceous calcareous mudstone (3 cm) and pale volcanic ash (2 cm) in the middle. 81 cm

13-1. Grayish black thin-bedded carbonaceous calcareous mudstone interbedded with pale volcanic ash to form two cycles. 21 cm

12. Yellowish green medium- to thin-bedded calcareous mudstone intercalated with pale volcanic ash and carbonaceous mudstone to form four cycles. 72 cm
11. Yellowish green medium- to thin-bedded calcareous mudstone in the upper part (77 cm); pale volcanic ash in the lower part (15 cm). Fragments of plant fossils and complete ammonoid fossils were abundantly preserved in the mudstone. 92 cm
10. Yellowish green medium-bedded calcareous mudstone in the upper (17 cm); yellow shale in the middle (8 cm); pale volcanic ash in the lower (6 cm). 31 cm
9. Grayish yellow thin-bedded calcareous mudstone. 105 cm
8. Black to grayish green thin-bedded manganous volcanic ash intercalated with pale thin-bedded calcareous mudstone. 20 cm

7. Light gray thin-bedded calcareous mudstone, brown manganous calcareous mudstone (3 cm) and pale volcanic ash (6 cm) in descending order. 36 cm
6. Light gray thin-bedded calcareous mudstone, with brown manganous calcareous mudstone at the base (5 cm). 35 cm
5. Light gray thin-bedded calcareous mudstone, with brown manganous calcareous mudstone at the base (5 cm). 45 cm
4. Two cycles were deposited and each cycle consists: Light gray thin-bedded calcareous mudstone in the upper; brown manganous calcareous mudstone in the lower. 113 cm
3. Light gray thin-bedded calcareous mudstone in the upper and brown manganous calcareous mudstone in the lower. 130 cm
2. Brown thin-bedded calcareous mudstone in the upper and grayish black manganous mudstone in the lower. 150 cm
1. Brown thin-bedded calcareous mudstone. 100 cm

Covered by the Quaternary downwards

3.31 Dongpan Section at Dongpan Village, Liuqiao Town

The Dongpan section (22° 16′ 07″ N, 107° 41′ 47″ E) outcrops along the 322 National Highway and is located about 6 km east of Liuqiao Town, Chongzuo City, southern Guangxi, South China (Fig. 3.72).

The sequence of the Dongpan section is composed of the Talung Formation and the conformably overlying Luolou Formation (Fig. 3.73). The Talung Formation is dominated by dark gray cherts, grayish green siliceous mudstones, intercalated with yellowish green calcareous mudstones and pale volcanic ashes or yellow tuff-bearing mudstones in the lower part, and grayish green calcareous siliceous mudstones intercalated with pale volcanic ashes in the upper part. There are also minor beds of manganous limestones in the lower part, laminations prevalent in the mudstones and Bouma sequences evident in the siliceous limestones of the Talung Formation. The basal part of the Luolou Formation is characterized by the silt-bearing mudstones intercalated with volcanic ashes, and upwards by the silt-bearing mudstones intercalated with argillaceous limestones.

Luolou Formation in the Lower Triassic
Covered by the Quaternary at the top

16. Brownish yellow thin-bedded silt-bearing mudstone intercalated with dark gray thin-bedded argillaceous limestone in the upper part; brownish yellow thin-bedded calcareous mudstone in the lower part. The **ammonoid** *Ophiceras* sp. and numerous **bivalves** *Claraia* cf. *wangi*, *C. guizhouensis* and *C. griesbachi* were recorded in the bed. 100 cm

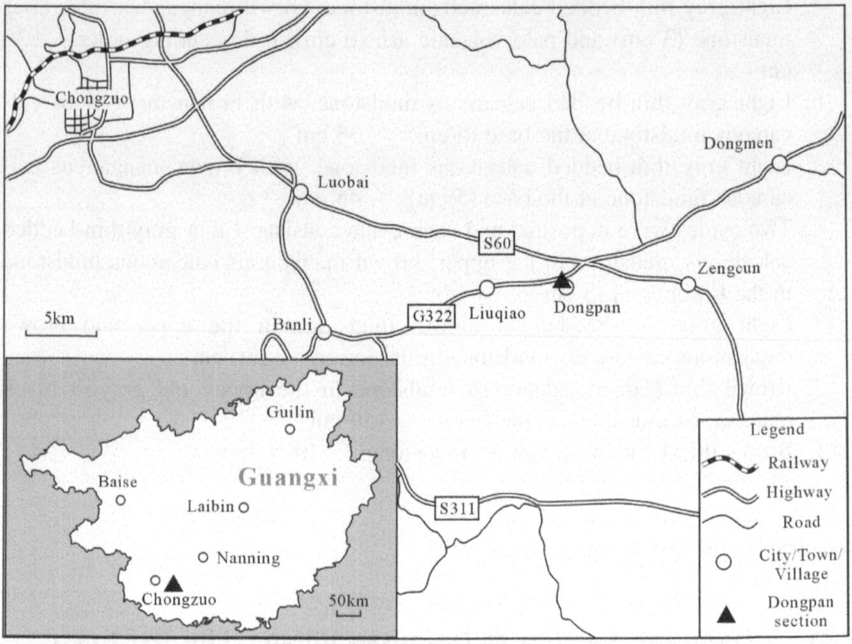

Fig. 3.72 Geographic location of the Dongpan section

15. Brownish yellow thin-bedded silt-bearing mudstone intercalated with dark gray thin-bedded argillaceous limestone which trends to become more abundant upwards. A few **bivalve** *Claraia* sp. was collected. 140 cm
14. Brownish yellow thin-bedded calcareous mudstone, occasionally intercalated with silt-bearing calcareous mudstone, with the **ammonoid** *Ophiceras* sp. and **bivalves** *C. griesbachi*, *C. hunanica*, *C. dieneri*, *C.* cf. *wangi*, *Entolium* sp. and *Bakevellia* sp.. 110 cm

13-9. Brownish yellow thin-bedded calcareous mudstone in the upper part; yellow to brownish yellow silt-bearing mudstone in the lower part. **Ammonoids** *Ophiceras* sp. and *Ophiceras tingi* and abundant **bivalves** *Entolium* sp., *Claraia* cf. *wangi*, *C. hunanica*, *C. dieneri*, *C. griesbachi*, C. sp. and *Bakevellia* sp. were collected. 60 cm

13-8. Brown thin-bedded marl, with the **bivalve** *Claraia dieneri* and the **ammonoid** *Ophiceras* sp.. 7 cm

13-7. Brownish yellow silt-bearing mudstone with abundant specimens of **ammonoids** *Ophiceras tingi* and *Ophiceras* sp.. 36 cm

13-6. Pale volcanic ash with abundant specimens of the **ammonoid** *Ophiceras tingi* Tien and a few specimens of **bivalve** *Claraia* sp.. 12 cm

13-5. Brownish yellow silt-bearing mudstone. 28 cm

13-4. Pale thin-bedded volcanic ash with a few specimens of the **ammonoid** *Ophiceras* sp. and the **bivalve** *Claraia dieneri*. 1 cm

13-3. Yellow thin-bedded silt-bearing mudstone with a few bivalves and the **ammonoid** *Ophiceras tingi* and *Lytophiceras* cf. *chamunda.* 8 cm

13-2. Pale thin-bedded volcanic ash. 5 cm

13-1. Yellow silt-bearing mudstone with a few specimens of **bivalves** *Palaeoneilo qinzhouensis* and *Claraia liuqiaoensis* and **ammonoids** in the upper part (16 cm thick); pale thin-bedded volcanic ash with a few **ammonoid** *Ophiceras* sp. in the lower part (2.5 cm thick). 18.5 cm

Conformity
Talung Formation in the Upper Permian

12-3. Grayish green thin-bedded calcareous siliceous mudstone, occasionally intercalated with silt-bearing calcareous mudstone, with laminations. A few **radiolarians** *Copicyntra* sp., *Trilonche crassispinosa*, *Paracopicyntra ziyunensis* and *Palaeolithocyclia pilata* and the **ammonoid** *Laibinoceras* cf. *compressum* were collected. 55 cm

12-2. Light grayish green volcanic ash. 1.5 cm

12-1. Grayish green thin-bedded calcareous siliceous mudstone, occasionally intercalated with silt-bearing calcareous mudstone. A few **radiolarian** Entactinaria and Spumullaria forms (details see Fig. 3.74), a few **bivalves** *Hunanopecten exilis*, *Lepdochondria intermedia*, *Hunanopecten qujiangensis*, *Euchondria fusuiensis*, *Posidonia* sp., *Streblochondria* sp., *Palaeoneilo qinzhouensis*, *Claraia liuqiaoensis* and *Promyalina*? sp., and **cephalopods** *Lopingoceras guangdeensis*, *Huananoceras* cf. *perornatum*, *Qianjiangoceras* sp. were recovered. 10 cm

11. Brownish yellow thin-bedded silt-bearing mudstone. 10 cm

10-3. Grayish green thin-bedded calcareous siliceous mudstone with **radiolarian** Entactinaria and Spumullaria forms (details see Fig. 3.74) and **ostracods** *Pseudobythocypris* aff. *procera*, *Spinomicrocheilinella anterocompressa*, *Discoidella xingyangense*, *Cavellina* cf. *nebruscensis*, *Waldekella*? *sphaerula*, *Fabalicypris* cf. *minuta*, *Macronotella*? sp. and *Microcheilinella aculeata.* 12 cm

10-2. Pale volcanic ash. 1 cm

10-1. Grayish green thin-bedded calcareous siliceous mudstone with a few **radiolarian** Entactinaria and Spumullaria forms (details see Fig. 3.74), some **brachiopods** *Tornquistia changhsingia*, *Attenuatella mengi*, *Martinia liuqiaoiensis*, *Costatumulus dongpanensis*, *Minutomarginifera semicircridge*, *Pygmochonetes jingxianensis*, *P.* sp., *Rugivestis elegans*, *Permorhipidomella ovatus*, *Enteletes subaequivalis*, *Micromartinia kwangsiella*, *Anidanthus parvimucronata*, *A. subquadratus* and *Rhipidomella parvula*, a few **bivalves** *Palaeoneilo* cf. *qiubeiensis*, *Hunanopecten exilis*, *Euchondria fusuiensis*, and the **ammonoid** *Stacheoceras* sp.. 40 cm

9-9. Pale volcanic ash. 2 cm

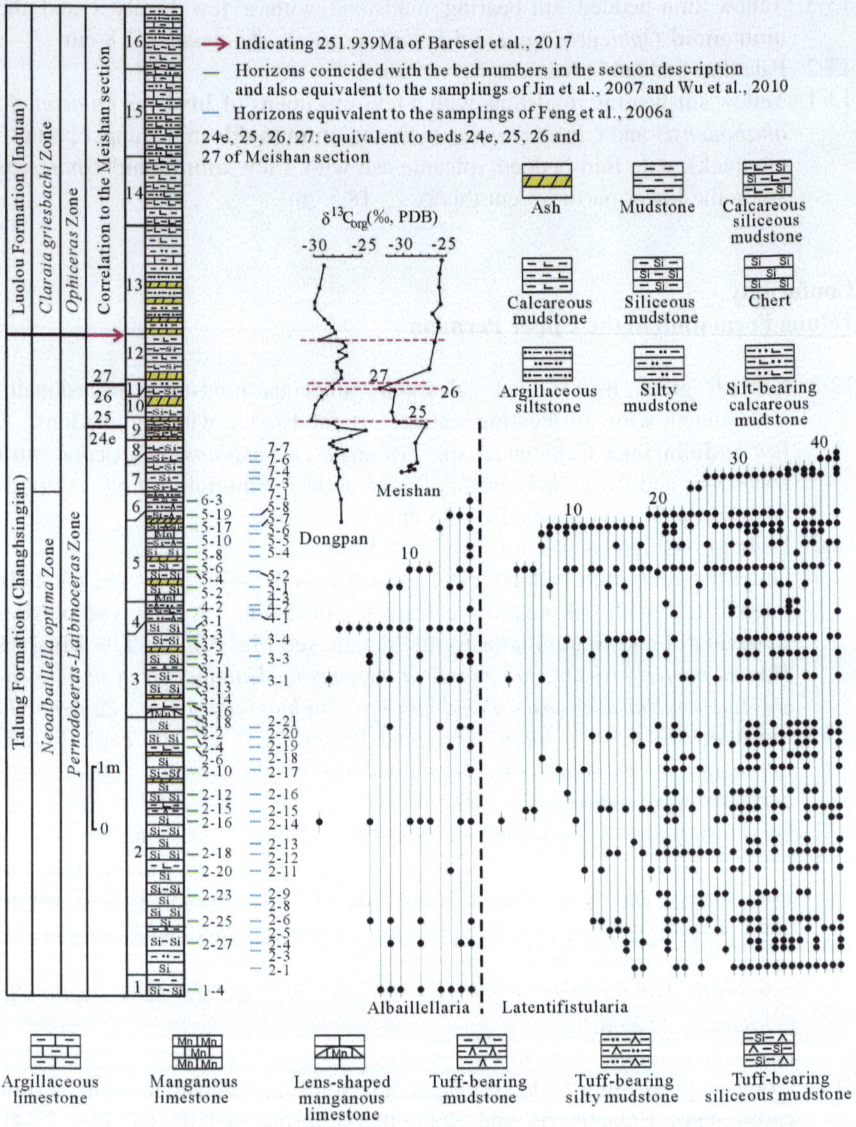

Indicating 251.939Ma of Baresel et al., 2017
Horizons coincided with the bed numbers in the section description and also equivalent to the samplings of Jin et al., 2007 and Wu et al., 2010
Horizons equivalent to the samplings of Feng et al., 2006a
24e, 25, 26, 27: equivalent to beds 24e, 25, 26 and 27 of Meishan section
Luolou Formation (Induan)
Claraia griesbachi Zone
Ophiceras Zone
Correlation to the Meishan section
Talung Formation (Changhsingian)
Neoalbaillella optima Zone
Pernodoceras-Laibinoceras Zone
δ13Corg(‰, PDB)
-30 -25 -30 -25
Dongpan
Meishan
Albaillellaria
Latentifistularia
Ash
Mudstone
Calcareous siliceous mudstone
Calcareous mudstone
Siliceous mudstone
Chert
Argillaceous siltstone
Silty mudstone
Silt-bearing calcareous mudstone
Argillaceous limestone
Manganous limestone
Lens-shaped manganous limestone
Tuff-bearing mudstone
Tuff-bearing silty mudstone
Tuff-bearing siliceous mudstone

◂**Fig. 3.73** Stratigraphic column and radiolarian occurrences (Albaillellaria, Latentifistularia) at the Dongpan section. *Note*: **Albaillellaria**, 1—*Albaillella* sp. 1, 2—*Neoalbaillella* sp. 3, 3—*Albaillella* sp. 2, 4—*Neoalbaillella* cf. *grypa*, 5—*Neoalbaillella* sp. 1, 6—*Albaillella* cf. *triangularis*, 7—*Neoalbaillella minuta*, 8—*Albaillella flabellata*, 9—*Albaillella fida*, 10—*Neoalbaillella* sp. 2, 11—*Albaillella yaoi yaoi*, 12—*Albaillella levis*, 13—*Albaillella yaoi longa*, 14—*Neoalbaillella optima*, 15—*Albaillella triangularis*, 16—*Neoalbaillella camarata*; **Latentifistularia**, 1—*Triplanospongos* sp., 2—*Cauletella* sp. 1, 3—*Octatormentum*? sp., 4—*Ishigaum* sp., 5—*Foremanhelena intermixta*, 6—*Cauletella porosa*, 7—*Areolicaudatus* sp., 8—*Cauletella* sp. 2, 9—*Polyfistula regularis*, 10—*Octatormentum*? *floriferum*, 11—*Ruzhencevispongus totundus*, 12—*Tormentum rotundatum*, 13—*Ishigaum fusinum*, 14—*Ormistonella robusta*, 15—*Ormistonella elegans*, 16—*Quadricaulis* cf. *femoris*, 17—*Latentifistula crux*, 18—*Ormistonella*? *cylindrata*, 19—*Ruzhencevispongus uralicus*, 20—*Shangella regularis*, 21—*Foremanhelena triangula*, 22—*Cauletella delicata*, 23—*Cauletella constricata*, 24—*Foremanhelena robusta*, 25—*Ishigaum craticula*, 26—*Ishigaum complanum*, 27—*Tetragregon longispinatum*, 28—*Grandetortura nipponica*, 29—*Cauletella manica*, 30—*Cauletella paradoxa*, 31—*Ishigaum trifustis*, 32—*Ishigaum longispina*, 33—*Ishigaum tristylum*, 34—*Foremanhelena circula*, 35—*Areolicaudatus semiglobosa*, 36—*Shangella longa*, 37—*Ishigaum obesum*, 38—*Triplanospongos musashiensis*, 39—*Nazarovella inflata*, 40—*Ormistonella adhaerens*, 41—*Nazarovella scalae*, 42—*Nazarovella gracilis*

9-8. Grayish green thin-bedded calcareous siliceous mudstone with laminations. **Radiolarian** Entactinaria and Spumullaria forms (details see Fig. 3.74), and a few **brachiopods** *Attenuatella mengi*, *Martinia liuqiaoiensis*, *Costatumulus dongpanensis*, *Minutomarginifera semicircridge*, *Pygmochonetes jingxianensis*, *Rugivestis elegans*, *Anidanthus parvimucronata* and *A. subquadratus* were collected. 3 cm

9-7. Yellowish green thin-bedded silt-bearing mudstone with radiolarians. 8 cm

9-6. Grayish green thin-bedded calcareous siliceous mudstone, intercalated with dark grayish green silty mudstone with laminations. **Radiolarian** Entactinaria and Spumullaria forms (details see Fig. 3.74), a few **brachiopods** *Tornquistia changhsingia*, *Attenuatella mengi*, *Costatumulus dongpanensis*, *Minutomarginifera semicircridge*, *Permorhipidomella ovatus*, *Anidanthus parvimucronata* and *Rhipidomella parvula*, and the **bivalve** *Lepdochondria intermedia* were found. 4 cm

9-5. Pale volcanic ash. 1 cm

9-4. Grayish green thin-bedded calcareous siliceous mudstone, with laminations at the base. **Radiolarian** Entactinaria and Spumullaria forms (details see Fig. 3.74), a few **brachiopods** *Attenuatella mengi*, *Martinia liuqiaoiensis*, *Costatumulus dongpanensis*, *Minutomarginifera semicircridge*, *Pygmochonetes jingxianensis*, *Anidanthus parvimucronata*, *A. subquadratus*, **bivalves** *Palaeoneilo* cf. *qiubeiensis*, *Hunanopecten exilis* and *Euchondria fusuiensis*, and bryozoans were discovered. 8 cm

9-3. Pale volcanic ash. 2 cm

9-2. Grayish green thin-bedded calcareous siliceous mudstone, with laminations in the lower part. A few radiolarians, brachiopods (same to that of bed 9-4) and ammonoids were collected. 4 cm

9-1. Pale volcanic ash. 2 cm

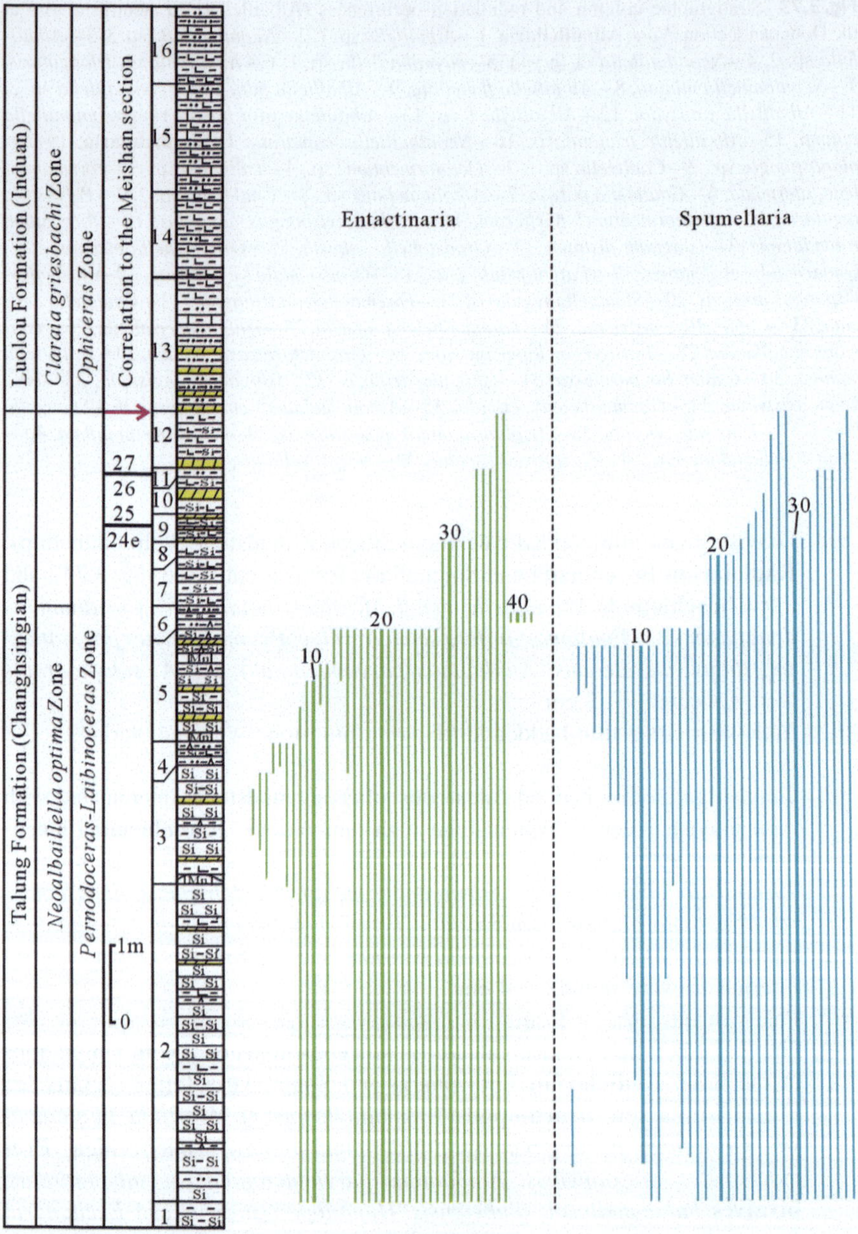
Luolou Formation (Induan)
Claraia griesbachi Zone
Ophiceras Zone
Correlation to the Meishan section
Talung Formation (Changhsingian)
Neoalbaillella optima Zone
Pernodoceras-Laibinoceras Zone
27
26
25
24e
1m
0
Entactinaria
Spumellaria
10
20
30
40

◂**Fig. 3.74** Stratigraphic column and radiolarian occurrences (Entactinaria, Spumellaria) at the Dongpan section. Legends same to Fig. 3.73. *Note* **Entactinaria**, 1—*Triaenosphaera* sp. 5, 2—*Triaenosphaera* sp. 1, 3—*Triaenosphaera* sp. 2, 4—*Triaenosphaera* sp. 4, 5—Genus and species indet., 6—*Entactinia* sp. 1, 7—*Uberinterna virgispinosum*, 8—*Triaenosphaera* sp. 3, 9—*Provisocyntra ormistoni*, 10—*Wuyia dongpanica*, 11—*Entactinia meishanensis*, 12—*Paracopicyntra longispina*, 13—*Entactinia* sp. 2, 14—*Polyentactinia* sp., 15—*Spongentactinia*? sp., 16—*Entactinia itsukaichiensis*, 17—*Entactinia reticulate*, 18—*Entactinia modesta*, 19—*Entactinia wangi*, 20—*Entactinia minuta*, 21—*Trilonche textilis*, 22—*Trilonche*? sp. 1, 23—*Trilonche*? sp. 2, 24—*Kashiwara magna*, 25—*Triaenosphaera minutus*, 26—*Triaenosphaera variabilis*, 27—*Wuyia permica*, 28—*Paracopicyntra simplex*, 29—*Trilonche cimelia*, 30—*Trilonche pseudocimelia*, 31—*Trilonche brevispinosa*, 32—*Triaenosphaera megacantha*, 33—*Paracopicyntra akikawaensis*, 34—*Entactinia sashidai*, 35—*Provisocyntra densa*, 36—*Hegleria mammilla*, 37—*Trilonche crassispinosa*, 38—*Paracopicyntra ziyunensis*, 39—*Megaporus jini*, 40—*Megaporus yini*, 41—*Megaporus yangi*, 42—*Megaporus* sp.; **Spumellaria**, 1—*Archaeospongoprunum* cf. *bispinosum*, 2—*Copicyntra*? sp. 1, 3—*Copicyntra*? sp. 2, 4—*Copicyntroides* sp., 5—*Archaeospongoprunum* sp., 6—*Paurinella mesotriassica*, 7—*Tetraspongodiscus tetramorpha*, 8—*Tetraspongodiscus*? *inaequispinosus*, 9—*Tamonella multispinosa*, 10—*Yujingella triacantha*, 11—*Ellipsocopicytra acantha*, 12—*Archaeospongoprunum chiangdaoensis*, 13—*Tetrapaurinella discoidalis*, 14—*Paroertlispongus chinensis*, 15—*Tamonella aculeata*, 16—*Archaeospongoprunum mengi*, 17—*Tetraspongodiscus stauracanthus*, 18—*Copicyntroides parvulus*, 19—*Paurinella aequispinosa*, 20—*Copiconulus acanthicus*, 21—*Paroertlispongus fontainei*, 22—*Tetrapaurinella nanjingensis*, 23—*Copiellintra elongata*, 24—*Bistarkum martiali*, 25—*Lithelius sinensis*, 26—*Ellipsocopicytra wani*, 27—*Copiconulus solidus*, 28—*Copicyntra* sp. 3, 29—*Copicyntra* sp. 4, 30—*Klaengspongus formosus*, 31—*Paramphibrachium woni*, 32—*Palaeolithocyclia platta*, 33—*Klaengspongus umbilicatus*, 34—*Sinosphaera spinosa*, 35—*Trigonosphaera calvata*, 36—*Klaengspongus spinosus*, 37—*Palaeolithocyclia pilata*

8-7. Yellowish green thin-bedded calcareous siliceous mudstone. 5 cm
8-6. Yellow mudstone. 1.5 cm
8-5. Yellowish green thin-bedded calcareous siliceous mudstone. 2.5 cm
8-4. Yellow mudstone. 2.5 cm
8-3. Yellowish green thin-bedded calcareous siliceous mudstone. 1 cm
8-2. Pale volcanic ash. 1 cm
8-1. Grayish green thin-bedded silt-bearing calcareous siliceous mudstone. 5 cm

Some **brachiopods** *Orthotichia* sp., *Tornquistia changhsingia*, *Attenuatella mengi*, *Martinia liuqiaoiensis*, *Costatumulus dongpanensis*, *Minutomarginifera semicircridge*, *Rugivestis elegans*, *Permorhipidomella ovatus*, *Enteletes subaequivalis*, *Micromartinia kwangsiella*, *Anidanthus parvimucronata*, *A. subquadratus* and *Rhipidomella parvula*, **bivalves** *Palaeoneilo fangi*, *P.* cf. *qiubeiensis*, *P. qinzhouensis*, *Parallelodon changhsingensis*, *Nuculupsis guangxiensis*, *N.* sp., *Pernopecten*? sp., *Tambanella*? *shaodongensis*, *Etheripecten* sp., *Euchondria fusuiensis*, *E.* sp., *Hunanopecten exilis*, *H. qujiangensis*, *Lepdochondria intermedia*, *Claraia liuqiaoensis* and *Bakevellia* sp., a few **foraminifers** *Lasiodiscus*? sp., *Glomospirella*? *Mamilla*, *Hemigordius*? *yini* and *Geinitzinita changhsingensis*, **radiolarian** Entactinaria and Spumullaria forms (details see Fig. 3.74) and the **ammonoids** *Stacheoceras* sp., *Pernodoceras robustum* and *Dushanoceras* cf. *rotolarium* were collected from bed 8 (not subdivided when collecting).

7. Grayish green thin-bedded calcareous siliceous mudstone intercalated with siliceous mudstone. Abundant **brachiopods** *Tornquistia changhsingia, Attenuatella mengi, Martinia liuqiaoiensis, Costatumulus dongpanensis, Minutomarginifera semicircridge, Pygmochonetes jingxianensis, Rugivestis elegans, Permorhipidomella ovatus, Enteletes subaequivalis, Micromartinia kwangsiella, Anidanthus parvimucronata, A. subquadratus* and *Rhipidomella parvula*, **radiolarians** *Tetragregnon longispinatum, Grandetortura nipponica, Shangella longa, Cauletella manica, C. paradoxa, Ishiguam trifustis, I. obesum, I. craticula, I. longispina, I. complanum, I. tristylum, Foremanhelena circula, F. robusta, Triplanospongos musashiensis, Nazarovella gracilis, N. scalae, N. inflata, Ormistonella adhaerens* and *Areohlicaudatus semiglobosa*, **bivalves** *Streblopteria* sp., *Tambanella*? gujoensis, *Palaeoneilo fangi, Hunanopecten exilis, Lepdochondria intermedia, Euchondria fusuiensis* and *Claraia liuqiaoensis*, **foraminifers** *Lasiodiscus*? sp., *Glomospirella*? *mamilla, Glomospirella*? *curva* and *Hemigordius*? *yini*, **ostracods** *Pseudobythocypris* sp. and *Spinomicrocheilinella anterocompressa*, and **cephalopods** *Pernodoceras robustum, Lopingoceras guangdeensis* and *Dushanoceras* cf. *rotolarium* were found. 60 cm
6. Yellow thin-bedded tuff-bearing silty mudstone at the top (4 cm); grayish green siliceous mudstone in the upper part (2 cm); yellow thin-bedded tuff-bearing silty mudstone in the middle and lower parts. A few **radiolarian** *Albaillella triangularis* and **foraminifers** *Lasiodiscus*? sp. and *Glomospirella*? *mamilla* were recorded in the siliceous mudstone. 25 cm

5-19. Dark gray thin-bedded tuff-bearing siliceous mudstone, with a large number of **radiolarians** *Albaillella triangularis, A. yaoi longa, Neoalbaillella optima, Tetragregon longispinatum, Ruzhencevispongus uralicus, Shangella regularis, Cauletella manica, C. paradoxa, C. delicata, C. constricata, Ishigaum trifustis, I. obesum, I. craticula, I. longispina, I. complanum, I. tristylum, Foremanhelena triangula, F. circula, F. robusta, Triplanospongos musashiensis, Nazarovella gracilis; N. inflata* and *Areolicaudatus semiglobosa*, a few **ostracods** *Pseudobythocypris* cf. *contrails, Spinomicrocheilinella* sp., *Spinomicrocheilinella anterocompressa, Waldekella*? *sphaerula* and *Nodokirkbya*? cf. *striatoreticulata*, and the **bivalve** *Posidonia* sp.. 6 cm

5-18. Pale volcanic ash. 4 cm

5-17. Dark gray thin-bedded tuff-bearing siliceous mudstone with laminations. A large number of **radiolarians** *Albaillella triangularis, A. levis, Tetragregon longispinatum, Octatormentum*? *floriferum, Grandetortura nipponica, Latentifistula crux, Ruzhencevispongus totundus, Tormentum rotundatum, Cauletella manica, C. paradoxa, C. delicata, C.* sp. 2, *Ishigaum trifustis, I. obesum, I. longispina, I. fusinum, I. tristylum, Foremanhelena triangula, F. circula, F. robusta, Triplanospongos musashiensis, Nazarovella gracilis; N. scalae, N. inflata, Ormistonella robusta, O. adhaerens, O. elegans, O.*? *cylindrata, Areolicaudatus semiglobosa, A.* sp., *Quadricaulis* cf. *femoris* and *Polyfistula regularis*, **ostracods** *Bairdiacypris* aff. *mirautaae, Monoceratina* sp., *Paraberounella*? cf. *laterospina, Bohemina* (*Pokornina*)? sp.,

Kirkbya sp., *Libumella*? *athabascensis*, *Bairdia dongpanensis*, *Bairdia*? sp., *Permoyoungiella*? sp. *Waldekella*? *sphaerula*, *Pseudobythocypris* cf. *contrails* and *Spinomicrocheilinella anterocompressa*, a few **brachiopods** *Orthotichia* sp., *Pygmochonetes* sp., *P. jingxianensis*, *Costatumulus dongpanensis*, *Enteletes subaequivalis*, *Tornquistia changhsingia*, *Attenuatella mengi*, *Martinia liuqiaoiensis* and *Micromartinia kwangsiella*, and **bivalves** *Hunanopecten exilis*, *Euchondria fusuiensis* and *Claraia liuqiaoensis* were recorded. 4 cm

5-16. Gray to brownish lens-shaped manganous limestone. 3 cm

5-15. Pale volcanic ash. 3 cm

5-14. Dark gray thin-bedded tuff-bearing siliceous mudstone with laminations. A few **radiolarians** *Cauletella porosa*, *C. delicata*, *Ishigaum trifustis*, *I. longispina*, *Foremanhelena triangula*, *F. circula*, *Triplanospongos musashiensis*, *Nazarovella inflata*, *Ormistonella adhaerens*, *Areolicaudatus semiglobosa*, *A.* sp. and *Polyfistula regularis*, **ostracods** *Spinomicrocheilinella* sp., *Healdiopsis thuringensis thuringensis* and *Spinomicrocheilinella anterocompressa*, and **cephalopods** *Dushanoceras* cf. *rotolarium*, *Lopingoceras guangdeensis*, *Agathiceras* sp. and *Stacheoceras* sp. were discovered. 9 cm

5-13. Pale volcanic ash. 1 cm

5-12. Dark gray thin-bedded chert, rich in radiolarians and brachiopods. 7 cm

5-11. Pale volcanic ash. 6 cm

5-10. Dark gray thin-bedded chert, with a large number of **radiolarians** *Albaillella triangularis*, *Cauletella manica*, *C. delicata*, *C. constricata*, *C. porosa*, *Ishigaum trifustis*, *Ishigaum complanum*, *Foremanhelena circula*, *Triplanospongos musashiensis*, *Nazarovella gracilis*, *N. scalae*, *N. inflata*, *Ormistonella elegans*, *O.*? *cylindrata* and *Areolicaudatus semiglobosa*, and the **ostracod** *Pseudobythocypris* cf. *contrails.* 25 cm

5-9. Pale volcanic ash. 3 cm

5-8. Grayish green thin-bedded silicerous mudstone, with a few **radiolarians** *Albaillella triangularis*, *Ishigaum trifustis* and *I. obesum*, **ostracods** *Petasobairdia* sp., *Spinomicrocheilinella* aff. *dargenioi*, *Bairdia dongpanensis*, *Fabalicypris* cf. *minuta*, *Spinomicrocheilinella anterocompressa* and *Pseudobythocypris* aff. *procera*, **brachiopods** *Tornquistia changhsingia* and *Attenuatella mengi*, and **bivalves** *Leptochondria intermedia*, *Hunanopecten exilis* and *Hunanopecten qujiangensis.* 20cm

5-7. Yellowish green calcareous mudstone. 2 cm

5-6. Gray to yellowish green thin-bedded calcareous mudstone, with the **radiolarian** *Shangella regularis*, a few **ostracods** *Bairdia dongpanensis*, *Spinomicrocheilinella anterocompressa* and *Pseudobythocypris* aff. *procera*, and **brachiopods** *Costatumulus dongpanensis*, *Tornquistia changhsingia*, *Attenuatella mengi*, *Martinia liuqiaoiensis* and *Minutomarginifera semicircridge.* 20 cm

5-5. Pale volcanic ash. 1 cm

5-4. Dark gray thin-bedded tuff-bearing chert with laminations. A large number of **radiolarians** *Albaillella triangularis, A. yaoi yaoi, Neoalbaillella optima, N.* sp., *N. camarata, Tetragregon longispinatum, Octatormentum*? *floriferum, Grandetortura nipponica, Latentifistula crux, Ruzhencevispongus uralicus, Tormentum rotundatum, Cauletella manica, C. paradoxa, C. delicata, C. constricata, Ishigaum trifustis, I. obesum, I. longispina, I. fusinum, I. complanum, I. tristylum, I.* sp., *Foremanhelena triangula, F. circula, F. robusta, F. intermixta, Triplanospongos musashiensis, Nazarovella gracilis, N. inflata, Ormistonella adhaerens, O. elegans, O.*? *cylindrata, Areolicaudatus semiglobosa* and *Polyfistula regularis* were collected. 9 cm

5-3. Pale volcanic ash. 10 cm

5-2. Dark gray medium-bedded chert with abundant **radiolarians** *Albaillella flabellata, A. yaoi longa, A. fida, Neoalbaillella optima, Shangella longa, Ishigaum trifustis, I. craticula, I. tristylum, Triplanospongos musashiensis, Ormistonella robusta, Areolicaudatus semiglobosa, A.* sp. and *Quadricaulis* cf. *femoris*, **ostracods** *Macronotella*? sp., *Microcheilinella* cf. *elatus, Waldekella*? *sphaerula, Spinomicrocheilinella anterocompressa, S.* sp., *Pseudobythocypris* cf. *contrails* and *P.* sp., **brachiopods** *Anidanthus parvimucronata, Costatumulus dongpanensis* and *Tornquistia changhsingia*, and the **bivalve** *Hunanopecten qujiangensis.* 12 cm

5-1. Gray to brown lens-shaped manganous limestone (laterally varied in the thickness). Abundant **foraminifers** *Lasiodiscus*? sp., *Glomospirella*? *curva, Hemigordius*? *yini, Geinitzinita changhsingensis, Glomospirella*? *mamilla, Geinitzina*? sp., *Howchinella*? sp., *Dalongella fusiformis, Nodosaria*? sp., *Pachyphloides inflatus, Pseudolangella costata*, and *Ichthyolaria* sp. have been collected in bed 5 (not subdivided when sampling). 0–30 cm

4. Yellow thin-bedded tuff-bearing silty mudstone at the top (4 cm); grayish green calcareous and siliceous mudstone in the upper part (2 cm); yellow thin-bedded tuff-bearing silty mudstone in the middle and lower parts. Laterally varied in the thickness for the bed. A few **radiolarians** *Albaillella triangularis, Shangella longa, Cauletella manica, C. delicata, Ishigaum trifustis, I. obesum, I. tristylum, Foremanhelena circula, F. circula, F. triangula, Nazarovella gracilis* and *Areolicaudatus semiglobosa*, **ostracods** *Cooperuna* cf. *tenuis, Absina* cf. *unispinosa, Absina*? sp., *Healdia* sp., *Healdiopsis thuringensis thuringensis, Permoyoungiella*? sp. and *Spinomicrocheilinella* sp., and **foraminifers** *Hemigordius*? *yini, Geinitzinita changhsingensis, Howchinella*? sp. and *Neoendothyra*? *reicheli* were recorded in the siliceous mudstone. 27–36 cm

3-1. Dark gray thin-bedded chert. Abundant **radiolarians** *Albaillella triangularis, A.* cf. *triangularis, A. flabellata, A. yaoi longa, A. fida, A. levis, Neoalbaillella optima, N.* cf. *grypa, N. minuta* and *N. camarata*, and **ostracods** *Paraparchites*? sp., *Libumella* cf. *kargalensis, Healdia* sp., *Bairdia*? sp., *Spinomicrocheilinella* sp., *Pseudobythocypris* cf. *contrails, Spinomicrocheilinella anterocompressa* and *Pseudobythocypris* aff. *procera* were recovered. 17 cm

3-2. Pale volcanic ash. 4 cm

3-3. Dark gray thin-bedded chert. 10 cm

3-4. Pale volcanic ash. 4 cm

3-5. Dark gray thin-bedded chert, with abundant **radiolarians** *Latentifistula crux, Shangella longa, Ruzhencevispongus totundus, Cauletella paradoxa, C. delicata, Ishigaum trifustis, I. obesum, I. longispina, I. tristylum, Foremanhelena circula, Triplanospongos musashiensis, Nazarovella scalae, N. gracilis, N. inflata, Ormistonella robusta, O. elegans, O.? cylindrata, Areolicaudatus semiglobosa, Quadricaulis* cf. *femoris* and *Polyfistula regularis*, **ostracods** *Bohemina (Pokornina)*? sp., *Kirkbya* sp. and *Waldekella*? *sphaerula*, and the **bivalve** *Euchondria fusuiensis.* 8.5 cm

3-6. Pale volcanic ash. 4 cm

3-7. Dark gray thin-bedded chert, with a large number of **radiolarians** *Albaillella triangularis, A.* cf. *triangularis, A. flabellata, A. yaoi yaoi, Neoalbaillella optima, Shangella longa, S. regularis, Ruzhencevispongus uralicus, Tormentum rotundatum, Cauletella paradoxa, Ishigaum trifustis, I. obesum, I. longispina, Foremanhelena triangula, F. circula, Triplanospongos musashiensis, Nazarovella gracilis, N. inflata, Areolicaudatus semiglobosa* and *Quadricaulis* cf. *femoris*, and a few **cephalopods** *Stacheoceras* sp. and *Lopingoceras guangdeensis.* 9.5 cm

3-8. Light grayish green volcanic ash. 2 cm

3-9. Dark gray thin-bedded chert. 6.5 cm

3-10. Light grayish green volcanic ash. 2 cm

3-11. Dark gray thin-bedded chert, with abundant **ostracods** *Bairdia dongpanensis, B.* aff. *birinae, B.* sp., *Polycope* aff. *baudi, Libumella* cf. *kargalensis, Microcheilinella* cf. *elatus, Healdiopsis thuringensis thuringensis, ?Libumella athabascensis, Permoyoungiella*? sp., *Waldekella*? *sphaerula, Spinomicrocheilinella anterocompressa, S.* sp., *Cavellina* cf. *nebrascensis* and *Pseudobythocypris* aff. *procera*, a few **radiolarians** *Albaillella triangularis* and *A.* cf. *triangularis*, **brachiopods** *Orthotichia* sp., *Pygmochonetes jingxianensis, P.* sp., *Costatumulus dongpanensis, Permorhipidomella ovatus, Enteletes subaequivalis, Tornquistia changhsingia, Attenuatella mengi, Martinia liuqiaoiensis, Minutomarginifera semicircridge, Rugivestis elegans* and *Micromartinia kwangsiella*, and the **bivalve** *Nuculopsis guangxiensis.* 5 cm

3-12. Pale volcanic ash. 2 cm

3-13. Dark gray thin-bedded chert altered with grayish yellow calcareous mudstone, with a few **bivalves** *Tambanella*? *shaodongensis, Nuculopsis*? sp., *Paralleloden changhsingensis, Hunanopecten exilis, Hunanopecten qujiangensis, Claraia liuqiaoensis* and *Palaeoneilo qinzhouensis* in the chert. 25 cm

3-14. Dark gray thin-bedded chert, with abundant **radiolarians** *Albaillella flabellata, A. yaoi yaoi, A.yaoi longa, A. levis, Neoalbaillella optima, N. minuta, N.* sp.3, *Tetragregon longispinatum, Shangella longa, Cauletella paradoxa, C.*

constricata, *C. porosa*, *C.* sp., *Ishigaum trifustis*, *I. longispina*, *Foremanhelena circula*, *F. robusta*, *Triplanospongos musashiensis*, *Nazarovella gracilis*, *N. inflata* and *Areolicaudatus semiglobosa*, a few **brachiopods** *Tornquistia changhsingia*, *Attenuatella mengi* and *Micromartinia kwangsiella*, and **bivalves** *Pernopecten*? sp., *Hunanopecten exilis*, *Euchondria fusuiensis* and *Hunanopecten qujiangensis*. 7 cm

3-15. Grayish yellow calcareous mudstone intercalated with chert. 5 cm

3-16. Dark gray thin-bedded chert with the **brachiopod** *Martinia liuqiaoiensis* and **bivalves** *Palaeoneilo leiyangensis*, *Nuculopsis guangxiensis*, *Paralleloden changhsingensis* and *Euchondria fusuiensis*. 10 cm

3-17. Grayish yellow thin-bedded calcareous mudstone. 15 cm

3-18. Gray to brown lens-shaped manganous limestone, with brown laminated calcareous mudstone at the base (laterally varied in the thickness for the limestone). A few **bivalves** *Nuculopsis guangxiensis*, *Hunanopecten exilis*, *H. qujiangensis* and *Euchondria fusuiensis* were collected in the mudstone. 5–17 cm

Additionally, **foraminifers** *Lasiodiscus*? sp., *Glomospirella*? *curva*, *G.*? *mamilla*, *Hemigordius*? *yini*, *Geinitzinita changhsingensis*, *Geinitzina ichnousa*, *Pachyphloides inflatus*, *Pseudolangella costata*, *Frondinodosaria fusuiensis* and *Ichthyolaria* sp. were recorded in bed 3 (not subdivided when sampling).

2-1. Light green to yellow thin-bedded tuff-bearing mudstone (laterally varied in the thickness). 0–15 cm

2-2. Dark gray thin-bedded chert with a few brachiopods, **radiolarians** *Albaillella triangularis*, *A. flabellata*, *A. levis*, *Neoalbaillella optima*, *Shangella longa*, *S. regularis*, *Cauletella delicata*, *Ishigaum trifustis*, *I. obesum*, *I. longispina*, *Foremanhelena circula*, *Triplanospongos musashiensis*, *Nazarovella gracilis* and *N. inflata*, and **cephalopods** *Pernodoceras robustum* and *Lopingoceras guangdeensis*. 15 cm

2-3. Light green to yellow thin-bedded tuff-bearing mudstone. 13 cm

2-4. Dark gray thin-bedded chert with a large number of **radiolarians** *Latentifistula crux*, *Shangella longa*, *Tormentum rotundatum*, *Cauletella delicata*, *C. constricata*, *Ishigaum trifustis*, *I. obesum*, *I. craticula*, *I. longispina*, *Triplanospongos musashiensis*, *Nazarovella gracilis*, *N. scalae*, *N. inflata*, *Ormistonella robusta* and *Areolicaudatus semiglobosa*, **ostracods** *Bairdia dongpanensis* and *Pseudobythocypris* sp., **brachiopod** *Attenuatella mengi*, and **bivalves** *Stutchburia*? sp., *Palaeoneilo leiyangensis*, *Streblopteria* sp., *Tambanella*? *gujoensis*, *Nuculopsis*? sp., *Paralleloden changhsingensis*, *Hunanopecten exilis*, *Euchondria fusuiensis* and *Palaeoneilo qinzhouensis*. 17 cm

2-5. Yellow thin-bedded tuff-bearing mudstone. 13 cm

2-6. Dark gray thin-bedded chert intercalated with calcareous mudstone. Abundant **radiolarians** *Albaillella triangularis*, *A. yaoi longa*, *Latentifistula crux*, *Shangella regularis*, *Cauletella manica*, *Ishigaum trifustis*, *I. obesum*, *I. longispina*, *I. tristylum*, *Foremanhelena circula*, *Nazarovella gracilis*, *N.*

inflata, *Areolicaudatus semiglobosa* and *Ormistonella*? *cylindrata*, **ostracods** *Spinobairdia* sp., *Pseudobythocypris* cf. *contrails*, *Pseudobythocypris* sp., *Spinomicrocheilinella anterocompressa* and *Pseudobythocypris* aff. *procera*, **brachiopods** *Attenuatella mengi* and *Micromartinia kwangsiella*, and **bivalves** *Tambanella*? *gujoensis*, *Euchondria fusuiensis* and *Palaeoneilo qinzhouensis* were collected. 17.5 cm

2-7. Yellow thin-bedded tuff-bearing mudstone. 2.5 cm

2-8. Dark gray thin-bedded chert, rich in **radiolarians** *Shangella longa*, *Cauletella constricata*, *C. porosa*, *Ishigaum trifustis*, *I. obesum*, *I. longispina*, *Foremanhelena triangula*, *F. circula*, *F. robusta*, *Triplanospongos musashiensis*, *Nazarovella gracilis* and *N. inflata*. 5 cm

2-9. Yellow tuff-bearing mudstone in the upper part; light green tuff-bearing mudstone in the lower part. 5 cm

2-10. Dark gray thin-bedded chert altered with gray thin-bedded tuff-bearing mudstone, with a few **radiolarians** *Albaillella triangularis*, *Shangella longa*, *Ruzhencevispongus totundus*, *Cauletella paradoxa*, *C. delicata*, *C. constricata*, *Ishigaum trifustis*, *I. craticula*, *I. longispina*, *Foremanhelena triangula*, *F. circula*, *Triplanospongos musashiensis*, *Nazarovella gracilis* and *Areolicaudatus semiglobosa*. 36 cm

2-11. Yellow mudstone. 12 cm

2-12. Yellowish gray chert with abundant **radiolarians** *Latentifistula crux*, *Shangella longa*, *Cauletella manica*, *C. paradoxa*, *C. delicata*, *Ishigaum trifustis*, *I. craticula*, *I. longispina*, *Foremanhelena triangula*, *F. circula*, *Triplanospongos musashiensis*, *Nazarovella gracilis*, *N. inflata* and *Polyfistula regularis*, **ostracod** *Spinomicrocheilinella anterocompressa*, and **bivalves** *Paralleloden changhsingensis* and *Euchondria fusuiensis*. 8.5 cm

2-13. Yellow tuff-bearing mudstone. 13 cm

2-14. Dark gray thin-bedded chert intercalated with mudstone, with radiolarians. 9.5 cm

2-15. Yellow to yellowish green thin-bedded tuff-bearing mudstone intercalated with chert. Abundant **radiolarians** *Ruzhencevispongus uralicus*, *Cauletella manica*, *C. paradoxa*, *C. delicata*, *Ishigaum trifustis*, *I. longispina*, *I. fusinum*, *I.* sp., *Foremanhelena triangula*, *F. circula*, *F. robusta*, *F. intermixta*, *Triplanospongos musashiensis*, *Nazarovella gracilis*, *N. scalae*, *N. inflata*, *Ormistonella robusta*, *O. adhaerens*, *O.*? *cylindrata*, *Areolicaudatus semiglobosa* and *Quadricaulis* cf. *femoris*, and a few **ostracods** *Petasobairdia* sp., *Aurikirkbya* sp., *Monoceratina*? sp., *Paraberounella* sp., *Spinomicrocheilinella* sp., *Pseudobythocypris* cf. *contrails*, *Discoidella xingyangensis* and *Pseudobythocypris* aff. *procera* were collected. 11 cm

2-16. Grayish green thin-bedded chert intercalated with yellow mudstone. A large number of **radiolarians** *Albaillella triangularis*, *A.* cf. *triangularis*, *A. yaoi yaoi*, *Neoalbaillella optima*, *N.* sp.2, *N. camarata*, *Tetragregon longispinatum*, *Octatormentum*? *floriferum*, *Grandetortura nipponica*, *Cauletella paradoxa*, *C. delicata*, *Ishigaum trifustis*, *I. longispina*, *Foremanhelena triangula*,

F. circula, *Triplanospongos musashiensis*, *Nazarovella scalae* and *N. gracilis* were examined in the chert. 20 cm

2-17. Yellow thin-bedded tuff-bearing mudstone intercalated with grayish green chert, with a few **radiolarians** *Latentifistula crux*, *Ishigaum trifustis*, *I. longispina*, *Foremanhelena triangula*, *F. circula*, *Triplanospongos musashiensis* and *Nazarovella gracilis.* 20 cm

2-18. Grayish green thin-bedded mudstone intercalated with chert, with **radiolarians** *Latentifistula crux*, *Shangella longa*, *S. regularis*, *Cauletella manica*, *C. paradoxa*, *C. delicata*, *C. constricata*, *Ishigaum trifustis*, *I. obesum*, *I. craticula*, *I. longispina*, *I. complanum*, *Foremanhelena triangula*, *F. circula*, *F. robusta*, *Triplanospongos musashiensis*, *Nazarovella gracilis*, *N. scalae*, *N. inflata*, *Ormistonella robusta*, *O. adhaerens*, *O. elegans* and *O.*? *cylindrata*, and the **bivalve** *Hunanopecten exilis.* 28 cm

2-19. Yellowish gray medium-bedded chert with radiolarians. 10 cm

2-20. Grayish green tuff-bearing mudstone intercalated with chert. Abundant **radiolarians** *Albaillella yaoi longa*, *Tetragregon longispinatum*, *Ruzhencevispongus uralicus*, *Shangella longa*, *S. regularis*, *Cauletella manica*, *C. paradoxa*, *Ishigaum obesum*, *I. longispina*, *I. fusinum*, *Foremanhelena triangula*, *F. robusta*, *Triplanospongos musashiensis* and *N. gracilis*, **ostracods** *Permoyoungiella*? sp. and *Pseudobythocypris* aff. *procera*, and **bivalves** *Paralleloden changhsingensis*, *Palaeolima fangi*, *Palaeoneilo* cf. *qinbeiensis*, *Hunanopecten exilis*, *Hunanopecten qujiangensis* and *Palaeoneilo qinzhouensis* were examined in the chert. 33 cm

2-21. Yellowish gray medium-bedded chert with a few radiolarians. 10 cm

2-22. Yellow to grayish green tuff-bearing mudstone with a few radiolarians and brachiopods. 11 cm

2-23. Grayish green thin-bedded chert with a few **radiolarians** *Tetragregon longispinatum*, *Grandetortura nipponica*, *Shangella regularis*, *Ishigaum obesum*, *I. craticula*, *I. longispina*, *I. tristylum*, *Foremanhelena triangula*, *F. circula*, *Triplanospongos musashiensis* and *T.* sp., **ostracods** *Bairdia*? sp., *Permoyoungiella*? sp. and *Pseudobythocypris* cf. *contrails*, and **brachiopods** *Orthotichia* sp., *Costatumulus dongpanensis*, *Permorhipidomella ovatus*, *Enteletes subaequivalis* and *Micromartinia kwangsiella.* 19 cm

2-24. Yellow medium-bedded mudstone with a few **radiolarians** *Cauletella paradoxa*, *Ishigaum trifustis*, *I. craticula*, *I. longispina* and *Foremanhelena triangula*, and ostracods *Kellettina* aff. *reticulata*, *Rectonaria* cf. *inclinata*, *Healdia* sp., *Spinomicrocheilinella anterocompressa*, *S.* aff. *dargenioi*, *Bairdia dongpanensis*, *B.*? sp., *Permoyoungiella*? sp., *Pseudobythocypris* cf. *contrails* and *P.* aff. *procera.* 15 cm

2-25. Yellowish green medium-bedded chert with a large number of **radiolarians** *Albaillella triangularis*, *A.* cf. *triangularis*, *A. flabellata*, *A. yaoi yaoi*, *Shangella longa*, *Tormentum rotundatum*, *Cauletella manica*, *C. paradoxa*, *C. delicata*, *C. constricata*, *Ishigaum obesum*, *I. craticula*, *I. longispina*, *I.*

fusinum, I. complanum, Foremanhelena triangula, F. circula, Triplanospongos musashiensis, Nazarovella gracilis, N. inflata and *Ormistonella elegans.* 15.5 cm

2-26. Yellow medium-bedded mudstone with a few **radiolarians** *Tetragregon longispinatum, Ishigaum trifustis, I. longispina, Foremanhelena triangula, F. circula, F. robusta, Triplanospongos musashiensis, Nazarovella gracilis, N. scalae, Ormistonella robusta, O. adhaerens, O. elegans, Areolicaudatus semiglobosa* and *Quadricaulis* cf. *femoris*, and **ostracods** *Polycope* sp. and *Bairdia dongpanensis.* 15 cm

2-27. Grayish green thin-bedded chert intercalated with silty mudstone. Abundant **radiolarians** *Albaillella triangularis, A. flabellata, A. yaoi yaoi, Neoalbaillella optima, Shangella longa, Ishigaum obesum, I. trifustis, I. longispina, Foremanhelena triangula, F. circula* and *Nazarovella gracilis* etc. were yielded in the upper part of the bed; abundant radriolarians *Latentifistula crux, Shangella longa, Cauletella manica, C. paradoxa, C. delicata, C. constricata, Ishigaum trifustis, I. obesum, I. craticula, I. longispina, I. tristylum, Foremanhelena circula, Triplanospongos musashiensis, Nazarovella scalae, N. gracilis, Megaporus jini, M. yini, Entactinia modesta* and *Provisocyntra densa*, **ostracods** *Bairdia dongpanensis, Bairdia*? sp., *Permoyoungiella*? sp., *Pseudobythocypris* aff. *procera, P.* cf. *contrails, P.* sp. and *Spinomicrocheilinella anterocompressa*, and **brachiopods** *Tornquistia changhsingia, Attenuatella mengi, Martinia liuqiaoiensis, Minutomarginifera semicircridge, Rugivestis elegans* and *Micromartinia kwangsiella* were collected in the lower part of the bed.
Additionally, **foraminifers** *Nodoinvolutaria bella, Neoendothyra* sp., *Ichthyofrondina*? *liuqiaoensis, Howchinella*? sp., *Geinitzinita kongdongshanensis, G. changhsingensis, Geinitzina ichnousa, Dalongella fusiformis, Pachyphloides inflatus, Pseudolangella costata, Glomospirella*? *curva, G.*? *mamilla, Hemigordius*? *yini* and *Lasiodiscus*? sp. were recorded in bed 2 (not subdivided when collecting).
The basal part (about 60 cm) of the bed is covered by the Quarternary sediments.

1. Grayish green thin-bedded chert intercalated with yellow mudstone, with a few **radiolarians** *Albaillella triangularis, A. flabellata, A. yaoi yaoi, A. yaoi longa, A. fida, Neoalbaillella optima, N. camarata* and *N. minuta* in the chert.

Unexposed at the base

Note: Data of radiolarian Albaillellaria is after Jin et al. (2007) and Wu et al. (2010); radiolarian Latentifistularia is after Feng et al. (2006a); radiolarian Entactinaria is after Feng et al. (2006b, 2007); radiolarian Spumellaria is based on Feng et al. (2004, 2009); ostracod lists are referred to Yuan et al. (2007); foraminifers are based on Gu et al. (2007); bivalves are based on He et al. (2007) and Yang et al. (2015); ammonoids are based on Bu et al. (2006); brachiopods after He et al. (2020) (Fig. 3.75); data of carbon isotope are based on Zhang et al. (2006).

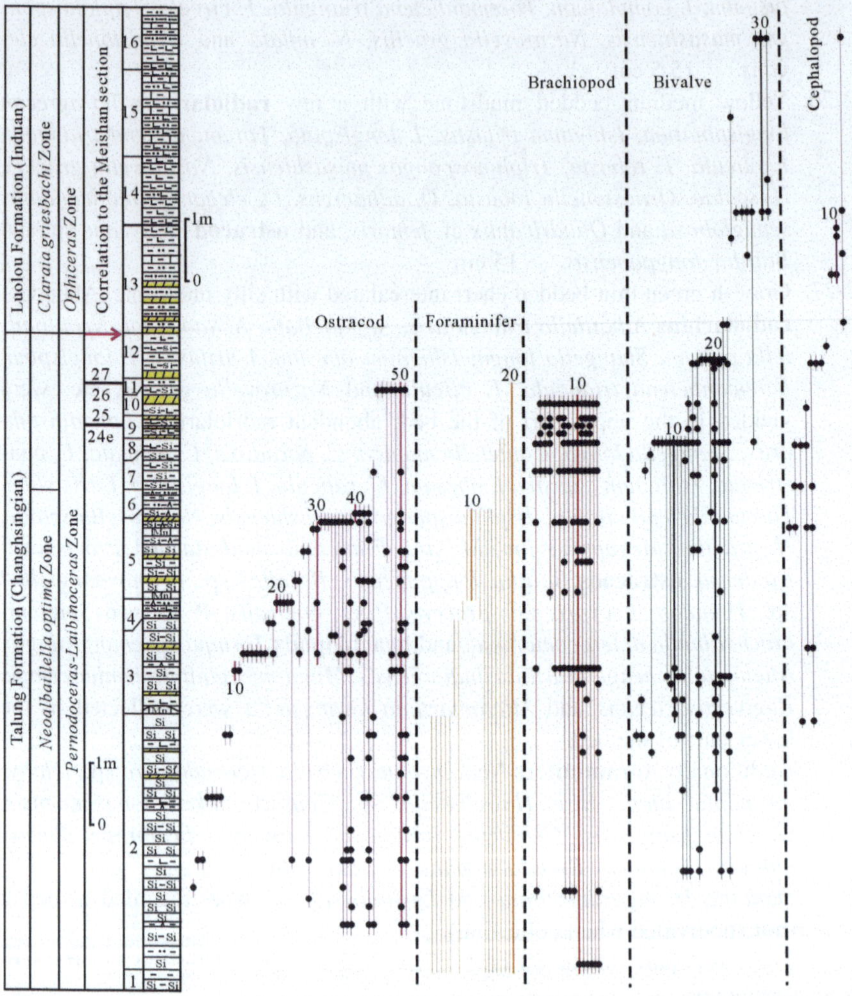
Luolou Formation (Induan)
Talung Formation (Changhsingian)
Claraia griesbachi Zone
Neoalbaillella optima Zone
Ophiceras Zone
Pernodoceras-Laibinoceras Zone
Correlation to the Meishan section
27
26
25
24e
1m
0
Ostracod
Foraminifer
Brachiopod
Bivalve
Cephalopod

◂**Fig. 3.75** Stratigraphic column and fossil occurrences (ostracod, foraminifer, brachiopod, bivalve and cephalopod) at the Dongpan section. Legends same to Fig. 3.73. *Note* **ostracod**, 1—*Polycope* sp. 1, 2—*Kellettina* aff. *reticulata*, 3—*Rectonaria* cf. *inclinata*, 4—*Petasobairdia* sp. 2, 5—*Aurikirkbya* sp. 1, 6—*Monoceratina*? sp. 2, 7—*Paraberounella* sp. 1, 8—*Spinobairdia* sp. 1, 9—Bythocytheridae gen. and sp. indet., 10—Rectonariidae gen. and sp. indet., 11—*Kirkbya* cf. sp. A, 12—*Bairdia* sp. 4, 13—*Bairdia* sp. 1, 14—*Bairdia* sp. 2, 15—*Bairdia* sp. 3, 16—*Bairdia* aff. *birinae*, 17—*Polycope* aff. *baudi*, 18—*Paraparchites*? sp. 1, 19—*Libumella* cf. *kargalensis*, 20—*Cooperuna* cf. *tenuis*, 21—*Absina* cf. *unispinosa*, 22—*Absina*? sp. 1, 23—*Absina*? sp. 2, 24—*Healdia* sp. 1, 25—*Macronotella*? sp. 2, 26—*Microcheilinella* cf. *elatus*, 27—*Petasobairdia* sp. 1, 28—*Spinomicrocheilinella* aff. *dargenioi*, 29—*Spinomicrocheilinella* sp. 2, 30—*Healdiopsis thuringensis thuringensis*, 31—*Bairdiacypris* aff. *mirautaae*, 32—*Monoceratina* sp. 1, 33—*Paraberounella*? cf. *laterospina*, 34—*Bohemina* (*Pokornina*)? sp. 1, 35—*Kirkbya* sp. 1, 36—?*Libumella athabascensis*, 37—*Bairdia dongpanensis*, 38—*Bairdia*? sp. 6, 39—*Permoyoungiella*? sp. 1, 40—*Nodokirkbya*? cf. *striatoreticulata*, 41—*Waldekella*? *sphaerula*, 42—*Spinomicrocheilinella* sp. 1, 43—*Pseudobythocypris* cf. *contrails*, 44—*Pseudobythocypris* sp. 1, 45—*Macronotella*? sp. 1, 46—*Microcheilinella aculeata*, 47—*Fabalicypris* cf. *minuta*, 48—*Cavellina* cf. *nebrascensis*, 49—*Discoidella xingyangensis*, 50—*Spinomicrocheilinella anterocompressa*, 51—*Pseudobythocypris* aff. *procera*; **foraminifer**, 1—*Nodoinvolutaria bella*, 2—*Neoendothyra* sp., 3—*Ichthyofrondina*? *liuqiaoensis*, 4—*Howchinella*? sp. B, 5—*Geinitzinita kongdongshanensis*, 6—*Frondinodosaria fusuiensis*, 7—*Geinitzina ichnousa*, 8—*Neoendothyra*? *reicheli*, 9—Nodosariidae gen. and sp. indet., 10—*Nodosaria*? sp., 11—*Geinitzina*? sp., 12—*Howchinella*? sp. A, 13—*Dalongella fusiformis*, 14—*Pachyphloides inflatus*, 15—*Pseudolangella costata*, 16—*Glomospirella*? *curva*, 17—*Hemigordius*? *yini*, 18—*Geinitzinita changhsingensis*, 19—*Glomospirella*? *mamilla*, 20—*Lasiodiscus*? sp.; **brachiopod**, 1—*Orthotichia* sp., 2—*Anidanthus subquadratus*, 3—*Rhipidomella parvula*, 4—*Anidanthus parvimucronata*, 5—*Pygmochonetes* sp., 6—*Pygmochonetes jingxianensis*, 7—*Costatumulus dongpanensis*, 8—*Permorhipidomella ovatus*, 9—*Enteletes subaequivalis*, 10—*Tornquistia changhsingia*, 11—*Attenuatella mengi*, 12—*Martinia liuqiaoiensis*, 13—*Minutomarginifera semicircridge*, 14—*Rugivestis elegans*, 15—*Micromartinia kwangsiella*; **bivalve**, 1—*Stutchburia*? sp., 2—*Palaeoneilo leiyangensis*, 3—*Streblopteria* sp., 4—*Tambanella*? *gujoensis*, 5—*Nuculopsis* sp., 6—*Etheripecten* sp., 7—*Euchondria* sp., 8—*Tambanella*? *shaodongensis*, 9—*Pernopecten*? sp., 10—*Nuculopsis guangxiensis*, 11—*Nuculopsis*? sp., 12—*Paralleloden changhsingensis*, 13—*Palaeolima fangi*, 14—*Palaeoneilo* cf. *qinbeiensis*, 15—*Leptochondria intermedia*, 16—*Hunanopecten exilis*, 17—*Streblochondria*? sp., 18-*Promyalina*? sp., 19—*Posidonia* sp., 20—*Euchondria fusuiensis*, 21—*Hunanopecten qujiangensis*, 22—*Claraia liuqiaoensis*, 23—*Palaeoneilo qinzhouensis*, 24—*Claraia* sp., 25—*Entolium* sp., 26—*Claraia hunanica*, 27—*Claraia dieneri*, 28—*Bakevellia* sp., 29—*Claraia guizhouensis*, 30—*Claraia* cf. *wangi*, 31—*Claraia griesbachi*; **cephalopod**, 1—*Agathiceras* sp., 2—*Dushanoceras* cf. *rotolarium*, 3—*Pernodoceras robustum*, 4—*Stacheoceras* sp., 5—*Lopingoceras guangdoeensis*, 6—*Qianjiangoceras* sp., 7—*Huananoceras* cf. *perornatum*, 8—*Laibinoceras* cf. *compressum*, 9—*Lytophiceras* cf. *chamunda*, 10—*Ophiceras tingi*, 11—*Ophiceras* sp.

3.32 Penglaitan Section in Laibin City

The Penglaitan section (23° 41′ 41″ N, 109° 18′ 39″ E) outcropped along the Hongshui River, 200 m east to the Hongtang Village in Laibin City, Guangxi, South China (Fig. 3.76).

The sequence of the Penglaitan section is composed of the Heshan, Linhao and Shipao formations in an ascending order. The Heshan Formation is mainly composed of limestones and siliceous limestones. The Linhao Formation includes the Ergou Member in the lower and the Penglaitan Member in the upper. The Ergou Member is characterized by siliceous rocks and silty shales, intercalated with limestones and tuffaceous siltstones. The Penglaitan Member is characterized

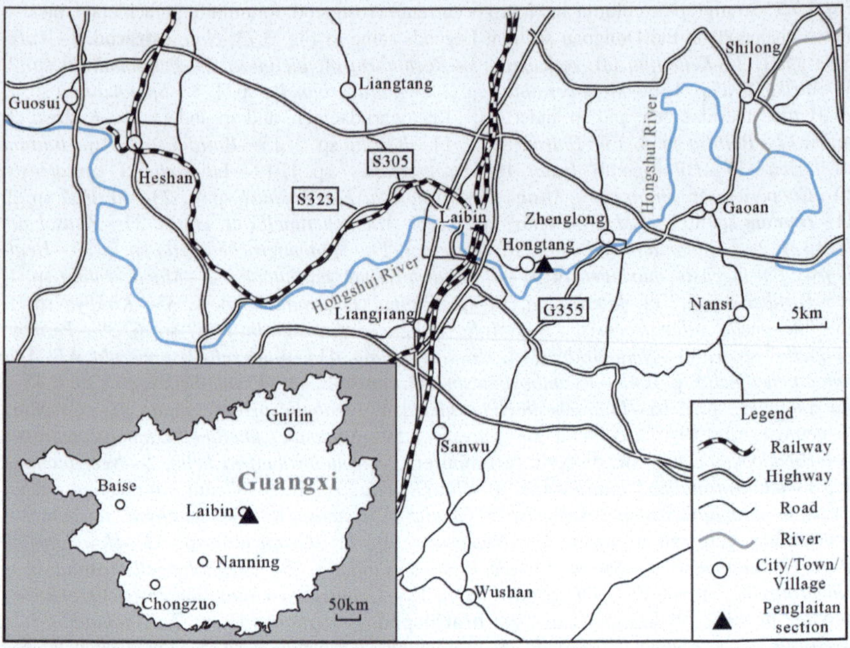

Fig. 3.76 Geographic location of the Penglaitan section

by tuffaceous sandstones, tuffaceous siltstones, siltstones and shales, intercalated with siliceous rocks and limestones. The Shipao Formation is characterized by argillaceous limestones intercalated with shales and several tuffaceous layers.

Shipao Formation in the Lower Triassic

143. Grayish green medium- to thin-bedded argillaceous limestone intercalated with shale. > 400 cm
142. Grayish black shale intercalated with grayish black thin-bedded argillaceous limestone, with two layers of grayish green volcanic ash (10 cm for each layer) within the basal part. Conodonts *Clarkina zhejiangensis*, *Hindeodus praeparvus* and *Hindeodus parvus*, the gastropod *Polygyrina depressa*, bivalves *Claraia* spp. and ostracods were recorded. 648 cm

Conformity

Linhao Formation in the Upper Permian
Penglaitan Member

141. Grayish black thick-bedded coarse-grained tuffaceous calcareous sandstone. Brachiopods *Anchorhynchia sarciniformis*, *Edriosteges poyangensis*, *Haydenella kiangsiensis*, *Leptodus nobilis*, *Meekella* sp., *Orthothetina ruber*,

Orthothetina regularis, *Orthothetina* sp., *Peltichia transversa*, *Cathaysia chonetoides*, *Laterispina parallela*, *Acosarina minuta*, *Paryphella orbicularis*, *Tropidelasma zhongliangshanensis*, *Hustedia* sp., *Spinomarginifera alpha*, *Spinomarginifera chenyaoyanensis*, *Fusichonetes pygmaea* and permianellids, the fusulinid *Colaniella parva*, non-fusulinid foraminifers *Palaeofusulina sinensis* and *Nankinella* sp., and cephalopods *Pseudotirolites* sp., *Pernodoceras* sp. and *Chaotianoceras* sp. were examinied. 228 cm

140. Grayish green thin-bedded tuffaceous mudstone intercalated with thin-bedded tuffaceous sandstone. Brachiopods *Anchorhynchia sarciniformis*, *Haydenella kiangsiensis*, *Orthothetina regularis*, *Laterispina parallela*, *Paryphella orbicularis* and *Spinomarginifera chenyaoyanensis*, fusulinid *Colaniella parva* and non-fusulinid foraminifer *Palaeofusulina sinensis* were recorded. 168 cm
139. Black siliceous rock intercalated with tuff in the upper part; grayish black thick- to medium-bedded siliceous rock intercalated with black thin-bedded mudstone and medium-bedded limestone in the lower part. The conodonts *Clarkina yini* and *Hindeodus praeparvus*, and brachiopods *Crurithyris* sp., *Oldhamina* sp., *Crenispirifer* sp., *Anchorhynchia sarciniformis* and *Leptodus nobilis* were found. 256 cm
138. Grayish green tuffaceous mudstone and lens-shaped limestone (mostly covered by the Quaternary sediments). The fusulinid *Colaniella parva* and non-fusulinid foraminifer *Palaeofusulina sinensis* were examined. 400 cm
137. Breccia-bearing tuffaceous sandstone in the upper part; medium- to thin-bedded limestone intercalated with tuffaceous sandstone and abundant pyrites in the middle part; medium- to thin-bedded limestone in the lower part (slightly covered by the Quaternary sediments). Brachiopods *Araxathyris sinensis*, *Araxathyris* sp., *Paryphella transversa*, *Paryphella sinuata*, *Pseudolabaia* sp., *Martinia* sp., *Anchorhynchia sarciniformis*, *Edriosteges poyangensis*, *Leptodus nobilis*, *Meekella* sp., *Orthothetina ruber*, *Orthothetina regularis*, *Peltichia transversa*, *Cathaysia chonetoides*, *Laterispina parallela*, *Acosarina minuta*, *Spinomarginifera alpha* and *Spinomarginifera chenyaoyanensis*, the coral *Lophophyllidium* sp., the non-fusulinid foraminifer *Geinitzina* sp., and the bivalve *Guizhoupecten regularis* were discovered. 575 cm
136. Gray thick-bedded limestone. The conodonts *Clarkina meishanensis*, *Clarkina yini*, *Hindeodus eurypyge* and *Hindeodus praeparvus*, the brachiopod *Schizodus* sp., the fusulinid *Colaniella parva* and the non-fusulinid foraminifer *Palaeofusulina sinensis* were recorded. 115 cm
135. Medium-bedded limestone intercalated with coal seams and with thin-bedded carbonaceous mudstone in the upper part; gray thick-bedded limestone with abundant pyrites in the lower part. The fusulinid *Palaeofusulina sinensis* was identified. 130 cm
134. Thin-bedded limestone, with coal seams at the top part (7 cm), lens-shaped limestone in the upper part, and with grayish green tuffaceous sandstone and purple to grayish black carbonaceous shale and siliceous rock at the basal part. Sponges *Amblysiphonella laibinensis*, *Amblysiphonella vesiculosa*,

Radiofibrosclera laibinensis, *Polycystocoelia* sp., *Colospingia* sp. and *Acoelia* sp. were examined. 220 cm

133. Gray thick-bedded limestone. The cephalopod *Changhsingoceras sichuanense*, conodonts *Clarkina meishanensis* and *Hindeodus praeparvus*, and the fusulinid *Palaeofusulina sinensis* were recovered. 110 cm
132. Grayish black medium-bedded calcareous tuff in the upper part; grayish green fine-grained tuffaceous sandstone in the lower part. 235 cm
131. Gray thick- to medium-bedded limestone intercalated with shale in the upper part; light gray thin-bedded limestone in the lower part (26 cm). Conodonts *Clarkina meishanensis*, *C. zhejiangensis*, *Clarkina yini* and *Hindeodus praeparvus*, and the fusulinid *Palaeofusulina sinensis* were recorded at the top part. The calcareous algaes and sponges were also discovered at the top part. 185 cm
130. Yellowish brown thick- to medium-bedded coarse- to medium-grained tuffaceous sandstone (fining upwards). Brachiopods *Spinomarginifera sichuanensis*, *Parapulchratia palliata*, *Neochonetes strophomenoides*, *Prelissorhynchia pseudoutah*, *Araxathyris bisulcata*, *Streptorhynchus* sp., *Phricodothyris guizhouensis*, *Martinia orbicularis*, *Paryphella transversa*, *Edriosteges poyangensis*, *Haydenella kiangsiensis*, *Leptodus nobilis*, *Meekella* sp., *Orthothetina ruber*, *Orthothetina regularis*, *Peltichia transversa*, *Cathaysia chonetoides*, *Laterispina parallela*, *Acosarina minuta*, *Fusichonetes pygmaea*, *Spinomarginifera alpha* and *Spinomarginifera chenyaoyanensis*, the bivalve *Guizhoupecten regularis*, and cephalopods *Pseudotirolites* sp., *Pernodoceras* sp. and *Huananoceras qingjiangensis* were recorded in the lower part. 170 cm
129. Tuffaceous, calcareous sandstone interbedded with tuffaceous siltstone, intercalated with volcanic ash in the upper part, with several layers of siliceous rock in the middle part, and with light gray thin-bedded limestone in the lower part. Cephalopods *Pleuronodoceras* sp. and *Rotodiscoceras* sp., brachiopods *Stenoscisma* sp. and *Anchorhynchia sarciniformis*, and the conodont *Clarkina yini* were examined in the lower part. 360 cm
128. Tuffaceous sandstone (mostly covered by the Quaternary sediments). 280 cm
127. Gray thin-bedded limestone intercalated with dark gray calcareous shale (deformed). 490 cm
126. Grayish black to gray medium-bedded limestone intercalated with mudstone. 60 cm
125. Gray thin-bedded mudstone intercalated with thin-bedded argillaceous limestone (limestones thickening upwards). 570 cm
124. Thin-bedded mudstone and calcareous mudstone in the upper part; yellow to grayish green thin-bedded tuff in the lower part. The bivalve *Clararia* sp. was collected at the basal part. 360 cm
123. Gray thick-bedded limestone with cherty nodules (or nodules of cherty limestone), with volcanic ash at the top. Conodonts *Clarkina meishanensis*, *Clarkina yini* and *Hindeodus praeparvus*, the brachiopod *Leptodus nobilis*, the

fusulinid *Colaniella parva* and the non-fusulinid foraminifer *Palaeofusulina sinensis* were recorded. 190 cm

122. Grayish green tuff, with fine-grained calcareous sandstone at the top and basal parts. The fusulinid *Colaniella parva* and the conodont *Clarkina yini* were identified. 60 cm
121. Grayish green tuff in the upper part; yellowish brown mudstone (10 cm) overlain by coal seams (3 cm) in the middle part; limestone in the lower part (10 cm). The non-fusulinid foraminifer *Colaniella parva* and the fusulinid *Palaeofusulina sinensis* were recovered. 95 cm
120. Gray thick-bedded limestone with chert nodules, with volcanic ash at the top. Conodonts *Clarkina yini, Hindeodus eurypyge*, and *Hindeodus praeparvus*, and the non-fusulinid foraminifer *Palaeofusulina sinensis* were recorded. 435 cm
119. Pale volcanic ash interbedded with coal seams. 10 cm
118. Brownish gray thick-bedded medium-grained sandstone. 100 cm
117. Dark gray shale intercalated with thin-bedded siltstone, with a layer of carbonaceous mudstone in the upper part. The plant fossil *Gigantonoclea guizhoueusis* was obtained in the lower part. 450 cm
116. Gray thick-bedded calcareous sandstone, with parallel beddings and cross beddings at the basal part. Gastropods and crinoids were found. 287 cm
115. Grayish green thin-bedded sandstone interbedded with thin-bedded mudstone in the upper part; gray to brownish gray medium- to thin-bedded argillaceous siltstone in the lower part. 158 cm
114. Dark gray silty shale. 452 cm
113. Grayish black mudstone and shale, intercalated with brownish gray thick- to medium-bedded calcareous siltstone, with ripple marks. 417 cm
112. Brownish gray to light gray thick-bedded medium-grained sandstone, with scour channel at the basal part. 122 cm
111. Gray shale intercalated with calcareous siltstone in the middle and upper parts; brownish gray to grayish yellow thick-bedded medium-grained sandstone in the lower part. Abundant charcoal fossils were found at the top part. 167 cm
110. Brownish gray shale intercalated with dark gray thick- to medium-bedded siltstone. Lens of sandstone were intercalated in the basal part (20 cm above the base). 950 cm
109. Gray thick- to medium-bedded medium-grained sandstone intercalated with silty shale, with ripple marks at the top and cross beddings and scour channel in the sandstone. 257 cm
108. Light gray thick- to medium-bedded fine-grained sandstone intercalated with shale in the middle and upper parts; brownish gray silty shale intercalated with siltstone in the lower part. Low-angled cross beddings and ripple marks in the middle and upper parts. 194 cm
107. Brownish yellow shale intercalated with coal seams. 176 cm
106. Brownish gray thick- to medium-bedded siltstone, with lens of sandstone and scour channel at the basal part. 31 cm

105. Silty shale in the upper part; black coal seams, shale (4 cm), volcanic ash (5 cm), coal seam (4 cm) in descending order in the lower part. Abundant plant fossils *Gigantonoclea guizhouensis*, *Gigantopteris dictyophylloides*, *Gigantonoclea* cf. *lagrelii*, *Gigantonoclea meridionalis*, *Szecladia multinervia*, *Asterophyllites* sp., *Paracalamites stenocostatus*, *Pecopteris* (*Asterotheca*) *guizhouensis*, *Pecopteris* cf. *lingulata*, *Rajahia rigida*, *Rajahia minor*, *Compsopteris contracta*, *Compsopteris impairs*, *Cladophlebis ozakii*, *Pterophyllum*? sp., *Taeniopteris* sp., *Fascipteris* sp., *Rhipidopsis panii* and *Phylladoderma*? sp. were recorded in the silty shale. 113 cm
104. Gray shale with a layer of thick-bedded argillaceous siltstone at the top. 94 cm
103. Dark gray thick-bedded medium-grained sandstone. 82 cm
102. Siltstone at the top part (50 cm); gray to brownish yellow shale in the upper part; brownish yellow fine-grained sandstone in the middle part; gray to brownish yellow shale in the lower part. 425 cm
101. Thick-bedded medium-grained sandstone with low-angled cross beddings at the top part; medium- to thin-bedded fine-grained sandstone interbedded with shale in the middle and upper parts; siltstone in the lower part (intercalated with shale upwards). 425 cm
100. Gray to brownish yellow shale. 205 cm
99. Gray thick- to medium-bedded siltstone in the upper part; dark gray shale interbedded with thin-bedded siltstone in the middle part; brownish gray thick-bedded siltstone in the lower part (60 cm). 171 cm
98. Brownish gray to gray shale. 148 cm
97. Three cycles were deposited and each cycle includes: brownish gray shale interbedded with thick- to medium-bedded fine-grained sandstone, with ripple marks. 307 cm
96. Dark gray siltstone in the upper part; thick- to medium-bedded fine-grained sandstone intercalated with shale in the middle part; grayish green thick-bedded medium-grained sandstone in the lower part. 360 cm
95. Brownish gray thick- to medium-bedded sandstone in the upper part (charcoal fossils discovered); brownish gray shale and siltstone in the lower part. 629 cm
94. Brownish gray silty shale with lens of sandstone. 164 cm
93. Grayish black silty mudstone intercalated with thin-bedded fine-grained calcareous sandstone. 450 cm
92. Gray thick-bedded medium-grained sandstone, occasionally intercalated with thin-bedded sandstone, with ripple marks. 964 cm
91. Gray medium-grained sandstone, with lens (20–30 cm in the diameter for each len) of sandstone. 96 cm
90. Light gray thin-bedded silty mudstone and sideritic siltstone, with straight-crested ripple marks at the top. 815 cm
89. Grayish yellow argillaceous siltstone (with lens of sandstone), with a layer of grayish yellow fine-grained sandstone at the base (10 cm), and with straight-crested ripple marks at the top. 258 cm

88. Brownish yellow lens-shaped fine-grained sandstone intercalated with dark gray shale. 156 cm
87. Gray shale intercalated with thin-bedded siltstone. 359 cm
86. Grayish yellow to brownish gray silty mudstone with nodules of siderite. 789 cm
85. Dark gray silty mudstone interbedded with siltstone. 2282 cm
84. Dark gray mudstone (mostly covered by the Quaternary sediments). 2565 cm
83. Dark gray shale intercalated with siltstone, with pale claystone at the base. 2274 cm
82. Light gray to brownish gray shale intercalated with siltstone, with a layer (1 cm) of medium-grained sandstone at the top part (30 cm below the top). 1755 cm
81. Gray shale intercalated with gray thin-bedded fine-grained calcareous sandstone. 397 cm
80. Gray to grayish yellow shale (with lens of sandstone) intercalated with thin-bedded fine-grained sandstone. 1175 cm
79. Gray shale intercalated with gray thin-bedded fine-grained sandstone. 1146 cm
78. Dark gray shale intercalated with thin-bedded siltstone. 1031 cm
77. Light gray shale intercalated with grayish yellow sideritic siltstone, with a layer of dark gray shale in the middle part and with lens of sandstone in the upper part. 1076 cm
76. Brownish yellow to light gray shale intercalated with thin-bedded fine-grained sandstone. 3358 cm
75. Grayish black to dark gray shale. 702 cm
74. Dark gray shale, with a layer of fine-grained calcareous sandstone at the top (20 cm). 855 cm
73. Dark gray shale intercalated with thin-bedded fine-grained sandstone (five cycles formed); lens of sandstone and abundant fragments of plant fossils at the top. 1877 cm
72. Dark gray shale intercalated with thin-bedded fine-grained calcareous sandstone, with shale interbedded with thin-bedded fine-grained sandstone at the top part (100 cm). 1145 cm
71. Grayish yellow thick- to medium-bedded fine-grained sandstone with graded beddings and cross beddings at the top part (sole mark on the top surface); dark gray shale intercalated with grayish yellow thin-bedded fine-grained sandstone in the upper part; dark gray shale intercalated with siltstone in the middle and lower parts. 428 cm
70. Dark gray shale interbedded with grayish yellow thin-bedded siltstone, with a layer of fine-grained sandstone (with graded beddings) at the top part (10 cm). 286 cm
69. Grayish yellow thick- to medium-bedded medium-grained sandstone, with ripple marks on the top surface. 114 cm

68. Dark gray shale interbedded with fine-grained sandstone in the upper part; thick- to medium-bedded fine-grained sandstone intercalated with shale in the middle part (with trace fossil *Zoophycos*); dark gray shale interbedded with fine-grained sandstone in the lower part. Isolated ripple marks were recorded in the fine-grained sandstone of the upper and lower parts. 86 cm
67. Dark gray shale intercalated with lens of sandstone in the upper part; dark gray shale interbedded with gray thin-bedded fine-grained sandstone in the middle and lower parts. 286 cm
66. Grayish brown thin-bedded fine-grained sandstone intercalated with shale, with a layer of fine-grained sandstone in the middle part. Abundant fragments of plant fossils and asymmetric ripple marks were discovered in the sandstone. 228 cm
65. Grayish yellow thin-, medium- to thick-bedded fine-grained sandstone, with a layer of dark gray mudstone (20 cm) in the middle part and abundant fragments of plant fossils at the base. 281 cm
64. Dark gray shale intercalated with grayish yellow siltstone and with fine-grained sandstone, with laminations. 767 cm
63. Grayish yellow shale. 118 cm
62. Dark gray silty shale. 107 cm
61. Thin-bedded siltstone in the upper part; brownish gray shale intercalated with two layers (5 cm for each layer) of fine-grained sandstone in the lower part. 780 cm
60. Thin-bedded siltstone in the upper part; brownish yellow shale in the lower part. 470 cm
59. Thin-bedded siltstone in the upper part (deformed); gray thick- to medium-bedded fine-grained calcareous sandstone interbedded with brownish gray shale in the lower part. 396 cm
58. Light gray to brownish yellow shale, intercalated with thin-bedded siltstone and fine-grained sandstone. 583 cm
57. Light gray shale intercalated with thin-bedded siltstone, with grayish yellow thick- to medium-bedded fine-grained sandstone at the top. 326 cm
56. Four cycles were deposited and each cycle includes: fine-grained sandstone in the upper part; thin-bedded siltstone in the middle part; brownish gray shale in the lower part. 289 cm
55. Fine-grained sandstone at the top part (15 cm); thin-bedded siltstone in the upper part; brownish gray shale in the lower part. 389 cm
54. Brownish gray shale and silty shale intercalated with three layers of grayish yellow medium- to fine-grained sandstone. 655 cm
53. Gray shale, occasionally intercalated with brownish yellow thick- to medium-bedded fine-grained tuffaceous calcareous sandstone, with a layer of brownish gray thick-bedded fine-grained sandstone at the top. 1487 cm
52. Brownish yellow thick- to medium-bedded medium-grained tuffaceous sandstone. 48 cm

51. Brownish gray shale, with thin-bedded fine-grained sandstone and siltstone at the top, and respectively with a layer of grayish yellow thick- to medium-bedded fine-grained sandstone in the middle and basal parts. 399 cm
50. Brownish gray shale, with thin-bedded siltstone at the top part. 357 cm
49. Grayish yellow medium-bedded medium-grained sandstone intercalated with brownish gray shale in the upper part; thick- to medium-bedded medium-grained sandstone intercalated with shale in the lower part. 387 cm
48. Brownish gray shale, with thin-bedded siltstone and fine-grained sandstone at the top part. 149 cm
47. Grayish yellow medium- to thin-bedded medium-grained sandstone (thickening upwards). 36 cm
46. Brownish gray shale intercalated with siltstone in the upper part; grayish yellow thin-bedded medium- to fine-grained sandstone intercalated with a layer of shale in the lower part. Abundant fragments of plant fossils were discovered in the siltstone. 47 cm
45. Thin-bedded silty mudstone and siltstone in the upper part; brownish gray shale intercalated with siltstone in the lower part. 448 cm
44. Brownish gray thick- to medium-bedded medium- to fine-grained sandstone (thickening upwards), intercalated with two layers of shale. 83 cm
43. Brownish gray shale and grayish yellow sandstone (mostly covered by the Quaternary sediments). 1426 cm
42. Grayish green thin-bedded fine-grained tuffaceous calcareous sandstone interbedded with shale, with fragments of plant fossils at the base (8 cm). 226 cm
41. Grayish green to brownish gray fine-grained tuffaceous calcareous sandstone interbedded with shale. 238 cm
40. Grayish green medium- to thin-bedded fine-grained tuffaceous calcareous sandstone interbedded with shale, with graded beddings in the sandstone and ripple marks on the top surface of layer. 164 cm
39. Dark gray shale intercalated with gray thin-bedded fine-grained tuffaceous calcareous sandstone. 74 cm
38. Gray thick- to medium-bedded fine-grained tuffaceous calcareous sandstone, intercalated with thin-bedded fine-grained calcareous sandstone and siltstone in the middle part. 84 cm
37. Gray thin-bedded calcareous tuffaceous sandstone interbedded with dark gray shale. 143 cm
36. Grayish black to dark gray mudstone intercalated with thin-bedded silty limestone. 587 cm
35. Gray thick- to medium-bedded fine-grained calcareous sandstone. 136 cm
34. Dark gray mudstone, with gray thin-bedded limestone in the upper part. 1183 cm
33. Dark gray thin-bedded siliceous rock. Cephalopods *Pleuronodoceras carinatum* and *Pseudotirolites* cf. *laibinensis* were found. 1850 cm
32. Grayish black siliceous rock interbedded with mudstone, intercalated with grayish yellow thin-bedded siltstone and fine-grained sandstone.

Cephalopods *Pseudogastrioceras* sp., *Pleuronodoceras carinatum, Pseudotirolites* cf. *laibinensis, Pseudotirolites* sp., *Huananoceras* sp., *Tapashanites* sp. and *Sinoceltites* sp. were examined. 300 cm

31. Grayish black thin-bedded siliceous rock intercalated with dark gray thin-bedded limestone. Cephalopods *Pseudotirolites laibinensis, Pseudogastrioceras* sp., *Pernodoceras kwangsiense, Pleuronodoceras carinatum, Pleuronodoceras mapingensis, Tapashanites costatus, Tapashanites mingyuanxiaensis, Pachydiscoceras flexoplicatum, Rotodiscoceras dushanensis* and *Rotodiscoceras* sp., and the bivalve *Guizhoupecten regularis* were recorded. 475 cm
30. Dark gray thick- to medium-bedded siliceous rock intercalated with volcanic ash, intercalated with two layers (5 cm for each layer) of limestone in the middle part. Cephalopods *Pseudotirolites regularis, Tapashanites mingyuexiaensis, Pleuronodoceras dushanense, Pseudogastrioceras jiangxiense* and *Pseudogastrioceras* sp., the brachiopod *Prelissorhynchia pseudoutah*, and the bivalve *Guizhoupecten regularis* were recovered. 250 cm
29. Dark gray thin-bedded siliceous rock intercalated with pale volcanic ash, with the ammonoid *Tapashanites costatus.* 130 cm
28. Gray siltstone in the upper part; volcanic ash in the middle part; grayish black siliceous mudstone in the lower part; volcanic ash at the base. Cephalopods *Tapashanites* sp. and *Pleuronodoceras* sp. were examined. 25 cm
27. Grayish green thick- to medium-bedded siltstone. 125 cm
26. Gray thick- to medium-bedded siltstone interbedded with grayish yellow fine-grained sandstone. 225 cm
25. Grayish yellow thick- to medium-bedded fine-grained tuffaceous sandstone. 195 cm
24. Grayish yellow thick- to medium-bedded medium- to fine-grained tuffaceous sandstone (fining upwards). 145 cm
23. Grayish yellow thick- to medium-bedded coarse- to medium-grained tuffaceous sandstone. Cephalopods *Huananoceras perornatum, Huananoceras* cf. *qianjiangense, Tapashanites* sp., *Pseudotirolites* cf. *laibinensis, Pseudotirolites* sp., *Pleuronodoceras carinatum, Pleuronodoceras* cf. *guiyangensis, Pleuronodoceras mapingensis, Pleuronodoceras multinodosum, Pleuronodoceras radiatum, Pleuronodoceras robustum, Pleuronodoceras* sp. and *Rotodiscoceras dushanensis* were discovered. 190 cm
22. Dark gray medium- to thin-bedded siliceous rock, intercalated with grayish green calcareous tuffaceous siltstone and tuff respectively in the middle and upper parts; dark gray thin-bedded limestone in the lower part; tuffaceous sandstone at the basal part. Cephalopods *Pleuronodoceras carinatum, Pseudotirolites asiaticus, Pseudotirolites qianjiangensis, Pseudotirolites laibinensis, Huananoceras qianjiangense, Huananoceras perornatum, Tapashanites* sp. and *Sinoceltites* sp. were found. 458 cm

Ergou Member

21. Grayish black siliceous mudstone, with dark gray thin-bedded limestone at the basal part. Cephalopods *Lopingoceras guangdeense*, *Pseudotirolites laibinensis*, *Pseudotirolites radiaplicatis*, *Pseudotirolites asiaticus*, *Pseudotirolites acutus*, *Pseudotirolites anshunensis*, *Pseudogastrioceras* sp., *Pleuronodoceras dushanense*, *Rotodiscoceras* sp., *Tapashanites tenuicostatus* and *Longmenshanoceras* sp., and the brachiopod *Paracrurithyris pygmaea* were examined. 219 cm
20. Grayish black thin-bedded calcareous tuffaceous siltstone interbedded with grayish black calcareous shale. 109 cm
19. Grayish black siliceous mudstone in the middle and upper parts; grayish black thick- to medium-bedded siliceous limestone in the lower part. 229 cm
18. Grayish black siliceous mudstone and silty mudstone, intercalated with grayish black thin-bedded siliceous rock. 1021 cm
17. Dark gray medium- to thin-bedded siliceous rock interbedded with thin-bedded silty shale. Cephalopods *Mingyuexiaceras* cf. *honghense*, *Penglaites* sp., *Penglaites laibinensis*, *Penglaites costatus*, *Tapashanites mingyuexiaensis*, *Tapashanites costatus*, *Tapashanites* cf. *floriformis*, *Pseudotirolites asiaticus*, *Pseudotirolites* cf. *laibinensis*, *Pseudogastrioceras* sp., *Changhsingoceras* sp., *Huananoceras perornatum*, *Huananoceras* cf. *qianjiangense*, *Pernodoceras multinodosum* and *Pleuronodoceras* sp., and brachiopods *Permophricodothyris grandis*, *Permophricodothyris indica* and *Fusichonetes* sp. were obtained. 249 cm
16. Grayish black thin-bedded siliceous rock interbedded with silty shale and grayish green tuffaceous siltstone. Cephalopods *Mingyuexiaceras hongheense*, *Mingyuexiaceras changxingensis*, *Mingyuexiaceras radiatum*, *Pseudotirolites uniformis*, *Tapashanites chaotianense* and *Sinoceltites kwangsiense* were recorded. 396 cm
15. Dark gray to grayish green tuffaceous siltstone interbedded with silty shale, respectively intercalated with a layer of medium-bedded limestone in the middle and top parts. Cephalopods *Penglaites laibinensis*, *Changhsingoceras* sp., *Huananoceras qianjiangense* and *Qianjiangoceras multiseptum* were examined. 220 cm

Conformity
Heshan Formation in the Upper Permian

14. Dark gray thick- to medium-bedded limestone interbedded with siliceous rock. The ammonoid *Penglaites* sp. was discovered. 401 cm

Not described downwards
Note: Data of fossils in beds 120–142 are based on Shen et al. (2019), but updated in taxonomy in this book (Figs. 3.77, 3.78).

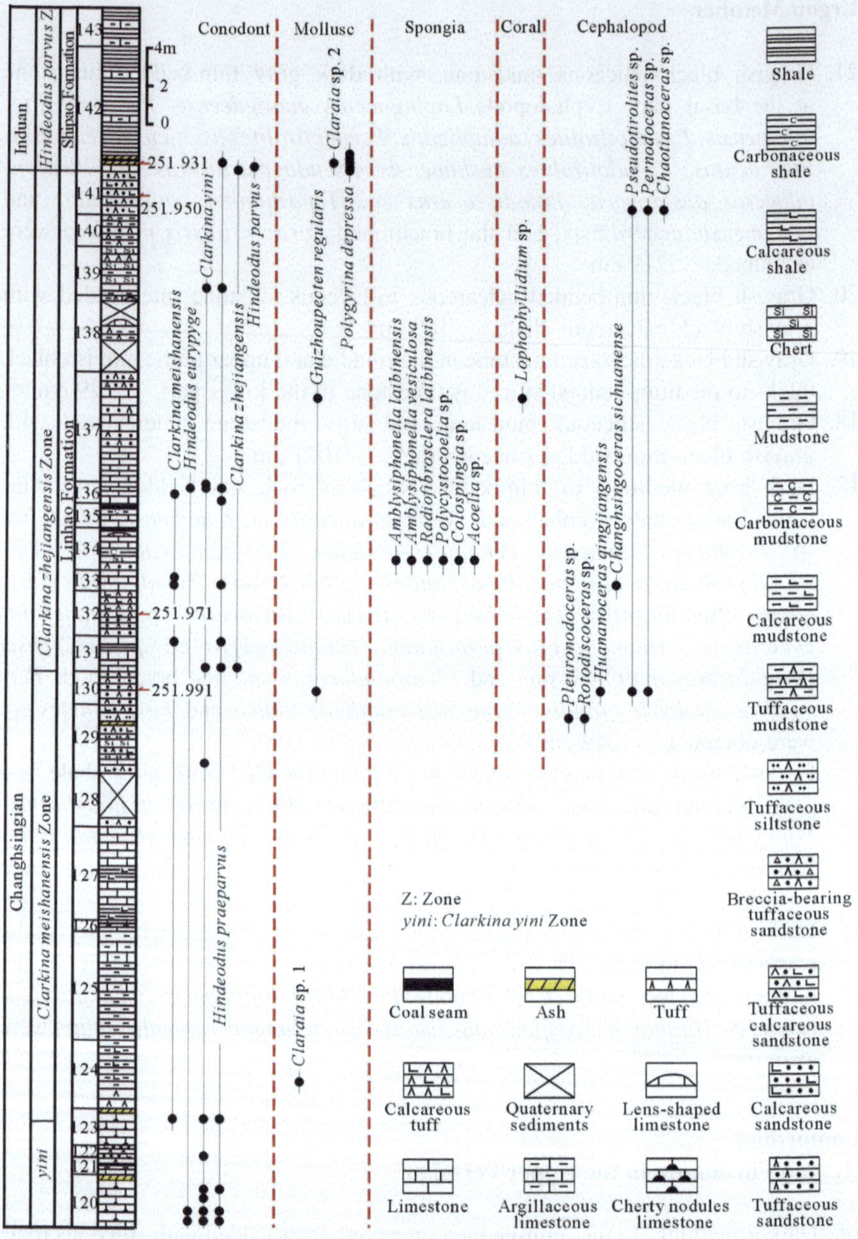

Fig. 3.77 Stratigraphic column and fossil occurrences (conodont, mollusca, spongia, coral, cephalopod) at the Penglaitan section (fossils and U–Pb date after Shen et al. 2019)

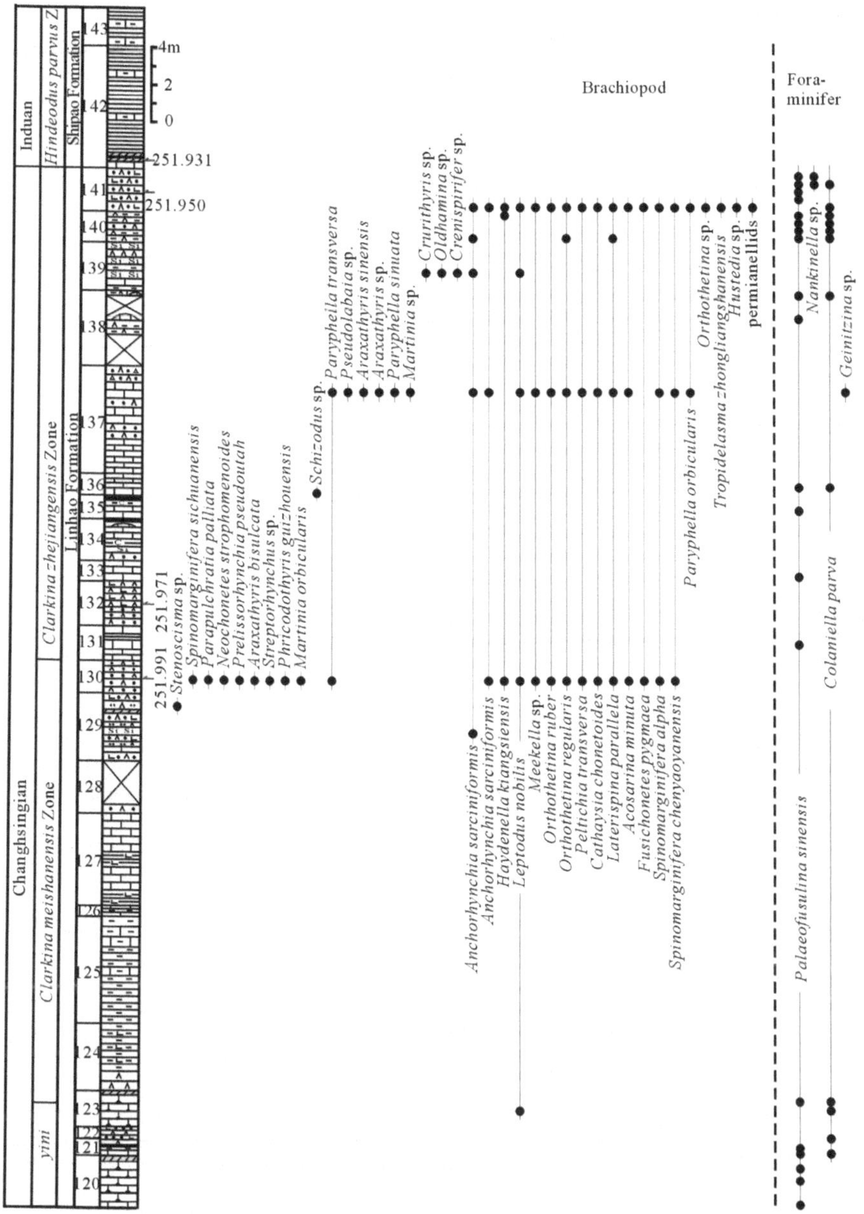

Fig. 3.78 Stratigraphic column and fossil occurrences (brachiopod, foraminifer) at the Penglaitan section. Note: fossils after Shen et al. (2019); legends and the abbreviation same to Fig. 3.77

References

Bu JJ, Wu SB, Zhang HL, Meng YY, Zhang F, Zhang LY (2006) Permian–Triassic cephalopods from Dongpan Section, Guangxi, and its geological significance. Geol Sci Technol Inf 25:47–51

Chen J, Henderson CM, Shen SZ (2008) Conodont succession around the Permian–Triassic Boundary at the Huangzhishan section, Zhejiang and its stratigraphic correlation. Acta Palaeontologica Sinica 47:91–114 [in English with Chinese abstract]

Chen ZQ, Tong JN, Zhang KX, Yang H, Liao ZT, Song HJ, Chen J (2009) Environmental and biotic turnover across the Permian–Triassic boundary on a shallow carbonate platform in western Zhejiang, South China. Aust J Earth Sci 56:775–797

Chu DL, Yu JX, Tong JN, Benton MJ, Song HJ, Huang YF, Song T, Tian L (2016) Biostratigraphic correlation and mass extinction during the Permian–Triassic transition in terrestrial–marine siliciclastic settings of South China. Glob Planet Change 146:67–88

Fang YH (2021) Late Permian to Early Triassic redox variations of South China and unusual biosedimentary responses. Doctoral Dissertation of China University of Geosciences, pp 1–145 [in Chinese with English abstract]

Feng QL, Gu SZ, Jiang ML, Jin YX (2004) Two new radiolarian genera from the uppermost Permian of southern China. Rev Micropaléontol 47:135–143

Feng QL, He WH, Zhang SX, Gu SZ (2006a) Taxonomy of order Latentifistularia (radiolaria) from the latest Permian in southern Guangxi, China. J Paleontol 80:826–848

Feng QL, Meng YY, He WH, Gu SZ (2006b) A new genus of Entactiniidae (Radiolaria) from the Upper Permian of South China. Eclogae Geol Helv 99:S67–S78

Feng QL, Gu SZ, He WH, Jin YX (2007) Latest Permian Entactinaria (Radiolaria) from southern Guangxi, China. J Micropalaeontol 26:19–37

Feng QL, Ye M, Crasquin S (2009) Latest Permian Palaeolithocycliidae (Radiolaria) from South China. Rev Micropaléontol 52:141–148

Gu SZ, Feng QL, He WH (2007) The last Permian deep-water fauna: latest Changhsingian small foraminifers from southwestern Guangxi, South China. Micropaleontology 53:311–330

Gui BW, Feng QL, Aihua Y (2009) Late Changhsingian (latest Permian) radiolarians from Chaohu, Anhui. J Earth Sci 20:797–810

He WH, Feng QL, Gu SZ, Jin YX (2005) Changxingian (Upper Permian) radiolarian fauna from Meishan D Section, Changxing, Zhejiang, China, and its possible paleoecological significance. J Paleontol 79:209–218

He WH, Feng QL, Weldon EA, Gu SZ, Meng YY, Zhang F, Wu SB (2007) A Late Permian to Early Triassic bivalve fauna from the Dongpan section, southern Guangxi, South China. J Paleontol 81:1009–1019

He WH, Zhang Y, Zheng YE, Zhang KX, Gui BW, Feng QL (2008) A late Changhsingian (latest Permian) radiolarian fauna from Chaohu, Anhui and a comparison with its contemporary faunas of South China. Alcheringa 32:199–222

He WH, Shi GR, Zhang Y, Yang TL, Zhang KX, Wu SB, Niu ZJ, Zhang ZY (2014) Changhsingian (latest Permian) deep-water brachiopod fauna from South China. J Syst Paleontol 12:907–960

He WH, Shi GR, Twitchett RJ, Zhang Y, Zhang KX, Song HJ, Yue ML, Wu SB, Wu HT, Yang TL, Xiao YF (2015) Late Permian marine ecosystem collapse began in deeper waters: evidence from brachiopod diversity and body size changes. Geobiology 13:123–138

He WH, Shi GR, Yang TL, Zhang KX, Yue ML, Xiao YF, Wu HT, Chen B, Wu SB (2016) Patterns of brachiopod faunal and body-size changes across the Permian–Triassic boundary: Evidence from the Daoduishan section in Meishan area, South China. Palaeogeogr Palaeoclimatol Palaeoecol 448:72–84

He WH, Shi GR, Zhang KX, Yang TL, Shen SZ, Zhang Y (2019) Brachiopods around the Permian–Triassic boundary of South China. Springer, Singapore, 261 pp

He WH, Weldon EA, Yang TL, Wang H, Xiao YF, Wu HT, Zhang KX, Wang YB, Wu SB (2020) The palaeoenvironmental and palaeobiogeographical significance of the Late Permian deep-water brachiopod fauna from Dongpan, South China, including descriptions of Micromartinia He & Weldon gen. nov. (Micromartiniidae He & Weldon fam. nov.) and Minutomarginifera nom. nov. J Syst Paleontol 18:885–909

Huang YF, He WH, Liao W, Wang YB, Yi ZX, Yang H (2022) Two pulses of increasing terrestrial input to marine environment during the Permian–Triassic transition. Palaeogeogr Palaeoclimatol Palaeoecol 586:110753. https://doi.org/10.1016/j.palaeo.2021.110753

Jiang HS, Lai XL, Luo GM, Aldridge RA, Zhang KX, Wignall PB (2007) Restudy of conodont zonation and evolution across the P/T boundary at Meishan section, Changxing, Zhejiang, China, environmental and biotic changes during the Paleozoic–Mesozoic transition. Glob Planet Change 55:39–55

Jiang HS, Lai XL, Sun YD, Wignall PB, Liu JB, Yan CB (2014) Permian–Triassic Conodonts from Dajiang (Guizhou, South China) and their implication for the age of microbialite deposition in the aftermath of the end-Permian mass extinction. J Earth Sci 25:413–430

Jin YX, Feng QL, Meng YY, He WH, Gu SZ (2007) Albaillellidae (radiolaria) from the latest Permian in southern Guangxi, China. J Paleontol 81:9–18

Li GS (2016) Fluctuations of redox conditions and ecological responses across the Permian–Triassic boundary in different paleogeographic settings of Middle & Lower Yangtze Region. Doctoral Dissertation of China University of Geosciences, pp 1–119 [in Chinese with English abstract]

Li ZJ, Zhu JN, Hu YF (1982) The new views on the correlation for the Late Permian coal-bearing strata in Junlian area, South Sichuan. J Stratigr 6:174–182 [in Chinese]

Li ZS, Zhan LP, Zhu XF, Zhang JH, Huang HQ, Xu DY, Yan Z, Li HM (1989) Study on the Permian–Triassic biostratigraphy and event stratigraphy of northern Sichuan and southern Shaanxi. Geological Publishing House, Beijing, 435 pp [in Chinese]

Liao W (2020) Palaeoenvironment change in shallow-marine carbonate platform across the Permian–Triassic boundary in South China and its possible cause. Doctoral Dissertation of China University of Geosciences, pp 1–92 [in Chinese with English abstract]

Liu H (2010) Ostracod fauna near the Permian–Triassic boundary in Chongyang, Hubei Province and its palaeoenvironmental significance. Dissertation submitted to China University of Geosciences for the Master Degree of Science, pp 1–52 [in Chinese with English abstract]

Mei SL, Zhang KX, Wardlaw BR (1998) A refined succession of Changhsingian and Griesbachian neogondolellid conodonts from the Meishan section, candidate of the global stratotype section and point of the Permian–Triassic boundary. Palaeogeogr Palaeoclimatol Palaeoecol 143:213–226

Shen SZ, He XL (1991) Brachiopod assemblages in the Changxing Formation of Zhongliangshan in Chongqing. J Stratigr 15:189–196 [in Chinese]

Shen SZ, Crowley JL, Wang Y, Bowring SA, Erwin DH, Sadler PM, Cao CQ, Rothman DH, Henderson CM, Ramezani J, Zhang H, Shen YA, Wang XD, Wang W, Mu L, Li WZ, Tang YG, Liu XL, Liu LJ, Zeng Y, Jiang YF, Jin YG (2011) Calibrating the end-Permian mass extinction. Science 334:1367–1372

Shen SZ, Ramezani J, Chen J, Cao CQ, Erwin DH, Zhang H, Xiang L, Schoepfer SD, Henderson CM, Zheng QF, Bowring SA, Wang Y, Li XH, Wang XD, Yuan DX, Zhang YC, Mu L, Wang J, Wu YS (2019) A sudden end-Permian mass extinction in South China. Geol Soc Am Bull 131:205–223

Song HJ, Tong JN, Chen ZQ (2009) Two episodes of foraminiferal extinction near the Permian–Triassic boundary at the Meishan section, South China. Aust J Earth Sci 56:765–773

Sun DY, Tong JN, Xiong YL, Tian L, Yin HF (2012) Conodont biostratigraphy and evolution across Permian–Triassic Boundary at Yangou Section, Leping, Jiangxi Province, South China. J Earth Sci 23:311–325

Sun X, Qiu XC, Tihelka E, Yang H, Sun DY, Tong JN, Tian L (2021) A diverse gastropod fauna from the shallow marine carbonate platform of the Yangou section (South China) in the immediate aftermath of the Permian–Triassic mass extinction. Geol J 56:6199–6209

Tian L, Tong JN, Sun DY, Xiong YL, Wang CG, Song HJ, Song HY, Huang YF (2014a) The microfacies and sedimentary responses to the mass extinction during the Permian–Triassic transition at Yangou Section, Jiangxi Province, South China. Sci China Earth Sci 57:2195–2207

Tian L, Tong JN, Song HJ, Liang L, Yang LR, Song HY, Wang CG, Zhao XM, Sun DY (2014b) Foraminferal evolution and formation of oolitic limestone near Permian–Triassic boundary at Yangou section, Jiangxi Province. Earth Sci J China Univ Geosci 39:1573–1586 [in Chinese with English abstract]

Wan JY (2021) Response of ostracod assemblages from microbialites in South China to the end-Permian mass extinction. Doctoral Dissertation of China University of Geosciences, pp 1–152 [in Chinese with English abstract]

Wan JY, Yuan AH, Crasquin S, Jiang HS, Yang H, Hu X (2019) High-resolution variation in ostracod assemblages from microbialites near the Permian–Triassic boundary at Zuodeng, Guangxi region, . Palaeogeogr Palaeoclimatol Palaeoecol 535.https://doi.org/10.1016/j.palaeo.2019.109349

Wang QX, Tong JN, Song HJ, Yang H (2009) Ecological evolution across the Permian/Triassic boundary at the Kangjiaping Section in Cili County, Hunan Province, China. Sci China Ser D Earth Sci 2009(52):797–806 [in Chinese with English abstract]

Wang T, Burne RV, Yuan AH, Wang YB, Yi ZX (2019) The evolution of microbialite forms during the Early Triassic transgression: a case study in Chongyang of Hubei Province, South China. Palaeogeogr Palaeoclimatol Palaeoecol 519:209–220

Wang H, He WH, Xiao YF, Yang TL, Zhang KX, Wu HT, Huang YF, Peng XF, Wu SB (2023) Stepwise collapse of biotic communities and its relations to oxygen depletion along the north margin of during the Permian–Triassic transition. Palaeogeogr Palaeoclimatol Palaeoecol.https://doi.org/10.1016/j.palaeo.2023.111569

Wu HT, He WH, Shi GR, Zhang KX, Yang TL, Zhang Y, Xiao YF, Chen B, Wu SB (2018) A new Permian–Triassic boundary brachiopod fauna from the Xinmin section, southwestern Guizhou, south China and its extinction patterns. Alcheringa 42:339–372

Wu J, Feng QL, Gui BW, Liu GC (2010) Some new radiolarian species and genus from Upper Permian in Guangxi Province, South China. J Paleontol 84:879–894

Yang H (2006) Calcimicrobialites after end-Permian mass extinction in South China and its implications of paleo-environment. These for Master Degree of China University of Geosciences, pp 1–61 [in Chinese with English abstract]

Yang H, Zhang SX, Jiang HS, Wang YB (2006) Age and general characteristics of calcimicrobialite near the Permian–Triassic Boundary in Chongyang, Hubei Province. Earth Sci J China Univ Geosci 31:165–170 [in Chinese with English abstract]

Yang LR, Song HJ, Tong JN, Chu DL, Tian L (2013) The extinction pattern of fusulinids during the Permian–Triassic crisis at the Kangjiaping section, Cili, Hunan Province. Micropalaeontol Sinica 30:353–366 [in Chinese with English abstract]

Yang TL, He WH, Zhang KX, Wu SB, Zhang Y, Yue ML, Wu HT, Xiao YF (2015) Palaeoecological insights into the Changhsingian–Induan (latest Permian–earliest Triassic) bivalve fauna at Dongpan, southern Guangxi, South China. Alcheringa 40:98–117

Yang ZY, Yin HF, Wu SB, Yang FQ, Ding MH, Xu GR (1987) Permian–Triassic boundary stratigraphy and fauna of South China. Geological Publishing House, Beijing, 378 pp [in Chinese with English abstract]

Yao A, Kuwahara K (1999) Middle–Late Permian Radiolarians from the Guangyuan–Shangsi Area, Sichuan Province, China. J Geosci Osaka City Univ 42:69–83

Yu JX, Broutin J, Lu ZS (2022) Plants and palynomorphs around the Permian–Triassic boundary of South China. Springer, Singapore, p 254

Yuan AH, Crasquin S, Feng QL, Gu SZ (2007) Latest Permian deep-water ostracods from southwestern Guangxi, South China. J Micropalaeontol 26:169–191

Yuan DX, Shen SZ (2011) Conodont succession across the Permian–Triassic boundary of the Liangfengya section, Chongqing, South China. Acta Palaeontologica Sinica 50:420–438 [in Chinese with English abstract]

Yuan DX, Shen SZ, Henderson CM, Chen J, Zhang H, Feng HZ (2014) Revised conodont-based integrated high-resolution timescale for the Changhsingian Stage and end-Permian extinction interval at the Meishan sections, South China. Lithos 204:220–245

Zhang F, Feng QL, He WH, Meng YY, Gu SZ (2006) Multidisciplinary stratigraphy across the Permian–Triassic boundary in deep-water environment of the Dongpan section, south China. Norw J Geol 86:125–131

Zhang KX, Tong JN, Shi GR, Lai XL, Yu JX, He WH, Peng YQ, Jin YL (2007) Early Triassic conodont-palynological biostratigraphy of the Meishan D Section in Changxing, Zhejiang Province, South China. Palaeogeogr Palaeoclimatol Palaeoecol 252:4–23

Zhang ZY, He WH, Zhang Y, Yang TL, Wu SB (2009) Late Permian–earliest Triassic ammonoid sequences from the Rencunping section, Sangzhi County, Hunan Province, South China and their regional correlation. Geol Sci Technol Inf 28:23–30 [in Chincse with English abstract]

Zhang N, Jiang HS, Zhong WL, Huang HH, Xia WC (2014a) Conodont biostratigraphy across the Permian–Triassic boundary at the Xinmin section, Guizhou, South China. J Earth Sci 25:779–786

Zhang Y, He WH, Shi GR, Zhang KX (2013) A new Changhsingian (Late Permian) Rugosochonetidae (Brachiopoda) fauna from the Zhongzhai section, southwestern Guizhou Province, South China. Alcheringa 37:223–247

Zhang Y, Shi GR, He WH, Zhang KX, Wu HT (2014b) A new Changhsingian (Late Permian) brachiopod fauna from the Zhongzhai section (South China), Part 2: Lingulida, Orthida, Orthotetida and Spiriferida. Alcheringa 38:480–503

Zhang Y, Zhang KX, Shi GR, He WH, Yuan DX, Yue ML, Yang TL (2014c) Restudy of conodont biostratigraphy of the Permian–Triassic boundary section in Zhongzhai, southwestern Guizhou Province, South China. J Asian Earth Sci 80:75–83

Zhang Y, He WH, Shi GR, Zhang KX, Wu HT (2015) A new Changhsingian (Late Permian) brachiopod fauna from the Zhongzhai section (South China) Part 3: Productida. Alcheringa 39:295–314

Zhang H, Cao CQ, Liu XL, Mu L, Zheng QF, Liu F, Xiang L, Liu LJ, Shen SZ (2016) The terrestrial end-Permian mass extinction in South China. Palaeogeogr Palaeoclimatol Palaeoecol 448:108–124

Zhao JK, Sheng JZ, Yao ZQ, Liang XL, Chen CZ, Rui L, Liao ZT (1981) The Changhsingian and the Permian-Triassic Boundary in South China. Nanjing Inst Geol Palaeontol Bull 2:1–112 [in Chinese with English abstract]

Age Analysis and Correlation

4

Song-Zhu Gu, Hai-Shui Jiang, G. R. Shi, Wei-Hong He, and Ting-Lu Yang

4.1 A Summary on the Stratigraphic Correlation Near the Permian–Triassic Boundary at the Well-Studied Meishan and Chahe Sections

In order to deepen our understanding of the dynamics of diversity changes and mass extinction across the Permian–Triassic boundary (PTB), the age analysis and stratigraphic correlation of the studied sections in varied palaeogeographic settings is essential for the study. Aiming for the age correlation, the Changhsingian depositional settings of South China could be subdivided into the terrestrial facies,

S.-Z. Gu
School of Earth Sciences, China University of Geosciences, 388 Luma Road, Wuhan, Hubei, China
e-mail: bossgu@cug.edu.cn

H.-S. Jiang · W.-H. He (✉)
State Key Laboratory of Biogeology and Environmental Geology, School of Earth Sciences, China University of Geosciences, 388 Luma Road, Wuhan, Hubei, China
e-mail: whzhang@cug.edu.cn

H.-S. Jiang
e-mail: jiangliuis@163.com

G. R. Shi
School of Earth, Atmospheric and Life Sciences, University of Wollongong, New South Wales, Wollongong, NSW 2522, Australia
e-mail: guang@uow.edu.au

T.-L. Yang
Faculty of Geosciences, East China University of Technology, 418 Guanglan Avenue, Nanchang, China
e-mail: yang@geology.hk

© The Author(s), under exclusive license to Springer Nature Singapore Pte Ltd. 2023
W. He et al. (eds.), *Stratigraphy Around the Permian–Triassic Boundary of South China*, New Records of the Great Dying in South China,
https://doi.org/10.1007/978-981-99-9350-5_4

terrestrial–marine transitional facies (=coastal marine), shallow-water marine clastic shelf facies (=clastic shelf), carbonate platform facies (including ramp), Isolated platform and reef facies, deep-water basin and slope facies, and rift basin facies (Fig. 4.1).

Much advance on biostratigraphy, carbon-isotope chemostratigraphy and geochronology has been made either in marine sequences or terrestrial facies in recent years. In order to analyze the age of sequences and define and correlate the location of the PTB at the studied sections, a summary of the age and stratigraphic subdivisions is given below for the well-studied shallow-marine Meishan and terrestrial-facies Chahe sections.

4.1.1 Meishan Section

At Meishan, three conodont zones have been established in the lowest Triassic, including *Hindeodus parvus*, *Isarcicella staeschei* and *Isarcicella isarcica* Zones in an ascending order (Zhang et al. 2007, 2009a) (Fig. 4.2). Four conodont zones are now recognized for the upper Changhsingian, based on a re-evaluation of the conodont taxonomy and the first occurrence for the index species (Yuan et al. 2014), comprising *Clarkina changxingensis*, *Clarkina yini*, *Clarkina meishanensis*, and *Hindeodus changxingensis*–*Clarkina zhejiangensis* Zones in an ascending order (Fig. 4.2). The ammonoid *Ophiceras* Zone found in bed 29 is regarded as the index fossil for the Induan, *Hypophiceras* Zone occurs in beds 25–28 (**'transitional bed'**), while ammonoids *Pleuronodoceras*, *Rotodiscoceras*, *Pernodoceras* and *Pseudotirolites* (*Pleuronodoceras*/*Rotodiscoceras*/*Pernodoceras*/*Pseudotirolites* Zone) are commonly found in the upper Changhsingian (although *Pseudotirolites* occasionally and locally has also been found in the lower Changhsingian) (Fig. 4.2; see He et al. 2010, 2015, 2019; Jiang et al. 2018; Yang et al. 1987; Yin et al. 2001; Zhang et al. 2009b; Zhao et al. 1978). The bivalve *Claraia wangi* and *Claraia griesbachi* Zones have long been considered as index fossils for the Induan, as has been the *Hunanopecten exilis* Zone for the Late Permian (Fig. 4.2, see He et al. 2007; Yang et al. 1987, 2015; Yin 1985a, b).

Besides the important biostratigraphic indicators, the end-Permian mass extinction is featured by a pronounced and globally recognizable negative excursion of $\delta^{13}C_{carb}$ (or $\delta^{13}C_{org}$) (Cao et al. 2010; Chen et al. 2019; Grasby and Beauchamp 2008; Hermann et al. 2010; Shen et al. 2019a; Xie et al. 2007; Yang et al. 2011; Zhang et al. 2016), located at, respectively, bed 24e (for carbonate carbon isotope) and bed 25 (for organic carbon isotopes) (marked by the dotted line in Fig. 4.2). 18 cm above this conspicuous negative excursion of $\delta^{13}C_{carb}$, the actual PTB (at base of bed 27c) is also found coinciding with a modest shift in $\delta^{13}C$ from positive to negative values (marked by dash lines in Fig. 4.2).

At the Meishan section, two high-precision U–Pb geochronologic dates have been published, including an age of 251. 941±0.037 Ma for bed 25 and 251. 880±0.031 Ma for bed 28 (Fig. 4.2, see Shen et al. 2019b).

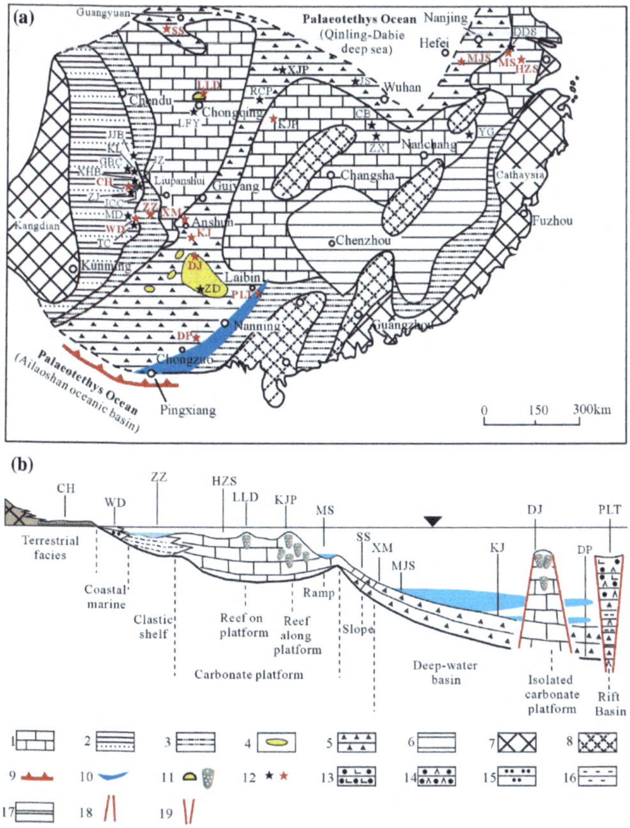

Fig. 4.1 Changhsingian lithofacies and depositional settings for the studied sections in South China. **a**—Index map for lithofacies palaeogeography of South China (Changhsingian) (revised after He et al. 2019, 2020); **b**—Cross section for the varied depositional settings. *Note* 1—Shallow-water marine carbonate platform (limestone); 2—Terrestrial facies (e.g., feldspathic sandstones, siltstones); 3—Shallow-water marine clastic shelf (e.g., siltstone and mudstone); 4—Isolated marine carbonate platform (limestone); 5—Deep-water marine siliceous basin; 6—Shallow-water restricted siliceous basin; 7—Ancient land mass; 8—Submerged rise; 9—Subduction of Palaeothys Ocean and possible subduction direction; 10—Rift basin; 11—Reef and microbialite; 12—Studied sections (abbreviation for section names sees Appendix A; red stars in (**a**) representing some typical sections marked in (**b**)); 13—Coarse-grained calcareous sandstone in rift basin; 14—Coarse-grained tuffaceous sandstone in rift basin; 15—Coastal marine silty facies; 16—Mudstone in rift basin; 17—Shale in rift basin; 18—Syndepositional faults; 19—Faults along margins of rift basin. Coastal marine = terrestrial − marine transitional facies in Chap. 2; ramp—a slope along the carbonate platform (shallow-water slope); slope—a slope downward to basin (deep-water slope); blue represents anoxic conditions and light blue represents dysoxic conditions (referred to Fig. 3.4 of He et al. 2019)

The lithological succession of the PTB interval at the Meishan section is also of note. In particular, the set of rock layers between bed 25 to bed 28 is highly distinctive and basically stable throughout South China, regardless of depositional

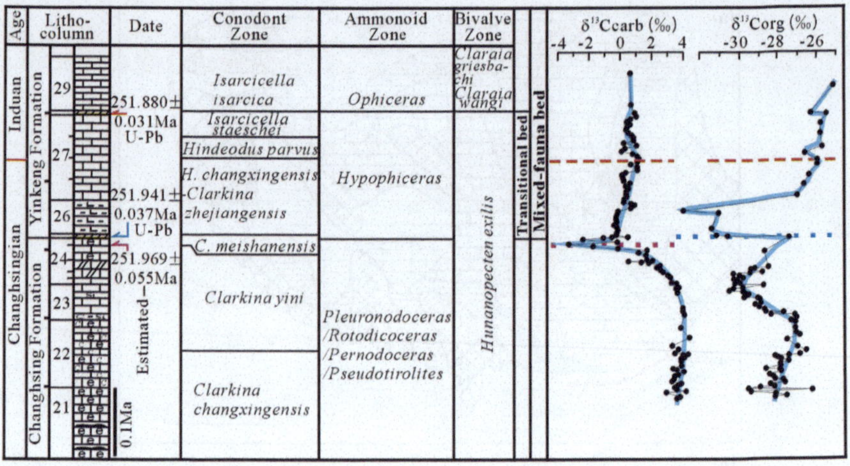

Fig. 4.2 Biostratigraphic zones, carbon isotopic profiles and U–Pb dates at the Meishan GSSP of the PTB. *Note* Conodont zones referring to Yuan et al. (2014) and Zhang et al. (2007, 2009a), carbon isotopes based on Shen et al. (2011), framework of stratigraphic column according to Shen et al. (2011), U–Pb date after Shen et al. (2019b); legends for lithologies same as in Fig. 3.32

facies. This set of rock layers was historically described as the **'transitional bed'** (Yang et al. 1991; Zhao et al. 1981), but later it was more formally defined and characterized and named as the 'Permian–Triassic Boundary Set' (PTBS) by Peng et al. (2001). In a similar vein, the interval from bed 25 to bed 32, spanning the *Clarkina meishanensis* Zone to the basal part of *Isarcicella isarcica* Zone, has been referred to as the **'mixed-fauna bed'** (e.g., Chen et al. 2005; Zhang et al. 2009a), and this informal stratigraphic term is also used here in this book for the sake of age correlation and consistency with past literature. If the definitions are followed, the upper boundary for the **'mixed-fauna bed'** is slightly younger than the counterpart for the **'transitional bed'**.

4.1.2 Chahe Section

The stratigraphic division across the PTB at the Chahe section has been better constrained compared to other sections of terrestrial facies, owing to the extensive biostratigraphic studies and recent advances in U–Pb geochronology and carbon-isotope chemostratigraphy (see below). The plant fossil *Tomiostrobus–Germaropteris* Zone associated with relics of gigantopterids (combined with data of the Tucheng section), could be regarded as the index fossils for the **'mixed-fauna bed'** (Chu et al. 2016; Feng et al. 2020a, b; Yu et al. 2007; Zhang et al. 2016). Besides, the conchostracan *Euestheria gutta – Palaeolimnadia xuanweiensis* Zone could broadly indicate the **'transitional bed'** in South China

(Chu et al. 2016, 2019; Scholze et al. 2020). Furthermore, the location of the **'transitional bed'** or PTBS at Chahe and its comparability with the Meishan section is supported by a negative excursion of the organic carbon isotopic curve (Fig. 4.3). The correlation of beds 68–71 of Chahe with bed 25 of Meishan is indicated by a rapid negative shift (marked by a dotted line) of organic isotope, as well as the U–Pb date of 252. 30 ± 0.07 Ma in bed 68 (Fig. 4.3; Shen et al. 2011). Bed 80 at Chahe is estimated to mark the PTB, based on a rapid positive shift (marked by a dash line) following a negative excursion of organic isotopes (Fig. 4.3). The underlying *Gigantonoclea*/*Gigantopteris*/*Pecopteris*/*Annularia*/*Compsopteris* Zones could be considered as the typical fossils for the Late Permian (Yu et al. 2007).

Additionally, in the terrestrial-marine transitional facies, the bivalve *Pteria* is found in the **'mixed-fauna bed'** and *Eumorphotis–Towapteria* Zone discovered in the Lower Triassic Induan in South China and can be adopted as biostratigraphic indicators for the correlation of PTB interval (Song et al. 2019).

4.2 Age Analysis and Correlation of Permian–Triassic Boundary Beds in the Other Studied Sections

Detailed fossil occurrences and the lithologic features of the other 30 PTB sections are already described in Chap. 3. Based on the index fossils, carbon isotope excursions, volcanic events and their comparisons with the well constrained Meishan and Chahe sections (see Sect. 4.1), a brief discussion is given below on age and correlation for each of these 30 sections.

4.2.1 Terrestrial Facies

At Tucheng, the plant fossil *Gigantonoclea–Annularia–Pecopteris* Zone is established in beds 1–16 and *Tomiostrobus–Germaropteris* Zone associated with relics of gigantopterids is recognized in beds 17–23 (Figs. 3.27, 4.4). The *Gigantonoclea–Annularia–Pecopteris* Zone is an index zone for the age of Changhsingian (see Fig. 4.3) and *Tomiostrobus–Germaropteris* Zone associated with relics of gigantopterids is typical of the **'mixed-fauna bed'** (Fig. 4.3). Thus, the Xuanwei Formation (beds 1–16) is assigned to the Changhsingian (Upper Permian) and the Kayitou Formation (beds 17–23) is roughly equivalent to the **'mixed-fauna bed'** of the Tucheng (Fig. 4.4).

At Mide, the plant fossil *Gigantonoclea–Gigantopteris* Zone is discovered in beds 1–18 and *Tomiostrobus–Germaropteris* Zone in beds 21 and 22 (Figs. 3.23, 4.4). The bivalve and ammonoid *Towapteria–Ophiceras* Zone is found in beds 23–28 (Figs. 3.22, 4.4). Therefore, the Xuanwei Formation (beds 1–20) basically belongs to the Changhsingian, while the basal part (beds 21 and 22) of the Kayitou Formaton may be equated to the lower part of the **'mixed-fauna bed'** or the

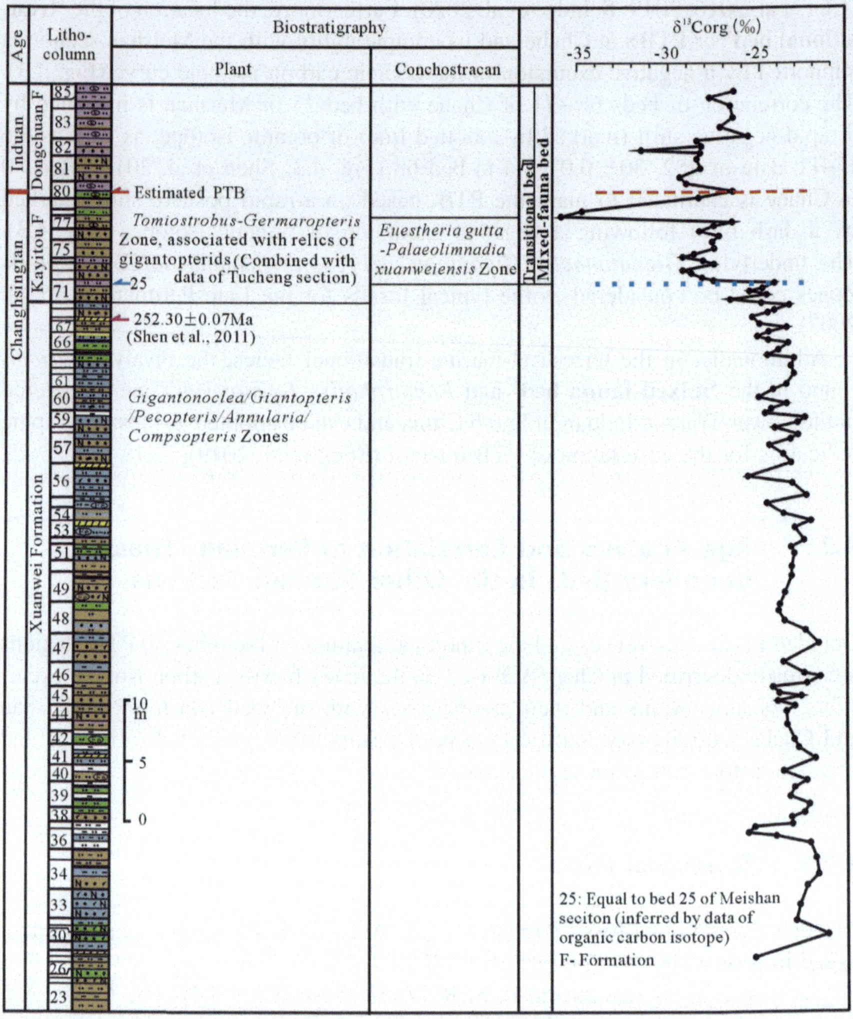

Fig. 4.3 Biostratigraphic and carbon-isotopic chemostratigraphic subdivisions at the Chahe section. *Note* Plant fossils from Chu et al. (2016) and Yu et al. (2007), conchostracan based on Chu et al. (2019) and Scholze et al. (2020), carbon isotopes from Zhang et al. (2016); other legends same as in Fig. 3.12

'transitional bed', and its other part (beds 23–28) to the Induan (Lower Triassic). Thus, the PTB is placed between beds 22 and 23 at Mide (Fig. 4.4).

At Zhejue, the plant fossil *Gigantonoclea–Gigantopteris–Pecopteris* Zone was yielded in the Xuanwei Formation and the basal part of the Kayitou Formation (Fig. 3.18), suggesting the age is of Changhsingian for the described sequence at this section (Fig. 4.4).

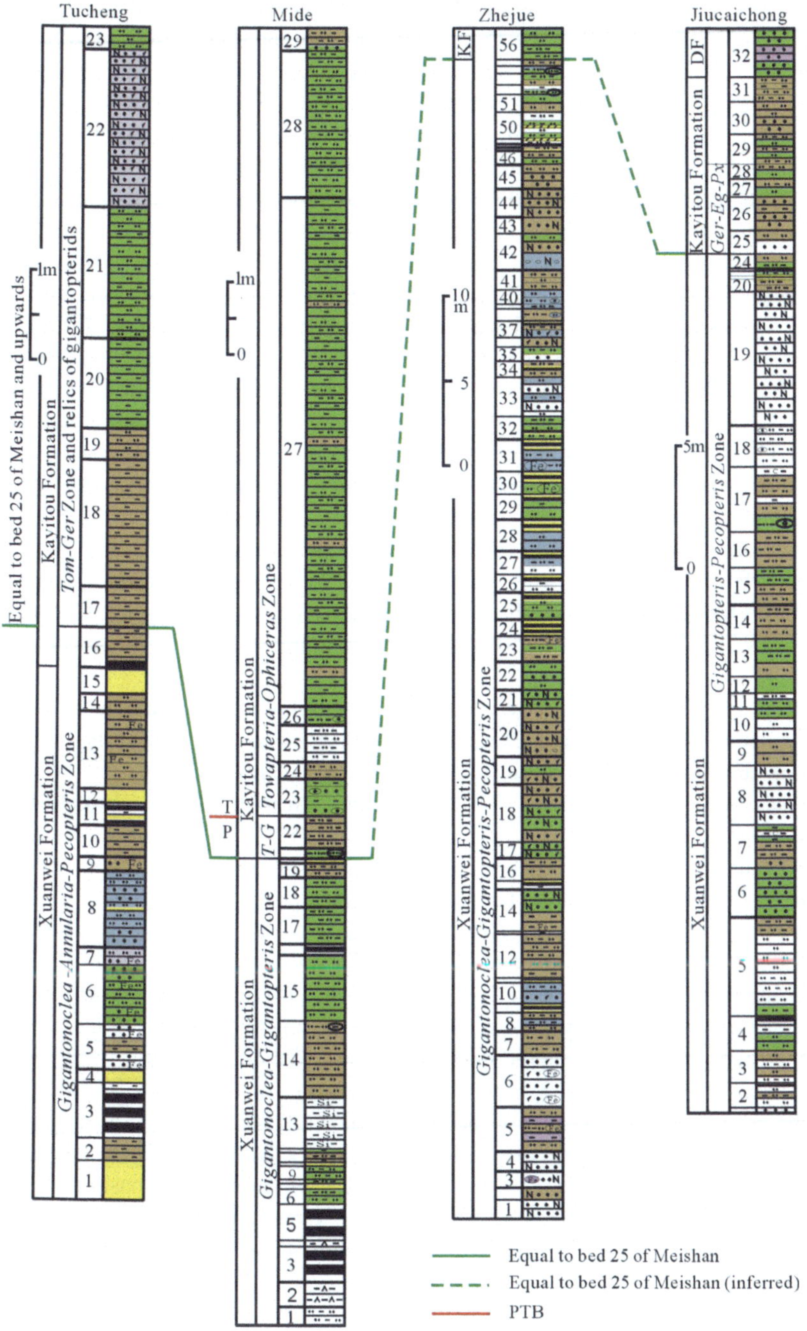

Fig. 4.4 Stratigraphic correlation of the studied sections in terrestrial facies (Tucheng, Mide, Zhejue, Jiucaichong sections). *Note* Columns of Tucheng, Mide, Zhejue and Jiucaichong sections respectively based on Figs. 3.27, 3.22, 3.17, 3.20; legends and abbreviations at Tucheng, Mide, Zhejue, Jiucaichong respectively same to Figs. 3.27, 3.22, 3.17, 3.20

At Jiucaichong, the plant fossil *Gigantopteris–Pecopteris* Zone is established in beds 1–24 (Figs. 3.20, 4.4). The combined plant and conchostracan fossil *Germaropteris–Euestheria gutta–Palaeolimnadia xuanweiensis* Zone are recognized in beds 25–28 (Figs. 3.20, 4.4). Thus, the Xuanwei Formation (beds 1–24) can be correlated with the Changhsingian while the lower part of the Kayitou Formation (beds 25–28) is equivalent to the **'transitional bed'** (Fig. 4.4).

At Jinzhong, the plant fossil *Gigantonoclea–Gigantopteris* Zone is discovered in beds 1–7, plant fossil *Tomiostrobus–Germaropteris* Zone in bed 8 and the lower part of bed 9, and the bivalve *Eumorphotis* Zone in the upper part of bed 9 to bed 15 (Figs. 3.15, 4.5). Therefore, the Xuanwei Formation (beds 1–7) is assigned to the Changhsingian, while the basal part of the Kayitou Formation (bed 8 and the lower part of bed 9) is correlated to the lower part of the **'mixed-fauna bed'** or the **'transitional bed'**, with the remainder of the Kayitou Formation (upper part of bed 9 to bed 17) assigned to the Induan. The PTB is placed in the middle part of bed 9 at Jinzhong (Fig. 4.5).

At Xiaohebian, the plant fossil *Gigantonoclea–Gigantopteris–Pecopteris* Zone is found in beds 1–9. The combined plant and conchostracan fossil *Germaropteris–Euestheria gutta–Palaeolimnadia xuanweiensis* Zone is recognized in beds 10–16 (Figs. 3.10, 4.5). As a result, the Xuanwei Formation together with the basal Kayitou Formation (beds 1–9) can be ascribed to the Changhsingian and the rest of the Kayitou Formation (beds 10–16) assigned to the **'transitional bed'** (Fig. 4.5).

At Kele, the plant fossil *Gigantonoclea–Gigantopteris–Compsopteris* Zone is recognized in beds 1–36 and *Tomiostrobus* is discovered in bed 37 (Figs. 3.8, 4.5). Hence, the Xuanwei Formation (beds 1–36) is assigned to the Changhsingian and the Kayitou Formation (beds 37–43) roughly corresponds to the **'mixed-fauna bed'** (Fig. 4.5).

At Guanbachong, the plant fossil *Gigantonoclea–Pecopteris* Zone occurs in beds 1–17 (Figs. 3.6, 4.5). The combined plant and conchostracan fossil *Germaropteris–Euestheria gutta–Palaeolimnadia xuanweiensis* Zone is established in beds 18 and 19 (Figs. 3.6, 4.5). Thus, the Xuanwei Formation (beds 1–17) can be correlated to the Changhsingian, and the lower part of the Kayitou Formation (beds 18 and 19) is roughly equivalent to the **'transitional bed'** (Fig. 4.5). Furthermore, based on the organic carbon isotope excursions, a rapid negative shift is evident at the top part of bed 19 (corresponding to bed 25 of Meishan), followed by a shift from positive to negative trend (relevant to the PTB at Meishan) detected in the basal part of bed 21 (Fig. 3.6; see Zhang et al. 2016), suggesting that the interval from the upper of the Kayitou Formation to the basal part of the Dongchuan Formation may be attributed to the **'transitional bed'** and the PTB is located in the basal part of bed 21 at the section (Fig. 4.5).

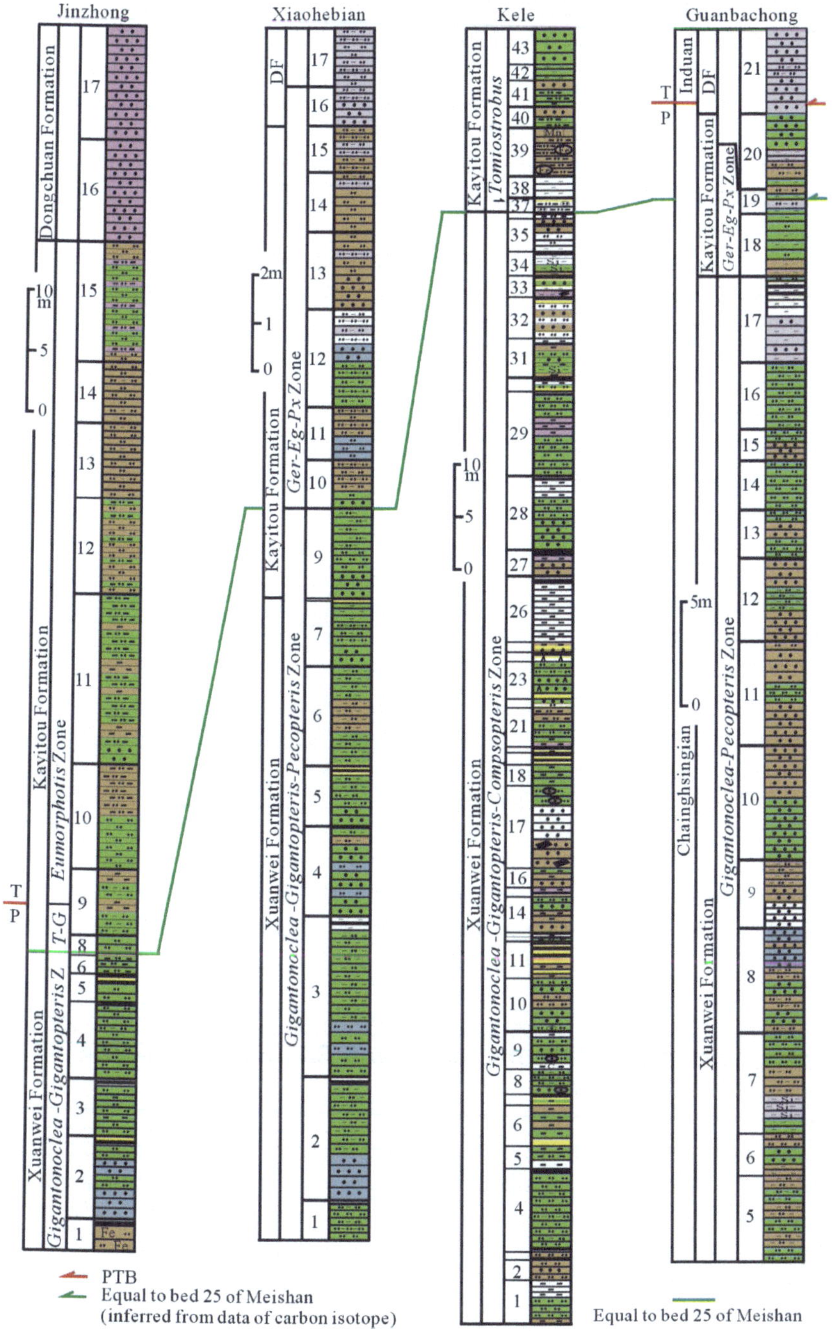

Fig. 4.5 Stratigraphic correlation of the studied sections in terrestrial facies (Jinzhong, Xiaohebian, Kele, Guanbachong sections). *Note* Columns of Jinzhong, Xiaohebian, Kele and Guanbachong sections respectively based on Figs. 3.15, 3.10, 3.8, 3.5; legends and abbreviations at Jinzhong, Xiaohebian, Kele, Guanbachong respectively same as in Figs. 3.15, 3.10, 3.8, 3.5

4.2.2 Terrestrial–Marine Transitional Facies and Shallow-Water Marine Clastic Shelf Facies

At Wadu, the plant fossil *Gigantonoclea–Gigantopteris–Pecopteris* Zone is recognized in beds 1–17, the bivalve *Pteria* Zone found in bed 18 to the lower part of bed 19, and *Pteria–Eumorphotis* Zone in the middle part of bed 19–22 (Figs. 3.25, 4.6). Thus the Xuanwei Formation (beds 1–17) is assignable to the Changhsingian, the basal part of the Feixianguan Formation (bed 18 to the lower part of bed 19) corresponds to the **'transitional bed'**, and the remainder of the Feixianguan Formation (middle part of bed 19 and upwards) is assigned to the Induan (Fig. 4.6). The PTB is placed in bed 19 at Wadu (Fig. 4.6).

At Jinjibang, the plant fossil *Gigantopteris–Pecopteris* Zone was discovered in beds 1–31 (Figs. 3.2, 3.3, 4.6). The combined plant and bivalve fossil *Tomiostrobus–Eumorphotis* Zone is recognized in beds 32–33 (Figs. 3.3, 4.6). Therefore, the Xuanwei Formation (beds 1–31) is correlated with the Changhsingian while the basal part of the Feixianguan Formation (beds 32 and 33) is assigned to the **'mixed-fauna bed'** or Induan (Fig. 4.6).

At Zhongzhai, the abundant presence of the brachiopod *Fusichonetes pygmaea* in beds 1–27 of the section, typical for the Changhsingian in South China (Fig. 3.30; He et al. 2019), suggests that these intervals should have been assigned to the Changhsingian. The U–Pb age in bed 29 of the section is of 252.24 ± 0.13 Ma and basically equivalent to the calibrated PTB age of 252.17 ± 0.06 Ma in bed 27c of Meishan (Fig. 4.6; Shen et al. 2011). Also, there is a similarity in lithological sequence and equivalence to the Permian–Triassic Boundary Stratigraphic Set (PTBS) (see Peng et al. 2001), which has been widely recognized and considered as an important indicator for the PTB correlation in South China. That is to say, bed 29 is a volcanic ash, bed 30 is limestone and the basal part of bed 31 is a volcanic ash at Zhongzhai (Fig. 4.6). The structure is basically matched with the counterpart (bed 25-ash, bed 27-limestone, bed 28-ash) at Meishan (Figs. 4.2, 4.6). Based on the correlation, the PTB is reliably placed in the middle part of limestone of bed 30 at Zhongzhai (Fig. 4.6).

4.2.3 Carbonate Platform Facies and Ramp of Carbonate Platform Facies

At Liangfengya, the conodont *Clarkina changxingensis*, *Clarkina yini* and *Clarkina meishanensis* Zones occur in ascending order in the Changhsing Formation, *Hindeodus praeparvus* was found in the upper part of bed 30, and *Hindeodus parvus* was discovered in bed 34 (Figs. 3.44, 4.7; see Yuan and Shen 2011). Hence, the Changhsing Formation (beds 1–32) is roughly assigned to the Changhsingian, the basal part of the Feixianguan Formation (beds 34–36) to the Induan, and the PTB is placed between beds 33 and 34 (Fig. 4.7).

At Chibi, the foraminferal *Colaniella* Zone, typical for the Changhsingian (Song et al. 2009a, 2009b), is known in beds 1–9 and the combined ammonoid

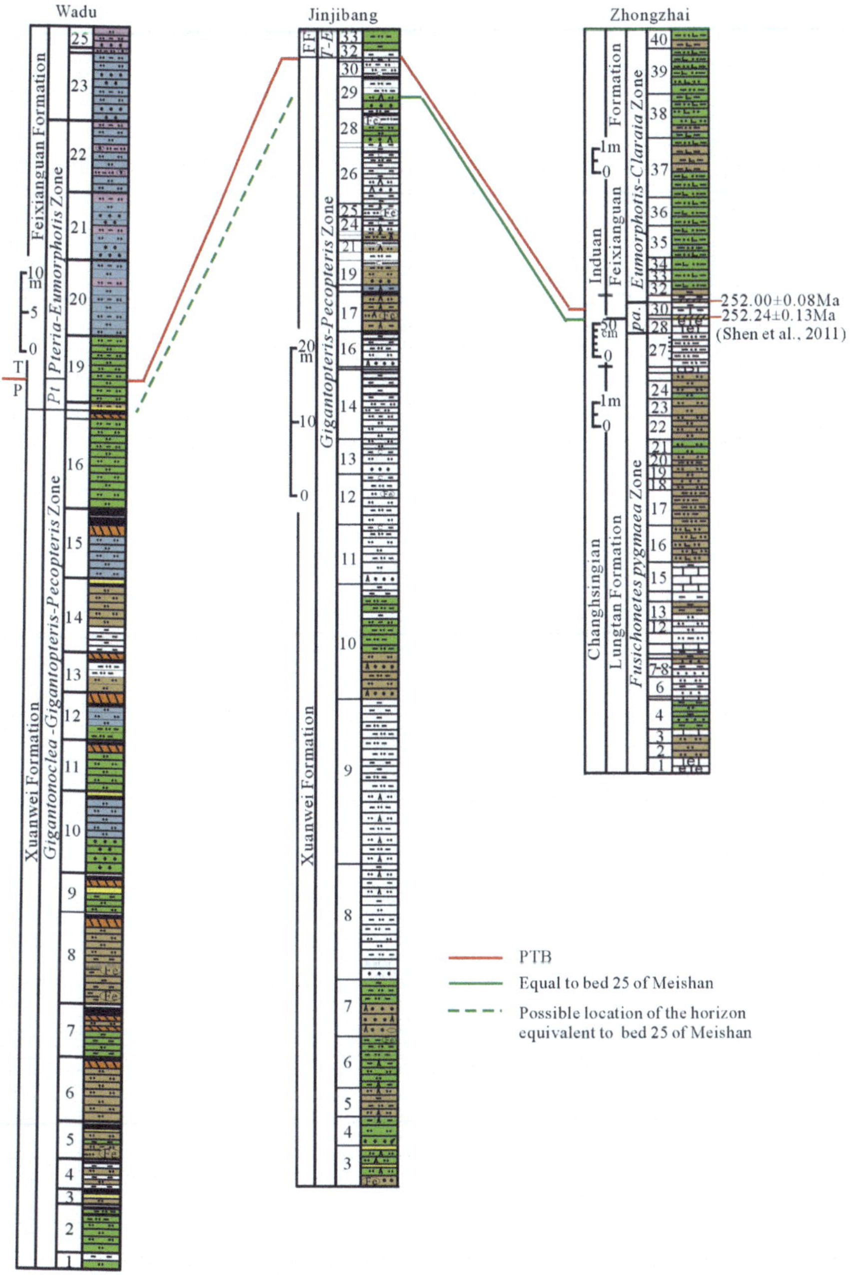

Fig. 4.6 Stratigraphic correlation of the studied sections in terrestrial–marine transitional facies (Wadu, Jinjibang) and shallow-water marine clastic shelf (Zhongzhai). *Note* Columns of Wadu, Jinjibang and Zhongzhai sections respectively based on Figs. 3.25, 3.2, 3.29; legends and abbreviations at Wadu, Jinjibang, Zhongzhai respectively same as in Figs. 3.25, 3.2, 3.29

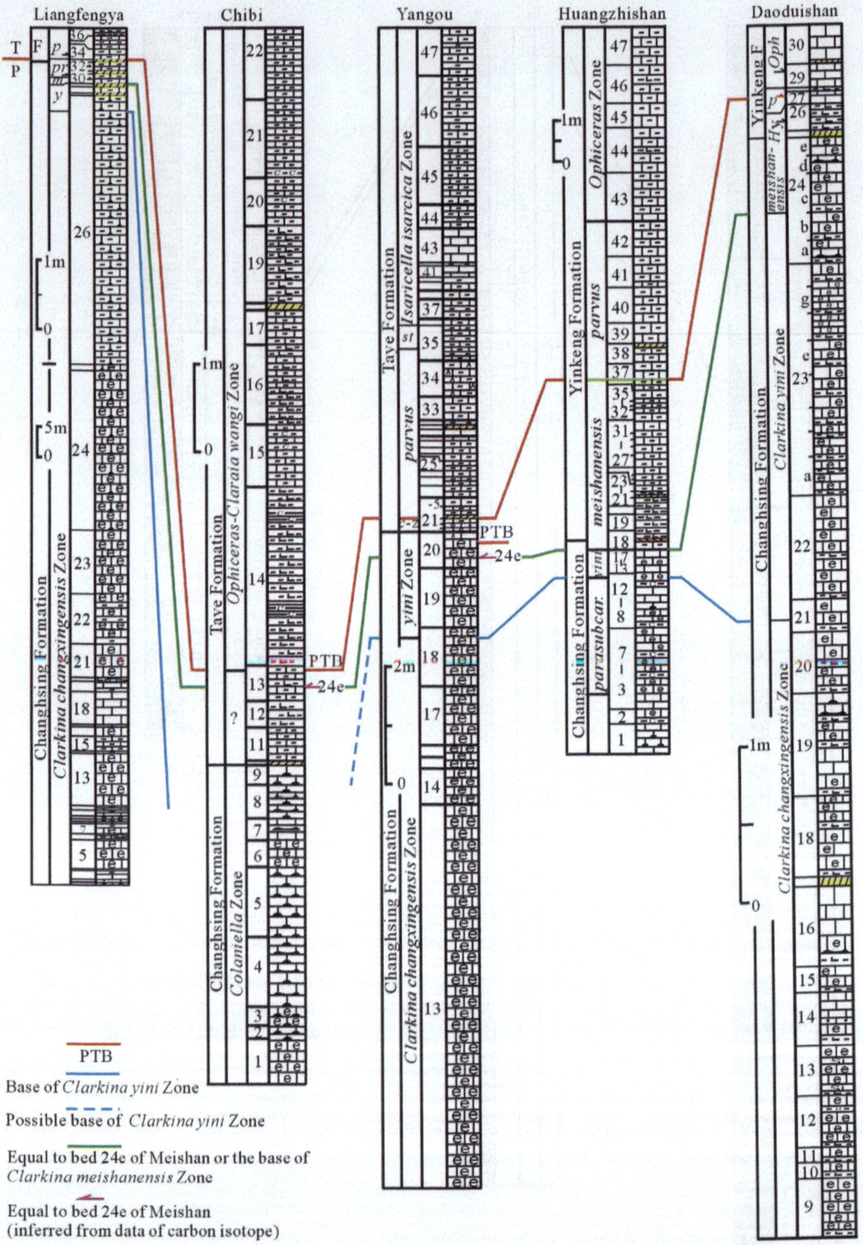

Fig. 4.7 Stratigraphic correlation of the studied sections in the carbonate platform facies. *Note* Columns of Liangfengya, Chibi, Yangou, Huangzhishan and Daoduishan sections respectively based on Figs. 3.44, 3.42, 3.39, 3.36, 3.33; legends and abbreviations at Liangfengya, Chibi, Yangou, Daoduishan, Huangzhishan respectively same as in Figs. 3.44, 3.42, 3.39, 3.36, 3.33

and bivalve *Ophiceras – Claraia wangi* Zone is found at the top part of bed 13 to bed 22 (Figs. 3.42, 4.7). Thus, the Changhsing Formation here is correlated with the Changhsingian, while most of the lower part of the Taye Formation (from the top part of bed 13 to bed 22) is assigned to the Induan (Fig. 4.7). Furthermore, the excursions of carbonate carbon isotope reveal that the upper part of bed 13 is equivalent to bed 24e of the Meishan section (Figs. 3.42, 4.7).

At Yangou, the conodont *Clarkina changxingensis* and *Clarkina yini* Zones were established in beds 13–20 ascendingly (Fig. 3.40). The conodont *Hindeodus changxingensis–Clarkina zhejiangensis* Zone is found in beds 21-1 to 21-3 (Figs. 3.40, 4.7). Conodonts *Hindeodus parvus, Isaricella staeschei* and *Isaricella isarcica* Zones were discovered from bed 21-4 to bed 39 in an ascending order (Figs. 3.40, 4.7; see Sun et al. 2012). Therefore, the Changhsing Formation (beds 13–20) can be assigned to the Changhsingian, the basal part of the Taye Formation (beds 21-1 to 21-3) is basically equated with the **'transitional bed'**, and other part of the Taye Formation is assigned to the Induan. The PTB is put between beds 21-3 and 21-4 (Fig. 4.7). This is despite that the $\delta^{13}C_{carb}$ excursions at this section suggest that the **'transitional bed'** here (in the middle part of bed 20) is slightly lower than the counterpart indicated by the biostratigraphic markers (Figs. 3.40, 4.7; see Tian et al. 2014). This preferential treatment in favour of biostratigraphic signals for this section is considered appropriate because (1) this book emphasizes the importance of biostratigraphic markers and (2) there are multiple excursions in the carbon isotope curves at the Yangon section (Fig. 3.40), rendering precise correlations with the Meishan section difficult.

At Huangzhishan, the conodonts *Clarkina yini* and *C. meishanensis* Zones were respectively found in the top part of the Changhsing Formation (beds 13–17) and in the basal part of the Yinkeng Formation (beds 18–36) (Figs. 3.37, 4.7; see Chen et al. 2008). *Hindeodus parvus* Zone (in beds 37–42), together with *Ophiceras* Zone (in bed 43 and upwards), was found in the lower Yinkeng Formation (Figs. 3.36, 3.37, 4.7; see He et al. 2015). Therefore, the PTB has been defined at the horizon between beds 36 and 37 (Fig. 4.7).

At Daoduishan, the conodonts *Clarkina changxingensis, C. yini* and *C. meishanensis* Zones were respectively found in bed 14 to base of bed 21, in the upper part of bed 21 to bed 24b and in beds 24c to 26 of the Changhsing Formation (Figs. 3.34, 4.7; see He et al. 2019). *Hindeodus parvus* was found in the middle of bed 27 and *Ophiceras* sp. in bed 29 (Figs. 3.34, 4.7; see He et al. 2019). Therefore, the PTB has been placed in the middle of bed 27 at Daoduishan (Fig. 4.7).

4.2.4 Isolated Platform and Reef Facies

At Zuodeng, the conodont *Hindeodus parvus* is discovered in the lower part of bed 21 (Figs. 3.54, 4.8). The $\delta^{13}C_{carb}$ excursions indicate bed 5 at Zuodeng could be correlated with bed 24e of Meishan and the PTB should therefore be defined in the

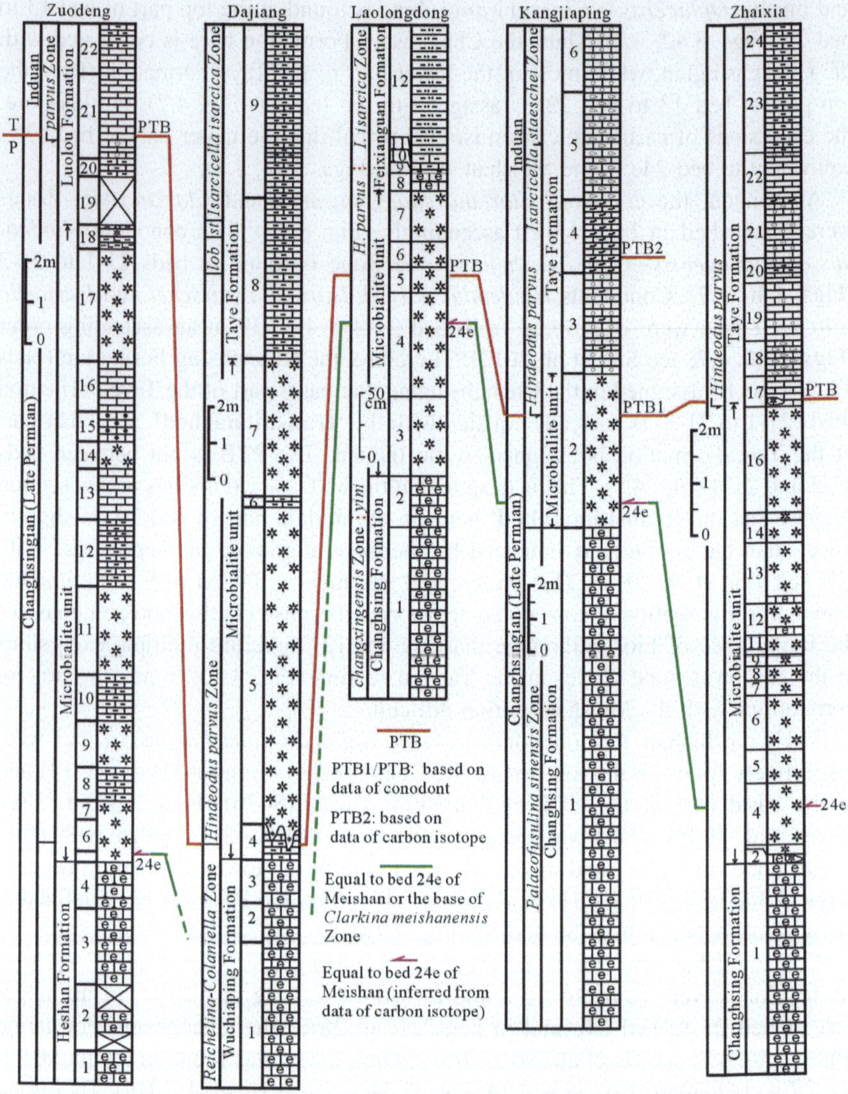

Fig. 4.8 Stratigraphic correlation of the studied sections from isolated platform and reef facies. *Note* Columns of Zuodeng, Dajiang, Laolongdong, Kangjiaping and Zhaixia sections respectively based on Figs. 3.54, 3.52, 3.50, 3.48, 3.46; legends and abbreviations at Zuodeng, Dajiang, Laolongdong, Kangjiaping, Zhaixia respectively same as in Figs. 3.54, 3.52, 3.50, 3.48, 3.46

lower part of bed 21 (Fig. 3.54), which is basically consistent with the first occurrence of *Hindeodus parvus* (Figs. 3.54, 4.8). The occurrence of the microbialite corresponds to bed 24e and upwards to below the PTB of Meishan (Fig. 4.8).

At Dajiang, the foraminiferal *Reichelina–Colaniella* Zone, typical for the Changhsingian (Song et al. 2009a, b), is found in bed 1 to the lower part of bed

4 (Figs. 3.52, 4.8). The conodont *Hindeodus parvus*, *Isarcicella lobata*, *Isarcicella staeschei* and *Isarcicella isarcica* Zones were discovered from the upper part of bed 4 to bed 9 in ascending order, suggesting the Induan (Figs. 3.52, 4.8; see Jiang et al. 2014). Thus, the PTB is put in the middle part of bed 4 at Dajiang (Fig. 4.8). The occurrence of the microbialite is roughly equivalent to the *Hindeodus parvus* Zone (earliest Triassic) (Fig. 4.8).

At Laolongdong, the conodont *Clarkina changxingensis* and *Clarkina yini* Zones were roughly established in the Changhsing Formation and the *Hindeodus parvus* Zone was found in the microbialite (basal part of Feixianguan Formation) (Figs. 3.50, 4.8; see Ezaki et al. 2003). The conodont *Isarcicella isarcica* Zone was found in the argillaceous limestone (overlying microbialite) of the Feixianguan Formation (Figs. 3.50, 4.8). Hence, the PTB here must be within the microbialite unit (Fig. 4.8). The $\delta^{13}C_{carb}$ excursions suggest that the upper part of bed 4 to bed 5 is basically correlated with bed 24e to the PTB of Meishan section, and the PTB is thus located in the upper part of bed 5 at Laolongdong (Figs. 3.50, 4.8; see Huang et al. 2022). Compared to the Meishan section and many other PTB sections in South China, the microbialite unit at Laolongdong seems to have been deposited over a longer period of time, beginning from a horizon most probably equivalent to bed 24 (i.e., below bed 24e) of Meishan to a level apparently above the PTB (Fig. 4.8).

At Kangjiaping, the fusulinid *Palaeofusulina sinensis* Zone, typical for the late Changhsingian (Shen et al. 2019a; Zhao et al. 1981), is established in the Changhsing Formation (Figs. 3.48, 4.8). The conodont *Hindeodus parvus* was found in the upper part of the microbialite unit (the basal part of Taye Formation) and *Isarcicella staeschei* in the calcirudite (bed 5) of the Taye Formation (younger than the microbialite) (Figs. 3.48, 4.8; see Wang et al. 2009). Therefore, the PTB should be placed within the microbialite unit (Fig. 4.8). Additionally, a rapid negative shift of the carbonate carbon isotope curves occurred in the basal part of bed 2 (i.e., the basal part of microbialite unit) and indicates a correlation of this horizon to bed 24e of Meishan (Fig. 3.48; see Wang et al. 2009). The PTB suggested by the carbon isotope excursion cannot be definitively decided at this section but probably located in the lower part of bed 4 (marked by PTB2), which is prominently younger than the PTB indicated by the first occurrence of *Hindeodus parvus* (marked by PTB1) (Fig. 4.8). As this book adopts the PTB as defined by the first occurrence of *Hindeodus parvus*, PTB1 is thus preferred. Consequently, the microbialite unit is considered to correspond to bed 24e to bed 27 of Meishan (Fig. 4.8).

At Zhaixia, the conodont *Hindeodus parvus* was found in the basal part of bed 17 (Figs. 3.46, 4.8). The excursions of carbonate carbon isotopes reveal a prominent negative shift in bed 4 and then an overturn from positive to negative shift in the basal part of bed 17 (Figs. 3.46, 4.8). This may suggest that the upper part of bed 4 could be correlated to bed 24e of Meishan while the basal part of bed 17 correlated to the PTB (Fig. 4.8). Consequently, the deposition of the microbialite unit at Zhaixia would have begun from the horizon slightly below bed 24e of Meishan and ended at the PTB (Fig. 4.8).

4.2.5 Deep-Water Facies

At Dongpan, the ammonoid *Pernodoceras–Laibinoceras* Zone, is defined in the Talung Formation (Figs. 3.75, 4.9). Radiolarian *Albaillella yaoi*, typical for the late Changhsingian (Wu et al. 2010; Zhang et al. 2018), was found in beds 1–5 in the Talung Formation (Fig. 3.73). The ammonoid *Ophiceras* sp. and bivalve *Claraia wangi* were commonly recorded in bed 13 (the basal part of the Luolou Formation) (Fig. 3.75; Bu et al. 2006; He et al. 2007; Yang et al. 2015). Therefore, the Talung Formation should be assigned to the late Changhsingian while the basal Luolou Formation to the Induan. The study on the organic carbon-isotope excursions suggests that the basal part of bed 9 at Dongpan is equivalent to bed 24e of Meishan and the basal part of bed 13 of Dongpan covers the PTB (Fig. 3.73; see Zhang et al. 2006). Additionally, the U–Pb age dated the volcanic ash in the basal part of bed 13 as 251.953 ± 0.038 Ma (Fig. 4.9; Baresel et al. 2017). Without doubt, the PTB is located at the basal part of bed 13 at the section.

At Xinmin, the combined conodont and ammonoid *Clarkina wangi–Pseudotirolites* Zone is established in beds 12–19, indicating that the basal part of Talung Formation is of early Changhsingian in age (Figs. 3.68, 4.9). The ammonoid *Rotodiscoceras–Pernodoceeas* Zone is found in beds 20–46, suggesting that most of the Talung Formation is of late Changhsingian in age (Figs. 3.68, 4.9). The ammonoid *Ophiceras* sp. was found in bed 53 and upwards and thus indicates that the basal part of the Taye Formation is roughly correlated with the Induan (Fig. 3.68). Additionally, the conodont *Clarkina meishanensis* and *Hindeodus changxingensis* Zones were recognized in the top part of bed 44 to the lower part of bed 48 in an ascending order (Figs. 3.68, 4.9; see Zhang et al. 2014). The conodont *Hindeodus parvus* was found in the upper part of bed 48 (Figs. 3.68, 4.9; see Zhang et al. 2014), signaling the base of the Triassic at this level (Fig. 4.9).

At Kejiao, the ammonoid *Tapashanites* and *Pleuronodoceras–Pseudotirolites* Zones are established respectively in beds 10–32 and beds 33–70 (Figs. 3.71, 4.9). The ammonoid *Hypophiceras* is found in beds 72–74 and *Ophiceras* Zone started from bed 74 and extended upwards (Figs. 3.71, 4.9). It suggests that the age of the Talung Formation is of Changhsingian and the Luolou Formation is of Induan (Triassic), beds 71–75 is basically equivalent to the **'transitional bed'**, and the PTB is placed between beds 73 and 74 (Fig. 4.9).

At Shangsi, the ammonoid *Tapashanites* Zone is defined in beds 11–18, the combined ammonoid and conodont *Rotodiscoceras–Clarkina yini* Zone in beds 19–26 (Figs. 3.62, 3.63, 4.9). The conodont *Clarkina meishanensis* Zone was recognized in beds 27 to 28-1, *Hindeodus changxingensis* Zone in beds 28-2 to 29-2, *Hindeodus parvus* in beds 29-3 to 30-1, *Isarcicella lobata* Zone in beds 30-2 to 31-1, and *Isarcicella isarcica* Zone in beds 31-2 and upwards (Figs. 3.62, 4.9; see Jiang et al. 2011). Additionally, the U–Pb age dated the volcanic ash in bed 28-3 as 252.16 ± 0.06Ma (Shen et al. 2011). Therefore, beds 27 to 29-2 at Shangsi are correlated to the **'transitional bed'** and the PTB is placed within bed 28-3 (Fig. 4.9).

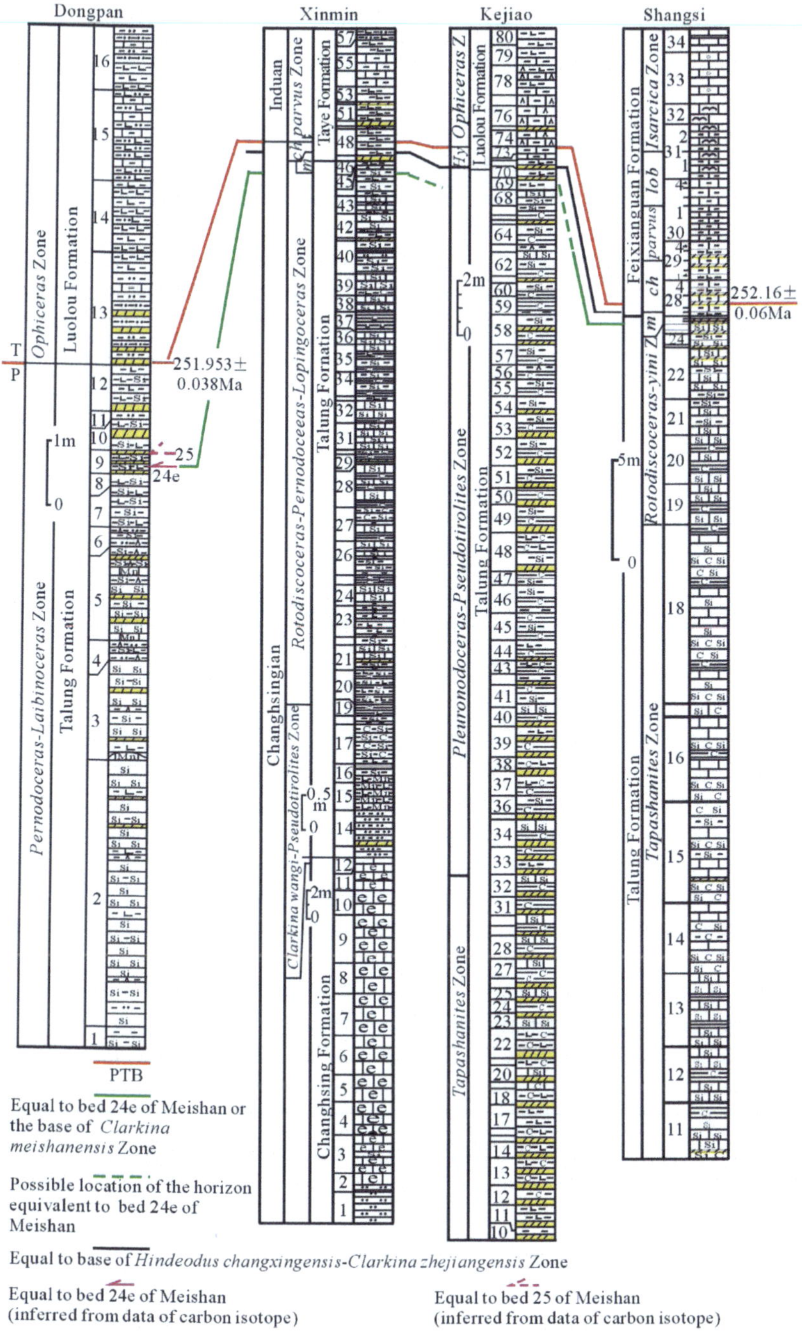

Fig. 4.9 Stratigraphic correlation of the studied sections from moderately deep- and deep-water facies (Dongpan, Xinmin, Kejiao, Shangsi). *Note* Columns of Dongpan, Xinmin, Kejiao and Shangsi sections respectively based on Figs. 3.73, 3.68, 3.71, 3.62; legends and abbreviations at Dongpan, Xinmin, Kejiao, Shangsi respectively same as in Figs. 3.73, 3.68, 3.71, 3.62

At Rencunping, the conodont *Clarkina changxingensis–Clarkina deflecta* and *Clarkina yini* Zones are respectively found from beds 18–21 and from bed 22 to the middle part of bed 23d (Figs. 3.66, 4.10). The conodont *Clarkina meishanensis* Zone is established in the upper part of bed 23d to bed 25. *Clarkina zhejiangensis* first appears in bed 26 (Figs. 3.66, 4.10). Additionally, ammonoids of the *Sinoceltites* Zone (equivalent to the *Tapashanites* Zone) and of the *Pleuronodoceras–Rotodiscoceras* Zone have been found, respectively, from bed 18 to the lower part of bed 20 and from the upper part of bed 21 to bed 24 at Rencunping (Figs. 3.66, 4.10; see Zhang et al. 2009b). Ammonoids of the *Sinoceltites* and *Pleuronodoceras–Rotodiscoceras* Zones, respectively, suggest ages of early and late Changhsingian (Yang et al. 1987). Ammonoids of the *Ophiceras* Zone, typical for the Induan (Yin et al. 2001), started from the upper part of bed 27 and upwards at Rencunping (Fig. 3.66; see Zhang et al. 2009b). Beds 18–24 at Rencunping, therefore, can be reliably assigned to the Changhsingian, and the upper part of bed 27 to bed 29 to the Induan (Triassic), with the PTB placed in the middle of Bed 27, as defined by the first appearance of *Ophiceras* sp. (Fig. 4.10).

At Xiejiaping, the ammonoid *Konglingites* Zone, typical for the Wuchiapingian (Yang et al. 1987; Zhao et al. 1978), was found in the Talung Formation (beds 2–9), suggesting a Wuchiapingian age for these beds (Figs. 3.60, 4.10). The ammonoid *Pleuronodoceras* Zone, typical for the late Changhsingian, was found in beds 17–24 in the Changhsing Formation of the Xiejiaping section (Figs. 3.60, 4.10). The combined ammonoid and conodont *Ophiceras–Isarcicella staeschei* Zone was found in beds 27–37 of the basal part of the Taye Formation (Figs. 3.60, 4.10). Additionally, the Late Permian ammonoid *Huananoceras* sp. and brachiopod *Fusichonetes pygmaea* were found in the Changhsing Formation (Fig. 3.60). These lines of evidence suggest that the middle and upper parts of the Changhsing Formation (beds 17–24) are assigned to the late Changhsingian and the lower part of the Changhsing Formation to the early Changhsingian (Fig. 3.60). The PTB is approximately placed at the base of bed 27 (Fig. 4.10).

At Jingshan, the combined bivalve and ammonoid *Claraia–Ophiceras* Zone is established in beds 35 and 36 and indicates the Taye Formation is of Induan in age (Figs. 3.58, 4.10). The trilobite *Pseudophillipsia subcircular* was discovered in bed 33 (Fig. 3.58, also see Li 2016).

The genus was regarded as one of the rare survivors into the latest Permian and became extinction in the end of the Permian (Fortey and Owens 1997). The top part of bed 34 is composed of chert, but chert disappeared prior to the Early Triassic generally. Based on the two lines of evidence it is inferred that beds 33 and 34 are of the latest Permian, not Triassic in age. Thus it is proper to put the PTB between beds 34 and 35 at Jingshan (Fig. 4.10).

At Majiashan, the ammonoid *Konglingites* Zone was found in bed 1, *Tapashanites* Zone in bed 3 to the basal part (20 cm) of bed 10, *Pseudotirolites–Rododiscoceras* Zone from the lower part of bed 10 to bed 16, and *Ophiceras* Zone in bed 20 (Figs. 3.56, 4.10; see He et al. 2015). The *Konglingites* Zone is typical for the Wuchiapingian, *Tapashanites* Zone for the early Changhsingian, and *Pseudotirolites–Rododiscoceras* Zone for the late Changhsingian (Yang et al.

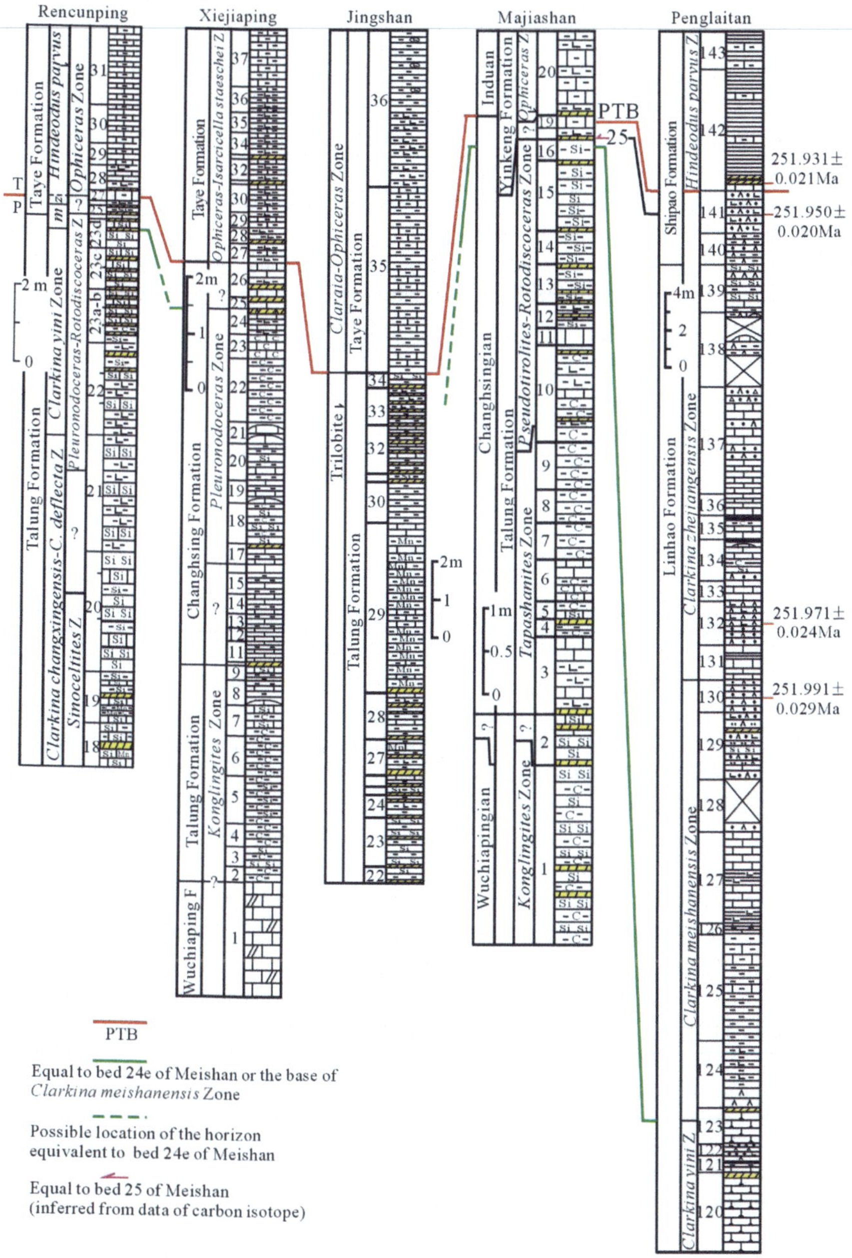

Fig. 4.10 Stratigraphic correlation of the studied sections from deep-water facies (Rencunping, Xiejiaping, Jingshan, Majiashan) and rift basin facies (Penglaitan). *Note* Columns of Rencunping, Xiejiaping, Jingshan, Majiashan and Penglaitan sections respectively based on Figs. 3.66, 3.60, 3.58, 3.56, 3.77; legends and abbreviations at Rencunping, Xiejiaping, Jingshan, Majiashan, Penglaitan respectively same as in Figs. 3.66, 3.60, 3.58, 3.56, 3.77

1987; Zhao et al. 1978). The *Ophiceras* Zone is an indicator for the Induan in age. The Talung Formation (beds 3–16), therefore, is assigned to the Changhsingian, and bed 20 to the Induan (Fig. 4.10). Although no index fossils have been found from beds 17–19, we place the PTB in the middle of bed 19, because of the similarity and equivalence to the PTBS (Peng et al. 2001), which has been precisely matched with the counterpart at Meishan (Fig. 4.10). Based on the correlation, bed 17 (pale volcanic ash) at Majiashan can be matched to bed 25 of Meishan (White clay). Likewise, bed 18 (dark green calcareous mudstones) at Majiashan can be correlated to bed 26 of Meishan (black calcareous mudstone), and bed 19 at Majiashan correlated to bed 27 of Meishan (limestone). The study of organic carbon isotopes well supports these correlations. A rapid negative shift occurred at the top of bed 16 to bed 18 and a reverse shift (from positive to negative excursion) took place in bed 19 (Fig. 3.56; see Fang 2021). The former indicates the horizon corresponding to bed 25 of Meishan and the latter suggests the location of the PTB at Majiashan (Fig. 4.10).

4.2.6 Rift Basin Facies

At Penglaitan, the conodont *Clarkina yini* Zone is established in bed 120 to the upper part of bed 123 and the *Clarkina meishanensis* Zone in the top part of bed 123 to bed 130 (Figs. 3.77, 4.10). The conodont *Clarkina zhejiangensis* Zone is discovered in beds 131–141 and the *Hindeodus parvus* Zone in bed 142 and upwards (Figs. 3.77, 4.10). The U–Pb geochronology dated bed 132 of Penglaitan as 251.971 ± 0.024 Ma, bed 141 as 251.950 ± 0.020 Ma, and bed 142 as 251.931 ± 0.021 Ma (Shen et al. 2019b). The first age is very close to the estimated age of bed 24e of Meishan (251.969 ± 0.055 Ma) (Fan et al. 2020). The latter two ages are close to the age of bed 25 of Meishan (251.941 ± 0.037 Ma) (Shen et al. 2019b). Therefore, the interval from beds 132–141 (or even basal part of bed 142) of Penglaitan is matched with beds 24e and 25 of Meishan (Fig. 4.10). This means that the thickness of the former (2437 cm or even thicker if considering the first occurrence of *Clarkina meishanensis*) is far greater than its counterpart at Meishan (19 cm), reflecting the difference in tectonic settings between the two areas. The PTB at Penglaitan lies at the level between beds 141 and 142, as defined by the first appearance of condont *Hindeodus parvus* (Fig. 4.10).

References

Baresel B, Bucher H, Brosse M, Cordey F, Kuang GD, Schaltegger U (2017) Precise age for the Permian–Triassic boundary in South China from high-precision U–Pb geochronology and Bayesian age–depth modeling. Solid Earth 8:361–378

Bu JJ, Wu SB, Zhang HL, Meng YY, Zhang F, Zhang LY (2006) Permian–Triassic Cephalopods from Dongpan Section, Guangxi, and its geological significance. Geol Sci Technol Inf 25:47–51 [in Chinese with English abstract]

Cao CQ, Yang YC, Shen SZ, Wang W, Zheng QF, Summons RE (2010) Pattern of δ^{13}Ccarb and implications for geological events during the Permian–Triassic transition in South China. Geol J 45:186–194

Chen ZQ, Kaiho K, George AD (2005) Survival strategies of brachiopod faunas from the end-Permian mass extinction. Palaeogeogr Palaeoclimatol Palaeoecol 224:232–269

Chen J, Henderson CM, Shen SZ (2008) Conodont succession around the Permian–Triassic boundary at the Huangzhishan section, Zhejiang and its stratigraphic correlation. Acta Palaeontologica Sinica 47:91–114 [in English with Chinese abstract]

Chen Y, Ye Q, Jiang HS, Wignall PB, Yuan JL (2019) Conodonts and carbon isotopes during the Permian–Triassic transition on the Napo Platform, South China. J Earth Sci 30:244–257

Chu DL, Yu JX, Tong JN, Benton MJ, Song HJ, Huang YF, Song T, Tian L (2016) Biostratigraphic correlation and mass extinction during the Permian–Triassic transition in terrestrial–marine siliciclastic settings of South China. Glob Planet Change 146:67–88

Chu DL, Tong JN, Benton MJ, Yu JX, Huang YF (2019) Mixed continental–marine biotas following the Permian–Triassic mass extinction in South and North China. Palaeogeogr Palaeoclimatol Palaeoecol 519:95–107

Ezaki Y, Liu JB, Adachi N (2003) Earliest Triassic microbialite micro- to mega structures in the Huaying area of Sichuan Province, South China: implications for the nature of oceanic conditions after the end-Permian extinction. Palaios 18:388–402

Fan JX, Shen SZ, Erwin DH, Sadler PM, MacLeod N, Cheng QM, Hou XD, Yang J, Wang DX, Wang Y, Zhang H, Chen X, Xiang G, Zhang YC, Shi YK, Yuan DX, Chen Q, Zhang LN, Li C, Zhao YY (2020) A high-resolution summary of Cambrian to Early Triassic marine invertebrate biodiversity. Science 367:272–277

Fang YH (2021) Late Permian to Early Triassic redox variations of South China and unusual biosedimentary responses. Doctoral Dissertation of China University of Geosciences, pp 1–145 [in Chinese with English abstract]

Feng Z, Wei HB, Guo Y, He XY, Sui Q, Zhou Y, Liu HY, Gou XD, Lv Y (2020a) From rainforest to herbland: new insights into land plant responses to the end-Permian mass extinction. Earth Sci Rev 204:103153

Feng Z, Wei HB, Ye RH, Sui Q, Gou XD, Guo Y, Liu LJ, Yang SL (2020b) Latest Permian peltasperm plant from southwest China and its paleoenvironmental implications. Front Earth Sci 8:559430. https://doi.org/10.3389/feart.2020.559430

Fortey RA, Owens, RM (1997) Evolutionary history. 249–288. In: Whittington HB, Chatterton BDE, Speyer SE, Fortey RA, Owens RM, Chang WT, Dean WT, Jell PA, Laurie JR, Palmer AR, Repina LN, Rushton AWA, Shergold JH, Clarkson ENK, Wilmot NV, Kelly SRA (1997) Treatise on invertebrate Paleontology, Part O, Trilobita (revised), volume 1: Introduction. The Geological Society of America, Inc. and the University of Kansas, Boulder, Colorado, Lawrence, Kansas

Grasby S, Beauchamp B (2008) Intrabasin variability of the carbon-isotope record across the Permian-Triassic transition, Sverdrup Basin, Arctic Canada. Chem Geol 253:141–150

He WH, Feng QL, Weldon EA, Gu SZ, Meng YY, Zhang F, Wu SB (2007) A Late Permian to Early Triassic bivalve fauna from the Dongpan section, southern Guangxi, South China. J Paleontol 81:1009–1019

He WH, Twitchett RJ, Zhang Y, Shi GR, Feng QL, Yu JX, Wu SB, Peng XF (2010) Controls on body size during the Late Permian mass extinction event. Geobiology 8:391–402

He WH, Shi GR, Twitchett RJ, Zhang Y, Zhang KX, Song HJ, Yue ML, Wu SB, Wu HT, Yang TL, Xiao YF (2015) Late Permian marine ecosystem collapse began in deeper waters: evidence from brachiopod diversity and body size changes. Geobiology 13:123–138

He WH, Shi GR, Zhang KX, Yang TL, Shen SZ, Zhang Y (2019) Brachiopods around the Permian–Triassic boundary of South China (261pp). Springer, Singapore

He WH, Weldon EA, Yang TL, Wang H, Xiao YF, Wu HT, Zhang KX, Wang YB, Wu SB (2020) The palaeoenvironmental and palaeobiogeographical significance of the Late Permian deep-water brachiopod fauna from Dongpan, South China, including descriptions of Micromartinia

He & Weldon gen. nov. (Micromartiniidae He & Weldon fam. nov.) and Minutomarginifera nom. nov. J Syst Paleontol 18:885–909
Hermann E, Hochuli PA, Bucher H, Vigran JO, Weissert H, Bernasconi SM (2010) A close-up view of the Permian–Triassic boundary based on expanded organic carbon isotope records from Norway (Trøndelag and Finnmark Platform). Glob Planet Change 74:156–167
Huang YF, He WH, Liao W, Wang YB, Yi ZX, Yang H, Li GS (2022) Two pulses of increasing terrestrial input to marine environment during the Permian–Triassic transition. Palaeogeogr Palaeoclimatol Palaeoecol 586.https://doi.org/10.1016/j.palaeo.2021.110753
Jiang HS, Lai XL, Yan CB, Aldridge RJ, Wignall PB, Sun YD (2011) Revised conodont zonation and conodont evolution across the Permian–Triassic boundary at the Shangsi section, Guangyuan, Sichuan, South China. Glob Planet Change 77:103–115
Jiang HS, Lai XL, Sun YD, Wignall PB, Liu JB, Yan CB (2014) Permian–Triassic conodonts from Dajiang (Guizhou, South China) and their implication for the age of microbialite deposition in the aftermath of the end-Permian mass extinction. J Earth Sci 25:413–430
Jiang Y, Wu HT, Xiao YF, Yang TL, He WH, Wang H, Sun X, Wu SB (2018) Ammonoid zonations of Permian–Triassic boundary strata from Kejiao section, Huishui county, southern Guizhou province, south China and regional correlation. Geol Sci Technol Inf 37:65–73 [in Chinese with English abstract]
Peng YQ, Tong JN, Shi GR, Hansen HJ (2001) The Permian–Triassic boundary set: characteristics and correlation. Newsl Stratigr 39:55–71
Scholze C, Shen SZ, Backer M, Wei HB, Hübner M, Cui YY, Feng Z, Schneider JW (2020) Reinvestigation of conchostracans (Crustacea: Branchiopoda) from the Permian–Triassic transition in Southwest China. Palaeoworld 29:368–390
Shen SZ, Crowley JL, Wang Y, Bowring SA, Erwin DH, Sadler PM, Cao CQ, Rothman DH, Henderson CM, Ramezani J, Zhang H, Shen YA, Wang XD, Wang W, Mu L, Li WZ, Tang YG, Liu XL, Liu LJ, Zeng Y, Jiang YF, Jin YG (2011) Calibrating the end-Permian mass extinction. Science 334:1367–1372
Shen SZ, Ramezani J, Chen J, Cao CQ, Erwin DH, Zhang H, Xiang L, Schoepfer SD, Henderson CM, Zheng QF, Bowring SA, Wang Y, Li XH, Wang XD, Yuan DX, Zhang YC, Mu L, Wang J, Wu YS (2019a) A sudden end-Permian mass extinction in South China. Geol Soc Am Bull 131:205–223
Shen SZ, Zhang H, Zhang YC, Yuan DX, Chen B, He WH, Mu L, Lin W, Wang WQ, Chen J, Wu Q, Cao CQ, Wang Y, Wang XD (2019b) Permian integrative stratigraphy and timescale of China. Sci China Earth Sci 62:154–188
Song HJ, Tong JN, Chen ZQ (2009a) Two episodes of foraminiferal extinction near the Permian–Triassic boundary at the Meishan section, South China, Australia. J Earth Sci 56:765–773
Song HJ, Tong JN, Chen ZQ, Yang H, Wang YB (2009b) End-Permian mass extinction of foraminifers in the Nanpanjiang Basin, South China. J Paleontol 83:718–738
Song T, Tong JN, Tian L, Chu DL, Huang YF (2019) Taxonomic and ecological variations of Permian–Triassic transitional bivalve communities from the littoral clastic facies in southwestern China. Palaeogeogr Palaeoclimatol Palaeoecol 519:108–123
Sun DY, Tong JN, Xiong YL, Tian L, Yin HF (2012) Conodont biostratigraphy and evolution across Permian–Triassic boundary at Yangou Section, Leping, Jiangxi Province, South China. J Earth Sci 23:311–325
Tian L, Tong JN, Sun DY, Xiong YL, Wang CG, Song HJ, Song HY, Huang YF (2014) The microfacies and sedimentary responses to the mass extinction during the Permian–Triassic transition at Yangou Section, Jiangxi Province, South China. Sci China Earth Sci 57:2195–2207
Wang QX, Tong JN, Song HJ, Yang H (2009) Ecological evolution across the Permian/Triassic boundary at the Kangjiaping Section in Cili County, Hunan Province, China. Sci China Ser D Earth Sci 52:797–806
Wu J, Feng QL, Gui BW, Liu GC (2010) Some new radiolarian species and genus from Upper Permian in Guangxi Province, South China. J Paleontol 84:879–894

Xie SC, Pancost RD, Huang JH, Wignall PB, Yu JX, Tang XY, Chen L, Huang XY, Lai XL (2007) Changes in the global carbon cycle occurred as two episodes during the Permian–Triassic crisis. Geology 35:1083–1086

Yang ZY, Yin HF, Wu SB, Yang FQ, Ding MH, Xu GR (1987) Permian–Triassic boundary stratigraphy and fauna of South China. Geological Publishing House, Beijing, 378 pp [in Chinese with English abstract]

Yang ZY, Wu SB, Yin HF, Xu GR, Zhang KX (1991) Permo–Triassic events of South China. Geological Publishing House, Beijing, 183 pp [in Chinese with English abstract]

Yang H, Chen ZQ, Wang YB, Tong JN, Song HJ, Chen J (2011) Composition and structure of microbialite ecosystems following the end-Permian mass extinction in South China. Palaeogeogr Palaeoclimatol Palaeoecol 208:111–128

Yang TL, He WH, Zhang KX, Wu SB, Zhang Y, Yue ML, Wu HT, Xiao YF (2015) Palaeoecological insights into the Changhsingian–Induan (latest Permian–earliest Triassic) bivalve fauna at Dongpan, southern Guangxi, South China. Alcheringa 40:98–117

Yin HF (1985a) Bivalves near the Permian—Triassic boundary in South China. J Paleontol 59:572–600

Yin HF (1985b) On *Hunanopecten*. Acta Palaeontologica Sinica 24:635–639, 1 p [in Chinese with English summary]

Yin HF, Zhang KX, Tong JN, Yang ZY, Wu SB (2001) The global stratotype section and point (GSSP) of the Permian–Triassic boundary. Episodes 24:102–114

Yu JX, Peng YQ, Zhang SX, Yang FQ, Zhao QM, Huang QS (2007) Terrestrial events across the Permian–Triassic boundary along the Yunnan–Guizhou border, SW China. Glob Planet Change 55:193–208

Yuan DX, Shen SZ (2011) Conodont succession across the Permian—Triassic boundary of the Liangfengya section, Chongqing, South China. Acta Palaeontologica Sinica 50:420–438 [in Chinese with English abstract]

Yuan DX, Shen SZ, Henderson CM, Chen J, Zhang H, Feng HZ (2014) Revised conodont-based integrated high-resolution timescale for the Changhsingian Stage and end-Permian extinction interval at the Meishan sections, South China. Lithos 204:220–245

Zhang F, Feng QL, He WH, Meng YY, Gu SZ (2006) Multidisciplinary stratigraphy across the Permian—Triassic boundary in deep-water environment of the Dongpan section, south China. Norw J Geol 86:125–131

Zhang KX, Tong JN, Shi GR, Lai XL, Yu JX, He WH, Peng YQ, Jin YL (2007) Early Triassic conodont–palynological biostratigraphy of the Meishan D Section in Changxing, Zhejiang Province, South China. Palaeogeogr Palaeoclimatol Palaeoecol 252:4–23

Zhang KX, Lai XL, Tong JN, Jiang HS (2009a) Progresses on study of conodont sequence for the GSSP section at Meishan, Changxing, Zhejiang Province, South China. Acta Palaeontologica Sinica 48:474–486 [in Chinese with English abstract]

Zhang ZY, He WH, Zhang Y, Yang TL, Wu SB (2009b) Late Permian–earliest Triassic ammonoid sequences from the Rencunping section, Sangzhi County, Hunan Province, South China and their regional correlation. Geol Sci Technol Inf 28:23–30 [in Chinese with English abstract]

Zhang N, Jiang HS, Zhong WL (2014) Conodont biostratigraphy across the Permian–Triassic boundary at the Xinmin Section, Guizhou, South China. J Earth Sci 25:779–786

Zhang H, Cao CQ, Liu XL, Mu L, Zheng QF, Liu F, Xiang L, Liu LJ, Shen SZ (2016) The terrestrial end-Permian mass extinction in South China. Palaeogeogr Palaeoclimatol Palaeoecol 448:108–124

Zhang L, Feng QL, He WH (2018) Permian radiolarian biostratigraphy. In: Lucas SG, Shen SZ (eds), The Permian timescale (vol 450, pp 143–163) London: Geological Society

Zhao JK, Liang XL, Zheng ZG (1978) Late Permian cephalopods of South China. Palaeontologia Sinica, Series B 12:1–194 [in Chinese with English abstract]

Zhao JK, Sheng JZ, Yao ZQ, Liang ZL, Chen CZ, Rui L, Liao ZT (1981) The Changhsingian and Permian—Triassic boundary of South China. Nanjing Inst Geol Palaeontol Bull 2:1–112 [in Chinese, with English abstract]

5 Temporal and Spatial Processes and Dynamics of the Permian–Triassic Boundary Mass Extinction (PTBME) in South China

Wei-Hong He, G. R. Shi, Ke-Xin Zhang, N. Suzuki, Han Wang, and Xiong-Hua Zhang

5.1 A Restudy of the Pattern of the Marine Mass Extinction Across the Permian–Triassic Transition

5.1.1 A Single- or Two-Pulsed Extinctions?

The end-Permian mass extinction is the most severe biodiversity crisis during the Phanerozoic (Alroy et al. 2008; Erwin 1993; Fan et al. 2020; Sepkoski 1981) and the extinction process has long been debated (e.g., Jin et al. 2000; Sepkoski 1981; Raup 1986; Shen et al. 2019a; Song et al. 2013). Typically, the extinction

W.-H. He (✉) · K.-X. Zhang
State Key Laboratory of Biogeology and Environmental Geology, School of Earth Sciences, China University of Geosciences, 388 Luma Road, Wuhan, Hubei, China
e-mail: whzhang@cug.edu.cn

K.-X. Zhang
e-mail: kx_zhang@cug.edu.cn

G. R. Shi
School of Earth, Atmospheric and Life Sciences, University of Wollongong, New South Wales, Wollongong, NSW 2522, Australia
e-mail: guang@uow.edu.au

N. Suzuki
Department of Earth Science, Graduate School of Science, Tohoku University, Sendai 980-8575, Japan
e-mail: noritoshi.suzuki.d3@tohoku.ac.jp

H. Wang · X.-H. Zhang
School of Earth Sciences, China University of Geosciences, 388 Luma Road, Wuhan, Hubei, China
e-mail: wh675515738@hotmail.com

© The Author(s), under exclusive license to Springer Nature Singapore Pte Ltd. 2023

W. He et al. (eds.), *Stratigraphy Around the Permian–Triassic Boundary of South China*, New Records of the Great Dying in South China,
https://doi.org/10.1007/978-981-99-9350-5_5

pattern has been described as either a single-pulse or two-pulse extinction event (Jin et al. 2000; Shen et al. 2011a, 2019a; Song et al. 2013; Yang et al. 1991). In both models, there is a consensus that the biodiversity dropped suddenly. The duration of the drop is approximately 30–60 kyr and equivalent either to bed 25 or beds 25–28 of the Meishan section, South China (Burgess et al. 2014; Shen et al. 2019a). A more recent study of composite species diversity using machine learning also supports a sudden nature of the mass extinction (over *ca* 63 kyr in 1-, 2- and 5-myr bins, see Fan et al. 2020).

5.1.2 A Review on the Pre-extinction Biotic Crisis

However, in contrast to the sudden extinction model, Fan et al. (2020) also presented a much earlier decline in biodiversity beginning some 780 kyrs prior to the final catastrophic breakdown (see page 274, Fig. 3 and Supplementary Materials of Fan et al. 2020). Before this, the study on the brachiopod diversity changes from different water depths spanning continental shelf to basinal facies in South China and even in the peri-Gondwanan region revealed that the ecosystem collapse began in deeper waters and then extended to shallow waters (He et al. 2015a). The process was considered to be possibly related to the formation and shoaling of a chemocline (within anoxic to euxinic waters) under a global warming and ocean stratification scenario (He et al. 2015a; Isozaki 2009; Jensen et al. 2009; Kump et al. 2005; Zopfi et al. 2001). For example, Late Permian radiolarian faunas in deep-water basinal facies of South China have been shown to have begun their extinctions earlier than benthos in the shallow water or ramp facies, probably due to the shoaling and expansion of the Oxygen Minimum Zone (OMZ) (Feng and Algeo 2014; Wang et al. 2023). In a similar vein, the Late Permian siliceous organisms (siliceous sponges and radiolarians) in the boreal pelagic environments became extinct at the base of the conodont *Clarkina hauschkei* Zone (equivalent to *Clarkina yini* Zone) (e.g., Algeo et al. 2012; Beauchamp and Baud 2002; Nabbefeld et al. 2010; Twitchett et al. 2001). These lines of evidence indicate that the evolution of the global ecosystem during the Late Permian involved a complex process, and its health and integrity were experiencing a rapid decline and deterioration in the lead up to the end-Permian mass extinction, possibly with some yet under-studied 'early warning signals' pointing to the approaching of a global ecosystem regime shift (tipping point) (see Wu et al. 2018a; Zhang et al. 2017).

It is true that evidence in support of a pre-extinction biotic crisis is still limited. There are several reasons for this. First, to date studies around the Permian–Triassic transition have mainly been focused on the diversity changes at or very close to the PTB, namely from the *Clarkina meishanensis* to *Isarcicella staeschei* Zones (basically equivalent to beds 25–28 of the Meishan section),

X.-H. Zhang
e-mail: 1065250198@qq.com

directly leading to a poorer understanding on the prelude of the mass extinction (e.g., in the *Clarkina yini* Zone). Second, until now, most data for biodiversity studies of the end-Permian extinction event have been collected from shallow-water settings in South China (e.g., Chen et al. 2009a; Jin et al. 2000; Shen et al. 2019a; Song et al. 2009, 2013; Wang et al. 2014). In contrast, only a few studies have systematically documented the less diverse, and less abundant fauna from the deep-water sections (with the exception of Dongpan) (Chen et al. 2006, 2009b; He et al. 2015a, 2019; Liao 1979, 1984; Wu et al. 2018b). Third, even rarer studies have investigated the temporal difference of species extinctions or extirpations among different organism groups according to their ecological or functional traits (e.g., categories of different life styles).

In this chapter, we attempt to decode the patterns and processes of the pre-extinction bioevents based on a comparison of timing in the changes of biodiversity between marine plankton (radiolarians) and benthos (brachiopods and foraminifers). We further explore the possible links both in timing and processes between marine and terrestrial extinctions with respect to marine invertebrates and terrestrial plant fossils.

5.1.3 Temporal Differences in Extinctions of Different Taxa in Varied Palaeogeographic Settings/Water Depths in South China

5.1.3.1 Selection of Sections for This Study

In order to investigate the prelude of the mass extinction, seven sections were chosen for the study to investigate the spatio-temporal extinction patterns of different kinds of organisms. The sections represent different palaeogeographic and palaeobathymetric settings, including deep-water basin (Dongpan and Rencunping), ramp along carbonate platform (Meishan), reef on carbonate platform (Laolongdong), restricted carbonate platform (Huangzhishan), and open carbonate platform (Liangfengya) (Fig. 2.2). Additionally, the age and key stratigraphic horizons (e.g., PTB, the bed(s) equivalent to beds 24e and/or 25) are well constrained. Data of fossil occurrences are systematically collected and the palaeogeographic condition has been more or less studied at the selected sections (see Chaps. 2 and 3). Specifically, the chosen sections all have yielded abundant fossils, e.g., planktonic radiolarians, benthic brachiopods and foraminifers. The study on the biodiversity changes of different taxonomic groups and ecological categories would provide insights into how different taxa and ecological traits responded to the mass extinction across different water depths and palaeoenvironments. This is so because the radiolarians dwelled in water columns at different depths while benthos inhabited the seafloor environments. A comparison of these two groups in their extinction timing and patterns would shed light on how the end-Permian mass extinction might have impacted the pelagic and benthic communities differently.

5.1.3.2 Palaeo-Bathymetric Reconstruction of South China in the Changhsingian

Generally, lithological feature (ratio of carbonate and chert/shale), sedimentary structures and ratio of benthos over plankton/nekton are basic indexes for determining palaeo-water depths. Carbonate deposits are considered to accumulate above the lysocline, whereas siliceous carbonate deposits formed between the lysocline and carbonate compensation depth (CCD), and opaline/siliceous sediments accumulate below the CCD (Dittert and Henrich 2000; Edmond and Huh 2003; Weber and Pisias 1999; Weber and von Stackelberg 2000). As the position of the CCD is deeper than the lysocline (Weber and Pisias 1999), siliceous mudstone or chert is considered to accumulate in deeper water environments than those for limestone and siliceous limestone. Additionally, fine-grained sediments (e.g., siliceous mudstone) with weak bioturbation could have accumulated in a setting from outer shelf to basin (at least > 120 m water depths, see Immenhauser 2009). The tempestite-related structures are generally formed near the storm wave base (50–250 m deep, see Immenhauser 2009). Besides, the presence/absence of radiolarians is a useful index for determining the palaeo-water depth in the Permian (He et al. 2008, 2011; Kozur 1993; Xiao et al. 2017). The absence of radiolarians is considered to suggest water depths of about (or <) 50 m. The presence of spherical radiolarians (orders of Entactinaria or Spumellaria) and lack of Latentifistularia (excluding *Ishigaum obesum*, *Quadricaulis inflata* and *Grandetortura*) and Albaillellaria forms, are generally taken to indicate water depths between 50 and 200 m. The presence of the order Latentifistularia (excluding *Ishigaum obesum* and *Quadricaulis inflata*) together with the absence of Albaillellaria forms suggests a water depth approximately 150 m. The presence (but not dominance) of Latentifistularia and Albaillellaria forms generally indicates water depths of 200 m or deeper. If Latentifistularia and Albaillellaria forms dominated a radiolarian fauna (especially the abundant presence of *Neoalbaillella* forms), it would indicate a water depth greater than 500 m.

Considering the above bathymetric indicators, the depositional depths of the chosen sections during the Changhsingian have been estimated (Fig. 5.1).

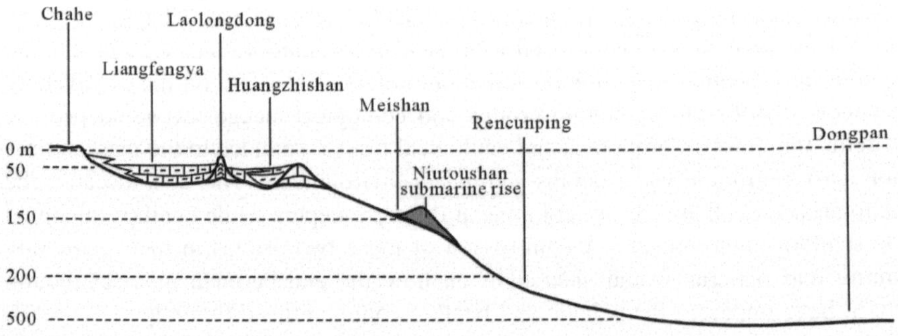

Fig. 5.1 Estimated palaeo-water depths for the chosen sections in the Changhsingian

At Dongpan, the Talung Formation is dominated by chert and siliceous mudstone and has yielded abundant radiolarian Latentifistularia and Albaillellaria forms (accounting for over 40%, see He et al. 2015b). Specifically, the radiolarian *Neoalbaillella* specimens are quite common. Also some cold-water brachiopod and ostracod genera were found in the Talung Formation. All the evidence would suggest that the water depth was, as a conservative estimate, deeper than 200 m (or even deeper than 500 m) through most of the Talung Formation.

At Rencunping, the Talung Formation is dominated by cherts, siliceous mudstones, siliceous limestones and carbonaceous mudstones. There were abundant presence of Entactinaria, Spumellaria and Latentifistularia elements and a few Albaillellaria elements (e.g., *Albaillella triangularis*, *Albaillella angusta*, ?*Neoalbaillella* sp.) in the Talung Formation. Therefore, the water depths of this section would have reached deeper than 200 m (or deep to 300 m) during the Changhsingian (Xiao et al. 2017), but were evidently shallower than those for the Dongpan section, because Albaillellaria specimens, indicative greater depths, are significantly less abundant at Rencunping (Figs. 3.66, 3.73) when compared to Dongpan.

At Meishan, a few spherical radiolarians are present, and those suggestive of moderately deep- or deep-water (Latentifistularia and Albaillellaria forms) are basically absent in the Changhsing Formation, suggesting the water depths were 50–200 m (or even < 150 m). On the other hand, benthic foraminifers are abundant in the bioclastic-dominated limestones of the Changhsing Formation, along with tempestite-related sedimentary structures (He et al. 2019), both of which corroborate a palaeo-water depth between 50 and 150 m.

At Laolongdong, the Changhsing Formation is dominated by reef limestones (calcareous sponge-dominated) and the basal part of the overlying Feixianguan Formation characterized by microbialites. Benthic foraminifers (including fusulinids), brachiopods, bryozoans, ostracods, crinoids and calcareous algae are abundant in the Changhsing Formation (Huang et al. 2022). Coccoidal cyanobacteria fossils and a few non-fusulinid foraminifers and ostracods have been recovered in the microbialites of the Feixianguan Formation (Huang et al. 2022). Generally, modern reefs are deposited in 10–20 m deep waters and modern microbialites or stromatolites accumulate in aquatic environments commonly less than 1 m deep with high salinity (see Dupraz and Visscher 2005; Immenhauser 2009; Petrash et al. 2012). Hence, the palaeo-water depths for the Laolongdong section are inferred to be shallower than 20 m during the Changhsingian.

At Huangzhishan, the top part of the Changhsing Formation is characterized by bioclastic limestone with abundant corals, brachiopods, crinoids, sponges, gastropods, bryozoans, ostracods, benthic foraminifers and calcareous algae (Chen et al. 2009a), but lacking radiolarians. The lithological and palaeontological features suggest that the formation was deposited nearby the fair-weather wave-base, about 50 mdeep (Immenhauser 2009). The basal part of the Yinkeng Formation is dominated by calcareous mudstones and marls, with exceptionally abundant brachiopods and absence of radiolarians; together they point to a shallow, low-energy setting probably just below the fair-weather wave-base (about or slightly > 50 mdeep) (He et al. 2017; Immenhauser 2009).

At Liangfengya, the Changhsing Formation is dominated by argillaceous limestones and bioclastic limestones, with abundant brachiopods and foraminifers but lacking any radiolarians. Thus the palaeo-water depths for the Changhsing Formation would be close to the fair-weather wave-base. Besides, the top part of the Changhsing Formation contained less mudstone (see the interval above bed 24e) at Liangfengya compared to its counterpart at Huangzhishan; therefore the palaeo-water depths for the interval should be slightly shallower than that at Huangzhishan in the end-Permian, probably shallower than 50 m.

5.1.3.3 Different Extinction/Extirpation Times of Planktonic and Benthic Organisms in Varied Water Depths

At Dongpan, the extinction/extirpation of radiolarians initiated from the middle part of bed 5 and mostly continued to bed 10 (although a few species survived to bed 12) (Fig. 5.2). The middle part of bed 5 of Dongpan is apparently below bed 24e of the Meishan section (Fig. 5.2). Furthermore, the extinction of radiolarians can be subdivided into two stages: deep-water Albaillellarian forms went extinct in the lower middle part of bed 5 (extinction rate 43%), then followed by a greater disappearance of radiolarians (moderately deep-water Latentifistularian and shallow-water Entactinarian and Spumellarian forms) in the top part of bed 5 and bed 6 (extinction rates respectively 33%, 62% and 31%) (Fig. 5.2). The extinction of brachiopods took place in the upper part of bed 10 (extinction rate 100%), which is equivalent to the horizon immediately above bed 25 of Meishan (Fig. 5.2). Therefore, the radiolarian extinction apparently began earlier than that of brachiopod at Dongpan (Fig. 5.2).

At Rencunping, the extinctions of radiolarians and brachiopods began in the upper part of bed 23c (extinction rates respectively 67% and 44%), which is equivalent to the strata below bed 24e of Meishan (Fig. 5.2).

At Meishan, the radiolarians disappeared mostly from the top part of bed 12 to the middle part of bed 19 (extinction rate 38% at the top part of bed 12) (Fig. 5.2). The extinction of brachiopods occurred in bed 24e–28 (extinction rate 50% in bed 24e; Fig. 5.2) (even continued to bed 30 according to data of the adjacent Daoduishan section). The extinction of foraminifers also took place from beds 25–28 (extinction rate 67% in bed 25) (Fig. 5.2).

At Huangzhishan, the extinction of foraminifers mostly took place in bed 18 (extinction rate 88%), which is equivalent to the strata above bed 24e of Meishan (Fig. 5.2). The extinction of brachiopods was gradual, beginning from bed 28 and continuing to bed 37 (extinction rate 71%), equivalent to the interval above bed 24e of Meishan and upwards across the PTB (Fig. 5.2).

At Laolongdong, the foraminifers and other benthos (e.g., brachiopods, bryozoans, ostracods, corals) became extinct at the basal part of bed 3 (extinction rate respectively 100% and 67%), which is equivalent to the strata below bed 24e of Meishan (Fig. 5.2).

At Liangfengya, the extinctions of foraminifers and brachiopods began in beds 28 and 29, which are correlated to the strata immediately below bed 24e of Meishan. The extinction rate is 91% for foraminifers from beds 28 to 32 and is 50% for brachiopods in bed 29 and 95% from beds 29 to 32 (Fig. 5.2).

5.1.3.4 Development of Biotic Crisis Across Varied Palaeogeographic Settings/Water Depths

As already outlined above (Sect. 5.1.3.2), the Dongpan and Rencunping sections are here considered as representing deep-water facies, Meishan as an example of a moderately deep-water facies, while the Huangzhishan, Laolongdong and Liangfengya sections all represent shallow-water settings. The discussion provided in the Section.

The analysis of extinction/extirpation timing of different taxa in varied water depths has revealed a number of significant features. First, as shown in Fig. 5.2, extinctions in deep waters took place evidently earlier than their counterparts in moderately deep and shallow water environments, except for the brachiopod extinction at the deep-water Dongpan section and the benthic fauna extinction in the shallow-water settings of the Laolongdong section. The brachiopod extinction of Dongpan occurred in the horizon above bed 25 of Meishan was slightly later than the onset of brachiopod extinctions took place immediately below bed 24e at Meishan and Liangfengya (Fig. 5.2). The extinction of benthic fauna at Laolongdong occurred probably in the beginning of bed 24 of Meishan (Huang et al. 2022), evidently earlier than the counterpart of shallow-water facies (Fig. 5.2; the reasons for the exceptions at Dongpan and Laolongdong see Sect. 5.3.3.2). Second, the extinction of benthic foraminifers occurred slightly earlier than the extinction of brachiopods in shallow-water carbonate-platform facies. Third, the timing for plant extinctions in terrestrial facies basically coincides with timing for marine invertebrate extinctions in shallow- and moderately deep-water sections (except the Laolongdong section) (Fig. 5.2).

For the sake of convenience in the following discussion, the extinctions that occurred below bed 24e of Meishan or equivalents are here considered as the **prelude** of the end-Permian mass extinction, while the extinctions enclosed from bed 24e/25 to PTB/bed 28 of Meishan (and equivalent beds in other sections) as the extinction **zenith**, and the extinction "younger" than bed 28 of Meishan as the **epilogue** of the end-Permian mass extinction (terms of prelude, zenith and epilogue referred to Huang et al. 2022; Yin et al. 2007).

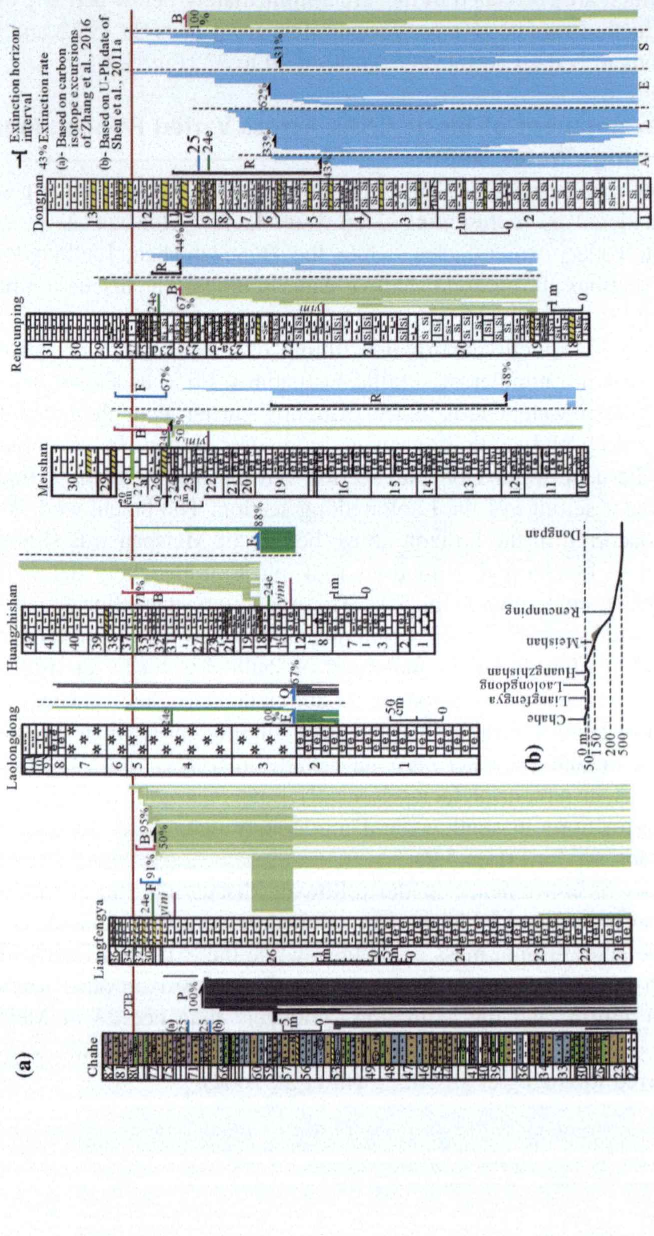
(a)
Chahe
Liangfengya
Laolongdong
Huangzhishan
Meishan
Rencunping
Dongpan
Extinction horizon/ interval
43% Extinction rate
(a)- Based on carbon isotope excursions of Zhang et al., 2016
(b)- Based on U-Pb date of Shen et al., 2011a
PTB
(b)

◀**Fig. 5.2** Horizons/intervals of extinctions and extinction rates for different taxa at the studied sections across different depositional settings. **a**—Stratigraphic ranges of species (marked by vertical lines), extinction horizons/intervals and extinction rates for different taxa. *Note* P—Plant; F—Foraminifer; B—Brachiopod; O—Other kinds of benthos excluding foraminifers; R—Radiolarian; A—Albaillellaria; L—Latentifistularia; E—Entactinaria; S—Spumellaria. 24e, 25—equivalent to bed 24e and 25 of Meishan (based on conodont zones or carbon isotope excursions, details see Chap. 4); *yini*—*Clarkina yini* Zone. Data of fossil stratigraphic ranges for (**a**): plant at Chahe from Figs. 3.12 and 3.13; brachiopod at Liangfengya from Fig. 3.44 and foraminifer extinction at Liangfengya from Liu et al. (2020); fossils at Laolongdong from Fig. 3.50; brachiopod at Huangzhishan from Fig. 3.37 and foraminifer at Huangzhishan from Fig. 3.36; brachiopod and radiolarian at Meishan from Fig. 3.32 and foraminifer at Meishan from Song et al. (2009); brachiopod and radiolarian at Rencunping from Fig. 3.66; Albaillellaria and Latentifistularia forms at Dongpan from Fig. 3.73, Entactinaria and Spumellaria forms at Dongpan from Fig. 3.74, and brachiopod species at Dongpan from Fig. 3.75. **b**—Estimated palaeo-water depth for each section in the Changhsingian (as a schematic figure simplified from Fig. 5.1, for the sake of comparing palaeo-water depths among sections)
Legends and abbreviations at Chahe, Liangfengya, Laolongdong, Huangzhishan, Meishan, Rencunping, Dongpan respectively same as in Figs. 3.12, 3.44, 3.50, 3.36, 3.32, 3.66, 3.73

5.2 Comments on the Global Terrestrial Extinction Across the Permian–Triassic Transition

In North China, the extinction rate of plants in terms of generic and species diversity was relatively high in the Wuchiapingian (corresponding to the basal part of Sunjiagou Formation) compared to that of Changhsingian, associated with a low origination rate in a regional scale (Xiong et al., 2021). Upward to the end of the Permian, almost all of the Permian plants became extinct with the exception of a few relics of gigantopterids and newly originated elements in the earliest Triassic.

In South China, the terrestrial end-Permian mass extinction is marked by the elimination of the *Gigantopteris* flora, followed immediately by a short-time survival of a few relics of gigantopterids associated with a few newly originated *Tomiostrobus* and *Germaropteris* elements in the earliest Triassic. For example, at Chahe, the flora became extinct in bed 70, which is comparable to bed 24e to bed 25 (Fig. 5.2). This terrestrial extinction was thus likely synchronous with the marine extinction zenith (Shen et al. 2011a; Zhang et al. 2016), or slightly earlier than the latter, based on the correlation of carbon isotope excursions between the two PTB sections. However, it is worthy noting that plants disappeared massively in the lower part of the *Gigantonoclea–Gigantopteris–Pecopteris* Zone which is probably equivalent to the early Changhsingian or the late Wuchiapingian (Fig. 3.12), signaling the earlier appearance of Late Permian terrestrial extirpation well before the end-Permian marine mass extinction.

Signals for an earlier decline of the Late Permian terrestrial ecosystem has also been documented in South Africa where the plant extinctions began also in the late Wuchiapingian, then followed by a gradual disappearing of land vertebrates (Retallack 2021). A similar stratigraphic pattern has also emerged from the Sydney Basin of eastern Australia where the extinction of glossopterid gymnosperms is found to have occurred in a horizon equivalent to bed 22 of Meishan and thus significantly predated the marine extinction zenith (Fielding et al. 2019; Mays et al. 2020). In Siberia, using high-resolution geochronology (Burgess et al. 2017; Gastaldo et al. 2020; Fielding et al. 2019), Davydov et al. (2021) also found the fresh-water bivalve faunas and gymnosperms becoming extinct *ca.* 820 kyrs prior to the marine extinction zenith.

Thus, in summary, the Late Permian global terrestrial extinctions, except South China, seem to significantly predate the end-Permian marine mass extinction zenith, and are characterized by a protracted decline from the Wuchiapingian to Changhsingian.

5.3 A Further Depiction for the Dynamics of the Permian–Triassic Mass Extinction and Palaeoenvironmental Crisis

5.3.1 A General Review on the Causes of the Permian–Triassic Mass Extinction

The potential causes for the Permian–Triassic mass extinction have been proposed to mainly include volcanism (e.g., Burgess et al. 2017; Davydov 2021; He et al. 2014; Isozaki et al. 2007; Joachimski et al. 2020; Sial et al. 2020; Yang et al. 1991; Zhang et al. 2021), global warming (e.g., Isozaki 1997; Joachimski et al. 2020; Kump 2018; Sun et al. 2012), anoxia (e.g., Hu et al. 2021; Li et al. 2016; Wignall and Twitchett 1996; Xie et al. 2017; Zhang et al. 2020), acidification (e.g., Clarkson et al. 2015; Garbelli et al. 2017; Jurikova et al. 2020; Silva-Tamayo et al. 2018), sea-level changes (e.g., Adachi et al. 2017; Baresel et al. 2017; Hallam and Wignall 1999; Wu 2017; Yin et al. 2014) and extraterrestrial impact (e.g., Becker et al. 2001; Burger et al. 2019; Farley and Mukhopadhyay 2001; Jin et al. 2000; Kaiho et al. 2001; Onoue et al. 2019; Yang et al. 1987). Among these hypotheses, volcanism and anoxia are the most debated factors at present and how they triggered the palaeoenvironmental crisis and mass extinction remain poorly known.

5.3.2 Influence of Volcanism on the Mass Extinction

5.3.2.1 Comment on the Relationship Between the Intrusion of Siberian Traps Large Igneous Province and the Mass Extinction

Large-scale volcanism has long been proposed as one of the factors which contributed to the Permian–Triassic mass extinction (e.g., Renne and Basu 1991; Yang et al. 1987; Zhou and Kyte 1988). Specifically in recent years, numerous works have linked the volcanism with the Siberian Traps Large Igneous Province (STLIP) (Joachimski et al. 2020; Regelous et al. 2020; Sial et al. 2020), aided by the availability of high-resolution and high-precision U/Pb ages for the STLIP (Burgess and Bowring 2015; Burgess et al. 2014, 2017). The evidence from these works is mainly based on the following aspects:

(1) Timing for the change in the emplacement style of STLIP from dominantly flood lavas to sill intrusions, overlaps with the timing of the Permian–Triassic mass extinction zenith (Burgess et al. 2017).
(2) Timing for the intrusive magmatism of STLIP was considered as the same as that of the negative $\delta^{13}C$ excursion (CIE) (Burgess et al. 2017). Magmatism and associated volcanism of large igneous provinces are generally thought to trigger or drive mass extinctions because of their ability to perturb the global carbon cycle and thus destabilize the global ecosystem. When magma

intruded into their surrounding shale and bitumen-bearing carbonate rocks, the hydrocarbon-bearing rocks would have been heated and metamorphosed, leading to a massive production and release of thermogenic gases (CO_2 and CH_4). These gases could then escape to the atmosphere through explosion pipes, increasing atmospheric CO_2 levels and also equilibrating with the dissolved inorganic carbon pool of oceanic surface waters. This process thus may have been responsible for the observed decrease in $\delta^{13}C$ (Cui et al. 2021; Retallack and Jahren 2008; Svensen et al. 2009).

(3) In recent years, an ever expanding body of literature reporting anomalies of abundance/isotopes of Hg, Zn, Pb, Cd, Sb, Ni, Te and Os have been published (e.g., Grasby et al. 2021; Kaiho et al. 2021; Regelous et al. 2020; Shen et al. 2019b, c; Sial et al. 2020), following the high-precision dates of the STLIP lava flows, sills, and explosively erupted rocks (Burgess and Bowring 2015). The anomalies have been generally considered as the proxy to indicate the intrusive magmatism of the STLIP (Regelous et al. 2020; Shen et al. 2023; Sial et al. 2020).

As for the second line of evidence (aforementioned), that argues the timing for the intrusive magmatism of STLIP to coincide with that of the CIE, we consider this argument to be contentious and highly debatable. This is because that the massive intrusion and lateral transport of the STLIP magmatism began from 251.9 Ma or even later in the Early Triassic to the Early Cretaceous (Burgess et al. 2017; Dovydov 2021; Dovydov et al. 2021), and if so, it would have lagged behind the CIE (specifically behind the onset of it, see Fig. 2 of Burgess et al. 2017). Likewise, we also question the validity of the third line of evidence whereby the anomalies of siderophile elemental abundance/isotope (e.g., Hg, Zn, Pb) were regarded as the proxy for the intrusive magmatism of STLIP which triggered the mass extinction. In the Meishan section, the anomaly of siderophile elements was found to appear from bed 24 (or even older) and extend upward to the extinction zenith horizons (or slightly younger) (Georgiev et al. 2020; Liu et al. 2017, 2020; Shen et al. 2019b, 2021). Hence, the timing for the onset of siderophile elements' anomalies is earlier than that for the massive intrusion of the STLIP, and therefore the mass extinction zenith cannot be linked to the STLIP intrusive magmatism.

In short, in our opinion it is difficult to consider the sill intrusion of STLIP as the trigger of the end-Permian mass extinction, but the long-term influence exerted by the early eruptions (release of CO_2), along with the late intrusion of STLIP (thermal metamorphism and release of CO_2 and CH_4), would have contributed to the crisis.

5.3.2.2 The Potential Influence of Volcanism on the Mass Extinction Owing to the Subduction of the Palaeotethys Ocean

Recently an ever expanding body of works in South China proposed that the arc volcanism along the convergent continental margin caused by the subduction of the Palaeotethys Ocean, was a trigger for the mass extinction (e.g., Wang et al. 2019; Zhao et al. 2019). The volcanism started from the early Changhsingian and continued to the Early Triassic (Shen et al. 2021). The influence of volcanism on organisms would therefore have to be a protracted process.

In the event of volcanism, ash generated from the volcano eruption brought some elements (e.g., Fe^{2+}, Mg^{2+}, NH_4^+, NO_3^-, PO_4^{3-}, MnO and SiO_2) into waters and provided nutrition for the explosive flourish of organisms (Frogner et al. 2001; Mahowald et al. 2018; Olsson et al. 2013). This proposition is supported by the evidence that a variety of microbial types have been discovered in the ashes dated as Changhsingian (Fang et al. 2017). Specifically, nearby the PTB, abundant ash layers have been discovered, along with widespread microbialites discovered in the shallow-water carbonate platform margin settings (Fig. 5.3). However, it is crucial to note that the initial occurrence of the microbialites is not synchronous across South China, with some starting from a time equal to or older than bed 24e of Meishan (Figs. 3.46, 3.48, 3.50, 3.54) and others first appearing from beds younger than bed 24e of Meishan (Fig. 3.52). The temporal difference is most likely related to the variation of water depths at different localities. In general, microbialites are better developed in shallow water depths where they are found in association with regressive sequences (microbial blooming also related to the salinity level, see the part of Sect. 5.1.3.2).

In addition to the proliferation of microbialites across the PTB, the influx of volcanic ash would also have benefited marine invertebrates, as evidenced by the presence of abundant brachiopod shells in tuffaceous rocks (e.g., in the Linhao Formation of Guman and Penglaitan of southern Guangxi) and abundant radiolarians discovered in tuff-bearing siliceous mudstones (e.g., the Talung Formation of Dongpan in southern Guangxi) in the Changhsingian.

On the other hand, the rapid concentration of nutrients in water bodies due to volcanism could also have been caused, triggering formation and expansion of the Oxygen Minimum Zone in both deep (e.g. the Panthalassa) and shallow waters (e.g., shelf) and leading to oxygen deficiency and anoxia hostile to the survival of marine organisms, as mentioned in the part of Sect. 5.3.3.

Furthermore, a large amount of toxic elements (e.g., Cd, Cu, Pb, Hg) have been detected from rocks spanning the Permian–Triassic transition in South China, believed to have been released by the volcanism (Chen and Xu 2019; Shen et al. 2019b). The presence of these toxic elements has been inferred to have also contributed to the end-Permian mass extinction (Hoffmann et al. 2012; Mahowald et al. 2018). This proposition is strongly evidenced by the widely reported temporal coincidence between the extinction zenith and the mercury spike (recorded by Hg/TOC) from marine to terrestrial facies and from deep Panthalassa to shallow shelf (Shen et al. 2019b, 2019c, 2021; Sial et al. 2020). However, it must be clarified that the anomalous inputs of Hg were not limited to just the Permian–Triassic transition, it actually began from the early Changhsingian in South China (Shen et al. 2021). Although the earlier episodes of Hg injection did not lead to significant extinctions in the early Changhsingian, the species diversity remained low throughout this interval (e.g., Majiashan and Shangsi, see Figs. 3.56, 3.62, 3.63, 3.64). With increasing research, there is little doubt that volcanisms intensified in

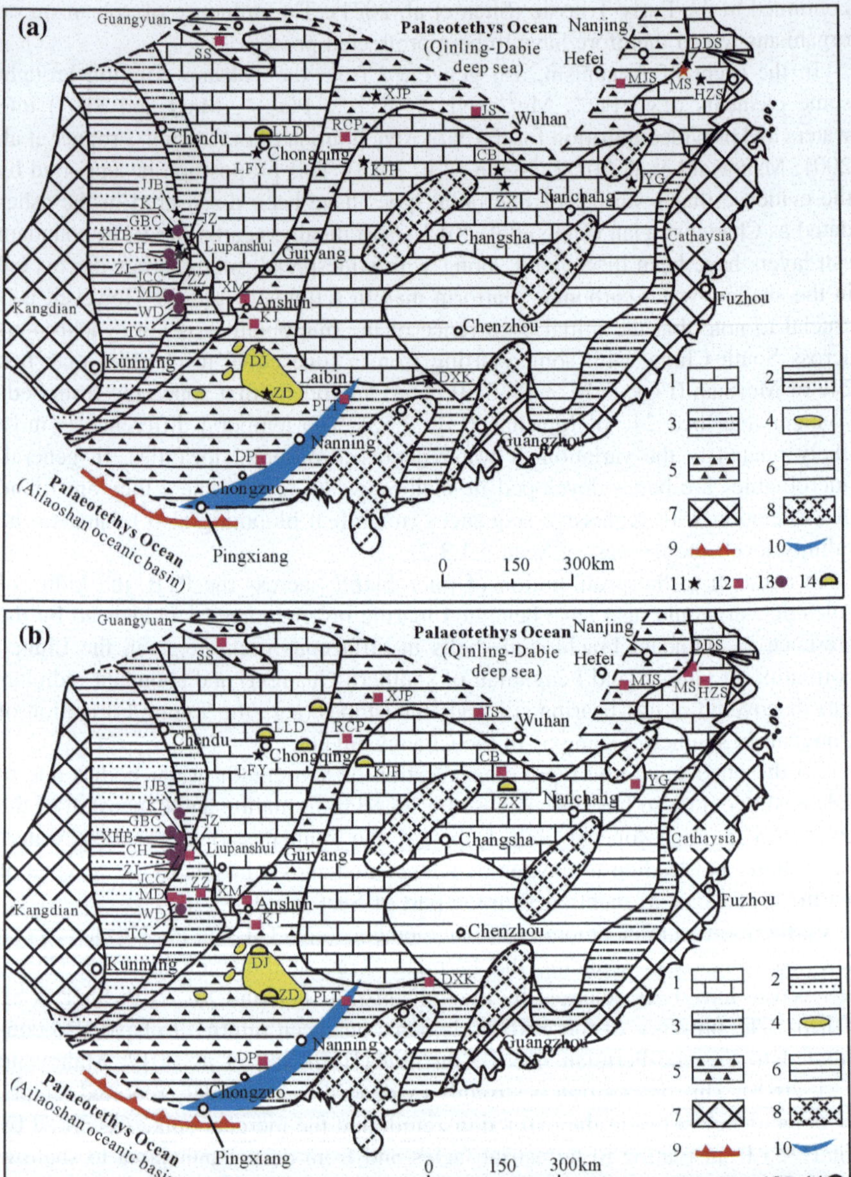

Fig. 5.3 Changes in the spatial distribution of airfall ash in two stages. **a**—Prior to the Permian–Triassic transition (palaeogeography revised after Fig. 2.2). **b**—During the Permian–Triassic transition (equivalent to *Clarkina meishanensis* to *Isarcicella staeschei* Zones) (palaeogeography revised after Fig. 2.3). 1—Shallow-water carbonate platform; 2—Terrestrial facies; 3—Shallow-water clastic shelf; 4—Isolated carbonate platform; 5—Deep-water siliceous basin; 6—Restricted siliceous basin; 7; Ancient land; 8—Rise under sea surface; 9—Subduction of Palaeotethys Ocean; 10—Rift related to the subduction; 11—Sections with rare/absent of volcanic ash; 12—Sections commonly with volcanic ashes; 13—Sections with tuffaceous clasts which were probably products of weathered Ermeishan basalts; 14—Microbialite

South China on approaching the PTB (Fig. 5.3), generating large amounts of poisonous elements into the atmosphere as well as into the oceans, thus precipitating an increasing hostile environment for the survival of organisms.

Finally, volcanic degassing would also have released huge amounts of CO_2 and CH_4 to accumulate in the air through the Changhsingian. This prospect of environmental change has been estimated to drive sea temperature increase by 5 °C in South China in the late Changhsingian (see that of bed 24 of Meishan, Joachimski et al. 2012). The elevation of temperature could have inhibited the circulation of sea waters and contributed to the development and/or expansion of seafloor and water column anoxia (discussed further below). At this time, the deep-water ecosystem is thought to begin collapsing (He et al. 2007, 2015a; Wu et al. 2018a) with the carbonate carbon isotope curve shifting to lower values (negative shift) (Shen et al. 2011a). Upwards to the Permian–Triassic transition (beds 25–28 of Meishan), more CO_2 and CH_4 were released into air with the intensifying of volcanic eruptions, intensifying the global warming trend and shifting the carbonate isotope curve to the lowest values.

To sum up, protracted and prevalent volcanism, generated by the subduction of the Palaeotethys Ocean underneath South China, was a prominent feature of the Changhsingian, and it became intensified toward the end Permian when it coincided with the extinction zenith. This coincidence would therefore support a highly possible cause and effect relationship between regional volcanism and at least a regional (South China) mass extinction.

5.3.3 Anoxia and Mass Extinction

Whether a prevalent and prolonged anoxia (or oxygen-restricted) event could be regarded as a trigger for the Permian–Triassic mass extinction has long been debated (e.g., Bond and Wignall 2010; Foster et al. 2020; Grasby et al. 2021; Grice et al. 2005; Hu et al. 2021; Li et al. 2016; Takahashi et al. 2021; Wignall and Hallam 1992; Wu et al. 2021; Xiang et al. 2022). There are a number of contentious aspects about this hypothesis. First, the duration and magnitude of this event is still not well defined, even with data collected from the same section (e.g., Meishan) but investigated by different workers (e.g., for the Meishan section, different interpretations exist among Chen et al. 2015; Li et al. 2016; Shen et al. 2011b). In order to clarify the debate, in the book we have performed an integrative and comparative study on when the anoxia began and how the onset and duration of anoxia in each studied section is related to the change of species diversity at each section, having regards to the specific palaeo-water depth of each section.

Seven PTB sections across a range of palaeo-bathymetric settings and depositional environments are chosen for this study (see Sect. 5.1.3.1). A crude and arbitrary classification scheme has been adopted to assign these sections to each of the following palaeo-water depths: deeper than 200 m (or even > 500 m), represented by the Dongpan section; between 200 and 500 m (Rencunping section, shallower than Dongpan); 50–150 m (Meishan section), 0–20 m (Laolongdong)

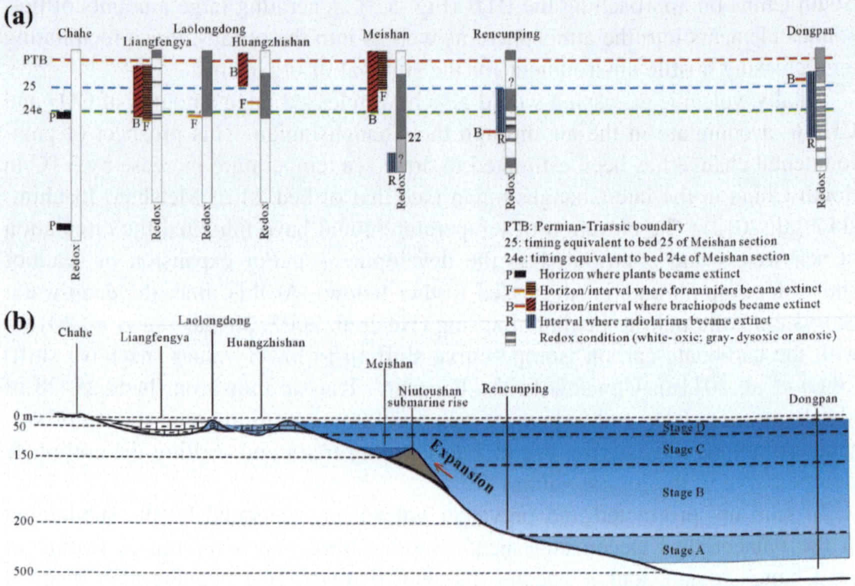

Fig. 5.4 Spatial (palaeoenvironmental) and temporal variations of extinction events of different organisms (pelagic and benthic forms) in relation to the proposed expansion dynamics of the Oxygen Minimum Zone (OMZ) in the latest Permian and across the Permian–Triassic transition. **a**—Comparison of timings in the extinction/deterioration of different taxa and change of redox conditions across the palaeo-bathymetric gradient of South China; **b**—Proposed scenario for understanding the dynamics of the OMZ during the latest Permian and its impacts on marine organisms. *Note* for Fig. 5.4a, information on extinctions is based on Fig. 5.2; data of redox condition at Dongpan after Feng and Algeo (2014), Shen et al. (2012), at Meishan after Li et al. (2016), Shen et al. (2011b), at Laolongdong after Huang et al. (2022), at Liangfengya after Li et al. (2021); for Fig. 5.4b, Stages A, B, C and D representing four stages successively influenced by anoxic/dysoxic conditions with the expansion of OMZ

or slightly > 50 m (Huangzhishan), or about 50 m (Liangfengya), and above the sea level (Chahe) (Figs. 5.1, 5.4).

5.3.3.1 Comparison of Anoxia and Extinction Timings Across the Palaeo-Bathymetric Gradient

At Dongpan, the redox conditions fluctuated between anoxia (or dysoxic) and oxygenation in the interval well below bed 24e of Meishan. This was then followed by more frequent fluctuations of anoxia/dysoxia in the interval equivalent close to bed 24e and upwards to the earliest Triassic (Shen et al. 2012; Feng and Algeo 2014; Fig. 5.4). The upper part of the Dongpan section broadly corresponds to the time when the water-column dwelling radiolarians gradually became extinct (Figs. 5.2, 5.4). In the horizon of bed 10, equivalent to immediately above bed 25 of Meishan, the benthos brachiopods abruptly became extinct (Figs. 5.2, 5.4). Thus here it is evident that the anoxic/dysoxic events first wiped out radiolarians and then brachiopods.

At Rencunping, the changing profile of its redox conditions is similar to that of Dongpan (Fig. 5.4). The interval where the disappearing of radiolarians and brachiopods took place is basically coincided with the duration when the anoxic/dysoxic events occurred more frequently (Fig. 5.4). Hence, the deterioration of both pelagic (radiolarians) and benthos (brachiopods) ecosystems can be linked to the intensifying of the oxygen-restricted condition. Here, the fact that brachiopods and radiolarians disappeared simultaneously is a significant observation though its cause remains unclear. This apparently synchronized extinction pattern between benthic and pelagic life forms is in contrast to the Meishan section where the radiolarian diversity started to decline earlier than the benthos (Figs. 5.2, 5.4), similar to that of Dongpan. Also, a comparison between Rencunping and Dongpan (Fig. 5.4) indicates that brachiopods became extinct earlier than their counterpart at Dongpan.

At Meishan, the study of sulfur isotopes and pyrite framboids revealed that it was anoxic from beds 22–26 (Shen et al. 2011b; Li et al. 2016). The color of rocks, grayish black to dark gray, indicates a dysoxic or anoxic setting in beds 13–19 although the study from the redox indicator is absent for this interval. These oxygen-deficient conditions could also have existed at Dongpan and Rencunping and is broadly comparable to each other (Fig. 5.4). The extinction of benthos (brachiopods and foraminifers) occurred in beds 24e–28 and the disappearing of radiolarians happened in beds 13–19 (below the conodont *Clarkina yini* Zone) at Meishan (Fig. 5.2). Hence, the extinction stage of brachiopods and foraminifers is coincided with the anoxia condition in beds 24e–26 while the disappearing of radiolarians is probably more or less related to oxygen constrictions of beds 13–19. Thus the mass extinction or ecosystem deterioration is basically relevant to the anoxia events at Meishan (Fig. 5.4), despite the times for brachiopod and foraminifer extinctions are not corresponding to each other sensu stricto and the time for benthic faunal extinctions lagged behind the termination of anoxic conditions. Besides, it is worthy to note that the benthos extinction is obviously later than that of Rencunping.

At Huangzhishan, the anoxic condition began from the interval equivalent to immediately below bed 24e of Meishan and continued to the horizon above bed 25 of Meishan (Fig. 5.4). The extinction of benthic foraminifers occurred in the horizon equivalent to beds 24e and 25 of Meishan, within an anoxic interval (Fig. 5.4). The extinction of brachiopods is more gradual and took place over an interval corresponding to bed 25 of Meishan to the earliest Triassic, also beginning in the anoxic interval despite the time for ending of the brachiopod extinction lagged behind the termination of anoxic conditions (Fig. 5.4). Therefore, the extinction of benthos is broadly related to the anoxia events at Huangzhishan.

At Laolongdong, the anoxia event happened in the interval beginning immediately below bed 24e of Meishan to the earliest Triassic (Fig. 5.4). The extinction of benthic foraminifers is synchronous with the onset of the anoxia event, reasonably indicating the extinction could have been related to the anoxia.

At Liangfengya, the interval equivalent to immediately below bed 24e of Meishan to the PTB was dominated by an oxic condition (albeit shortly interrupted by several dysoxic episodes). Thus, the extinction of benthos (brachiopods and foraminifers) does not seem to bear any significant relations to anoxia (Fig. 5.4).

In summary, the marine extinction or ecosystem deterioration is generally correlated, in timing, to oxygen deficient intervals in South China, as mentioned in our previous works (He et al. 2015a, 2017), despite a few exceptions (e.g., Shangsi and Penglaitan sections, see Xiang et al. 2016, 2022). As for the disagreement, there are possible explainations. The long-term anoxia in the Changhsingian at Shangsi hosted a low-diversity ecosystem. Even if the oxygen deficiency was temporarily improved but the improvement was offset by other ecological stressors (e.g., warming and toxification, see below), which may have worked together and amplified each other's negative impacts on an already highly stressed ecosystem and finally pushing it over a tipping point. The Penglaitan section was located in a rift basin, its sea level changed frequently in response to the shifts between extension and rapid basin filling. The presence of radiolarians in several horizons of the thick Changhsingian sequence (more than 560 m) may not indicate the presence of a deep-water setting through the section; instead, the basin could have largely remained a shallow-water condition, judging from the abundant presence of shallow-water foraminifers and brachiopods in some horizons of the section, especially in the end of the Permian. Thus, it would be reasonable to suggest that the Penglaitan section mostly remained shallow and oxygenated (Xiang et al. 2022) and therefore its end-Permian extinction would have little bearing on anoxia during the Permian–Triassic transition, a condition similar to that of Liangfengya.

5.3.3.2 How Had the Extinction/Ecosystem Deterioration Been Controlled by the Anoxic/Dysoxic Event

South China was a depositional basin across varied palaeogeographic settings with varied palaeo-water depths, e.g., from terrestrial facies, to coastal clastic shelf, carbonate platform, reef on/along the platform margin, and to deep-water basin. The deep-water basins were mainly located along the northern and southern margins of South China (Fig. 2.2). The areas were linked to the Palaeotethys and Panthalassa Oceans by ocean currents (Fig. 2.2; also see He et al. 2015b, 2020) and as such were conducive to the influence of upwelling systems and oxygen minimum zones (OMZs). Besides, in the southern deep-water basin, the palaeotopography was more complex, with several small isolated carbonate platforms scattered in the basin, which would have slowed the water circulations. The upwelling and restricted water circulation combinedly could have facilitated the enhancement of OMZ. Besides, the changes of dissolved oxygen concentration created by variations of upwellings, water circulations, and volcanism-related eutrophication, could have led to the expansion, contraction, or migration (upwards or downwards) of OMZ.

The comparison on the timing of anoxic events and extinctions (part of Sect. 5.3.3.1) across the palaeo-bathymetric gradient reveals that the deep-water setting was influenced by the oxygen-deficient events earlier compared to shallow water sections. This pattern recalls the effects of a vertically expanding OMZ. In the beginning (Stage A, Fig. 5.4b), the OMZ was formed and covered the water column, but would not have intercepted deep enough to reach the bottom depth of the Dongpan section (~500 m deep). Thus, at this stage the anoxia derived from the OMZ caused the extinction of radiolarians that dwelled in water columns but the benthos (e.g., brachiopods) were largely unaffected. In the next stage (Stage B, Fig. 5.4b), the OMZ expanded or migrated upward to a bathymetric level equivalent to the Rencunping section, where it first eliminated the brachiopods on the sea floor and then gradually wiped out the radiolarians in the water columns. Subsequent to this stage (Stage C, Fig. 5.4b), the OMZ continued to expand/migrate upward to shallower depths matched by the Meishan section, killing off most brachiopods and benthic foraminifers in these environments (a few benthic elements survived to the earliest Triassic probably because of their greater tolerance). The fact that radiolarians at Meishan were extirpated much earlier than brachiopods and benthic foraminifers remains an open question and may be explained by the appearance of a dysoxic episode led by the upward migration of OMZ, or by the presence of restricted circulation and oxygen deficiency due to the protection of the Niutoushan submarine rise (Fig. 5.4b; Wu et al. 1986; Zhang et al. 1997; He et al. 2017). In the final stage of the OMZ expansion (Stage D, Fig. 5.4b), coastal or shallow-water settings like the Huangzhishan section were affected by deteriorating anoxic/dysoxic conditions, decimating most benthic life inhabiting these habitats. Similar to Meishan, a few brachiopods survived this episode of the mass extinction and persisted to the earliest Triassic, presumably owing to their greater tolerance to oxygen fluctuations.

The scenario presented above invoking a gradual and staged expansion of the OMZ is consistent with the studies of water depth-dependent $\delta^{13}C_{carb}$, $\delta^{15}N$ and $\delta^{34}S$ gradients (Jiang et al. 2022; Wu et al. 2021), all of which corroborate a pattern of a gradual OMZ expansion eventually intercepting the intermediate water column (Levin 2003).

Under this model, it is conceivable that it would be difficult for the OMZ to intercept the very shallow environments (e.g., inter-tidal setting or above the fair-weather wave-base), because the dissolved oxygen concentration is higher than 0.5ml/1 (oxygen) in waters above the upper boundary of OMZ (Levin 2003). For example, at Liangfangya, it remained well oxygenated in the latest Permian and only became anoxic in the earliest Triassic as a result of a major transgression. The Laolongdong section was located in the middle part of a carbonate platform, very close to Liangfangya (but much shallower in water depths), and in theory it could have been also difficult to be influenced by any anoxia caused by an expanding OMZ. However, Laolongdong was evidently anoxic for the interval equivalent to immediately below bed 24e of Meishan to the earliest Triassic. How could we understand this paradox? As shown in Fig. 5.4b, Laolongdong was located on a raised carbonate platform linked to a short-lived reef builup structure. Considering this background, both the evaporation rate and salinity of this site would have been abnormally high, leading to the proliferation of microbials and the deposition of microbialite (see the part of Sect. 5.1.3.2) (Fig. 5.4). In turn, the accumulation of microbials would have brought about abundant organic materials, conducive to the formation of anoxic conditions. This interpretation is consistent with the notion that microbial blooming could have triggered oxygen depletion immediately after the end-Permian mass extinction (Chen et al. 2022).

5.3.4 A New Integrative Depiction for the Dynamics of the Permian–Triassic Mass Extinction

Although a pattern of a single- or two-pulse mass extinction is generally accepted, it should be noted that the collapse of the end-Permian palaeoecosystem had already begun to take place prior to the extinction zenith.

5.3.4.1 A Long-Term and Gradual Accumulation of Ecological Stresses Created by a Series of Environmental Changes from the Wuchiapingian to the Induan

The STLIP volcanism included three stages from the Wuchiapingian to Induan, namely pyroclastic eruption, lava eruption and intrusive magmatism (see Column 1 in Fig. 5.5; see Burgess et al. 2017; Davydov 2021; Shen et al. 2019a). In the first stage (pyroclastic eruption), a plenty of greenhouse gas (CO_2) was released into atmosphere and the emission could have been accumulated for a long time, at least beginning from the Wuchiapingian (Davydov 2021). In the second and third stages, it was characterized by the eruption and intrusion of basic magma in the late Changhsingian to the Induan, and thus the accumulation rate of released CO_2/CH_4 into atmosphere could have been relatively reduced but the amount of greenhouse gases could have been continuously increased.

The Late Permian climate (humidity) varied globally (Davydov et al. 2021; Fielding et al. 2019; Xiong et al. 2021). It was extremely arid in North China in the Wuchiapingian and Changhsingian, marked by widespread red beds. The aridity caused a prominent decline of plant generic and species diversity in the Wuchiapingian (Fig. 5.5, marked by A in Column 2), associated with a transition from cordaitalean (wet-adapted) to coniferalean (dry-adapted) dominance (Xiong et al. 2021). The decline trend in plant generic and species diversity remained in the Changhsingian and Induan (Xiong et al. 2021). Likewise, the aridity is also thought to have contributed to a decline of plant generic diversity in South Africa (Retallack 2021). In Siberia, a pronounced turnover of humid-adapted cordaitopsids to semi-dry- and dry-adapted plants occurred in the middle Changhsingian (*ca*, 253 Ma) (Fig. 5.5, marked by B in Column 2) (Davydov et al. 2021). Also, the ecosystem of glossopterids collapsed in the late Changhsingian (approximately equivalent to bed 20 of Meishan) in the Sydney Basin (Fig. 5.5, marked by C in Column 2) although the climate may not have been as dry as in other parts of the world at that time (Fielding et al. 2019).

The aforementioned massive disappearing of plants means a large amount of CO_2 could not have been sequestrated but released into the air. This is evidenced by the rapid elevation of CO_2 concentration during the Late Permian (Berner 2006; Song et al. 2019). The long-term accumulation of CO_2 in the air must have resulted in the rising of palaeotemperature both on land and in the sea (see Columns 3 and

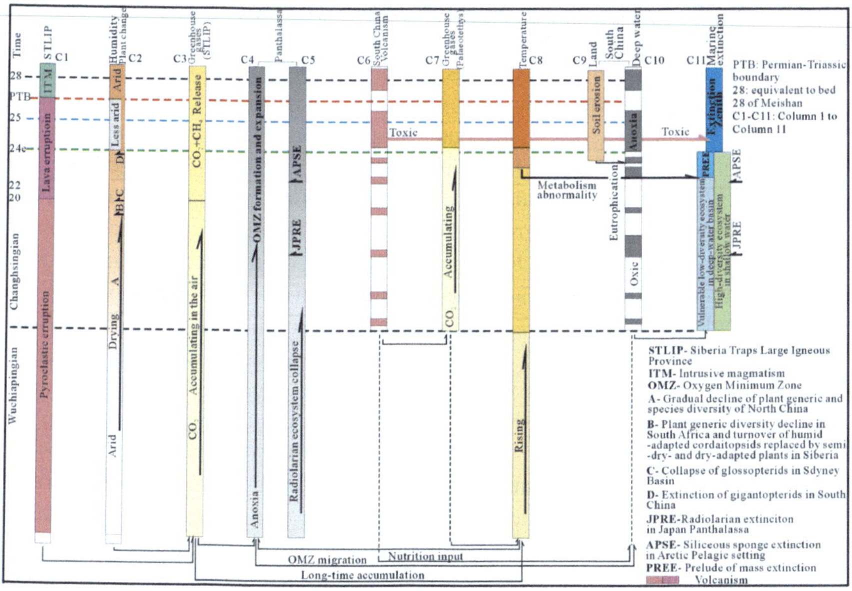

Fig. 5.5 Temporal correlation of postulated multiple killing factors and triggers for the Permian–Triassic mass extinction. Data of the STLP referred to Burgess et al. (2017), Shen et al. (2019a) and Davydov (2021); information on humidity change and plant deterioration in South Africa referred to Retallack (2021), in Siberia to Davydov et al.(2021), in Sydney Basin to Fielding et al. (2019), in North China to Xiong et al. (2021); information of greenhouse gases (STLIP) (Column 3) inferred from the column STLIP (Column 1); data of Panthalassa in Japan after Isozaki (1997), Isozaki (2009), Takahashi et al. (2013) and in Arctic pelagic setting after Algeo et al. (2012), Nabbefeld et al. (2010), Twitchett et al. (2001); data of temperature after Song et al. (2019), Sun et al. (2012), and Joachimski et al. (2020)

8 in Fig. 5.5; Song et al. 2019). Under this global warming background, the ocean current could have been sluggish to form a stratified ocean or even triggered the expansion of the OMZ in the Panthalassa (Column 4 in Fig. 5.5; Isozaki 2009; Takahashi et al. 2013). Generally, it was anoxic in deep waters of a stratified ocean or in the core zone of the OMZ. The anoxic setting was revealed by the presence of gray to black rocks yielding pyrite framboids and geochemical anomalies in the Late Permian to the Early Triassic in the Panthalassa Ocean (Isozaki 1997; Takahashi et al. 2019, 2021). The anoxic condition is believed to have led up to the collapse of the marine chert factory (e.g., radiolarians and siliceous sponges becoming extinct in deep sea of Japan and the Arctic region; Column 5 in Fig. 5.5) (Algeo et al. 2012; Beauchamp and Baud 2002; Isozaki 2009; Nabbefeld et al. 2010; Sano et al. 2012; Takahashi et al. 2009; Twitchett et al. 2001). Consequently, under all these deteriorating and deleterious environmental conditions, the global marine ecosystem must have been put under severe ecological stresses and become increasingly vulnerable and unstable from the Wuchiapingian.

5.3.4.2 Rapid Palaeoenvironmental Deterioration in the Changhsingian

Associated with the continued convergency of continents/blocks in the lead up to the final assembly of Pangea, massive volcanism related to the subduction of the Palaeotethys Ocean took place frequently around South China in the Changhsingian (see Column 6 in Fig. 5.5). A large quantity of CO_2 was released into the air (Column 7 in Fig. 5.5), adding to those emissions generated by the STLIP volcanism, which combinedly greatly boosted the Changhsingian global warming and the anoxic condition in the Panthalassa Ocean (Columns 4 and 8 in Fig. 5.5).

On the other hand, a great amount of nutrition associated with volcanic ash emissions could have been placed into the basin of South China. The Changhsingian nutrient influx boosted the flourishing of shallow-water organisms where a higher diversity was witnessed compared to that of the Wuchiapingian (Wang et al. 2014). That is to say, the nutrition was absorbed or diluted by organisms in the shallow-water setting. However, in the deep-water basin of South China, most of nutrition could have been accumulated (rather than being assimilated) because of the rarity of benthos, leading to the eutrophication and episodic anoxia events (Columns 6 and 10 in Fig. 5.5) as observed at several deep-water or moderately deep-water sections. The episodically oxygen-restricted setting in South China, together with the shoaling/migration of the OMZ from the Panthalassa, would have created the expansion of the OMZ into the deep-water basin of South China (see the part of Sect. 5.3.3.2), disrupting the deep-water marine ecosystem that was probably only suited to a vulnerable low-diversity benthic community in the Changhsingian (see the left part of Column 11 in Fig. 5.5).

5.3.4.3 The Temporal Coincidence of Multiple Killing Factors Triggered the Extinction Zenith in the Permian–Triassic Transition

Following the long-term magmatism from the STLIP and from the surrounding of Palaeotethys Ocean in the Wuchiapingian to the Changhsingian, the volcanism around South China could have been intensified, leading to the rising of influx of CO_2, CH_4, SO_2, Hg, Pb, Zn in an exponential manner across the PTB (equivalent to beds 25–28 of Meishan). That is to say, on the one hand, the intensifying of volcanism carried more and more greenhouse gases into the atmosphere rapidly and further triggered the global warming (Joachimski et al. 2020; Sun et al. 2012). It is important to note that the warming commenced in bed 24 (prior to bed 24e or 25/ onset of extinction zenith), based on oxygen isotope records of conodont apatites at Meishan (Joachimski et al. 2012, 2020). The warming might have caused the metabolism abnormality and even killed organisms (e.g., the extinction of benthos at Laolongdong exactly occurred at the beginning of bed 24 of Meishan section; also see Columns 8 and 11 in Fig. 5.5). On the other hand, the volcanism brought huge amount of toxic chemicals (e.g., Hg, Pb, Zn) that were deadly for animals (Columns 6 and 11 in Fig. 5.5). The abundance peaks of toxic elements (e.g., Hg, Zn, Pb) were recorded in the Permian–Triassic transition throughout the world (e.g., Ghosh et al. 2016; Grasby et al. 2017; Liu et al. 2017; Shen et al. 2021;

Sial et al. 2020). However, a question must be asked: why is there a lack of evidence for the poisoning of organisms caused by volcanism-introduced toxins recorded in rocks prior to the Permian–Triassic transition? One possibility would be that the volcanism-induced toxicity in the marine environments prior to the Permian–Triassic transition was low, therefore allowing organisms to have enough time to detoxify, but the speculation needs a further study.

The anoxic/dysoxic conditions caused by the upward migration of the OMZ had partly contributed to the extinction in the Permian–Triassic transition in settings where water depths were deeper than 50 m during the Changhsingian (see the part of Sect. 5.3.3.2).

The acidification putatively associated with the input of SO_2 during volcanisms could have also played a role in the end-Permian mass extinction (Clarkson et al. 2015; Foster et al. 2022; Garbelli et al. 2017; Hinojosa et al. 2012; Jurikova et al. 2020; Silva-Tamayo et al. 2018), but the extent of its influence and selectivity requires further studies.

In summary, this book emphasizes that during the Permian–Triassic transition, the coincidence of several factors (e.g., global warming, regional toxification related to the volcanism, regional anoxia, together with acidification) would have allowed the synergy and feedback systems of these factors to have worked together and amplified their deadly impacts of one another on the Late Permian global ecosystem and, finally, triggered and accelerated the crisis in the latest Permian (Column 11 in Fig. 5.5).

5.4 Outlook for the Future Studies on the Great Dying

5.4.1 Volcanism

As mentioned above, volcanism is believed to be the primary factor which had created several other palaeoenvironmental disasters (e.g., global warming, anoxia, toxication). However, how different is the influence on organisms exerted by between the long-term protracted volcanism initiated during the Wuchiapingian through to the late Changhsingian (prior to the extinction zenith) and the rapidly intensified volcanism in the Permian–Triassic transition? How much was the volcanism intensified and accelerated in both space and in eruption frequency across the PTB?

5.4.2 Anoxia

Although this book presented how the anoxic condition was developed and influenced the organisms from the Panthalassa Ocean, through deep-water basins, to shallow-water carbonate platforms, further studies using more sections from varied water depths or different palaeogeographic settings are essential for us to reach an agreement on the formation and development of the OMZ. Many factors could have contributed to oxygen restriction boosting the formation of OMZ, but what were these factors remain to be investigated.

5.4.3 The Process of Ecosystem Deterioration and Collapse

This book broadly presented a process of ecosystem deterioration and collapse beginning from the deep sea or deep-water basin and then spreading to the shallow waters. However, a deeper study is required to gain a higher resolution age framework to decipher further how exactly both regional and global ecosystems collapsed in the wake of both long-term protracted and short-lived intensified volcanism.

References

Adachi N, Asada Y, Ezaki Y, Liu JB (2017) Stromatolites near the Permian−Triassic boundary in Chongyang, Hubei Province, South China: a geobiological window into palaeo-oceanic fluctuations following the end-Permian extinction. Palaeogeogr Palaeoclimatol Palaeoecol 475:55–69

Algeo TJ, Henderson CM, Ellwood B, Rowe H, Elswick E, Bates S, Lyons T, Hower JC, Smith C, Maynard B, Hays LE, Summons R, Fulton J, Freeman KH (2012) Evidence for a diachronous Late Permian marine crisis from the Canadian Arctic region. Geol Soc Am Bull 124:1424–1448

Alroy J, Aberhan M, Bottjer DJ, Foote M, Fursich F, Harries PJ, Hendy AJW, Holland SM, Ivany LC, Kiessling W, Kosnik MA, Marshall CR, McGowan AJ, Miller AI, Olszewski T, Patzkowsky DME, Peters SE, Villier L, Wagner PJ, Bonuso N, Borkow PS, Brenneis B, Clapham ME, Fall LM, Ferguson CA, Hanson VL, Krug AZ, Layou KM, Leckey EH, Nurnberg S, Powers CM, Sessa JA, Simpson C, Tomasovych A, Visaggi CC (2008) Phanerozoic trends in the global diversity of marine invertebrates. Science 321:97–100

Baresel B, Bucher H, Bagherpour B, Brosse M, Kuang GD, Schaltegger U (2017) Timing of global regression and microbial bloom linked with the Permian−Triassic boundary mass extinction: implications for driving mechanisms. Sci Rep 7:43630.https://doi.org/10.1038/srep43630

Beauchamp B, Baud A (2002) Growth and demise of Permian biogenic chert along northwest Pangea: evidence for end-Permian collapse of thermohaline circulation. Palaeogeogr Palaeoclimatol Palaeoecol 184:37–63

Becker L, Poreda RJ, Hunt AG, Bunch TE, Rampino M (2001) Permian−Triassic boundary: evidence from extraterrestrial noble gases in fullerenes. Science 291:1530–1533

Berner RA (2006) GEOCARBSULF: a combined model for Phanerozoic atmospheric O_2 and CO_2. Geochim Cosmochim Acta 70:5653–5664

Bond DPG, Wignall PB (2010) Pyrite framboid study of marine Permian-Triassic boundary sections: a complex anoxic event and its relationship to contemporaneous mass extinction. Geol Soc Am Bull 122:1265–1279

Burger BJ, Estrada MV, Gustin MS (2019) What caused Earth's largest mass extinction event? New evidence from the Permian−Triassic boundary in northeastern Utah. Glob Planet Change 177:81–100

Burgess SD, Bowring S (2015) High-precision geochronology confirms voluminous magmatism before, during, and after earth's most severe extinction. Sci Adv 1:e1500470

Burgess SD, Bowring S, Shen SZ (2014) High-precision timeline for earth's most severe extinction. Proc Natl Acad Sci 111:3316–3321

Burgess SD, Muirhead JD, Bowring SA (2017) Initial pulse of Siberian Traps sills as the trigger of the end-Permian mass extinction. Nat Commun 8:164.https://doi.org/10.1038/s41467-017-00083-9

Chen ZQ, Shi GR, Yang FQ, Gao YQ, Tong JN, Peng YQ (2006) An ecologically mixed brachiopod fauna from Changhsingian deep-water basin of South China: consequence of end-Permian global warming. Lethaia 39:79–90

Chen ZQ, Shi GR, Gao YQ, Tong JN, Yang FQ, Peng YQ (2009a) A late Changhsingian (latest Permian) deep-water brachiopod fauna from Guizhou, South China. Alcheringa 33:163–183

Chen ZQ, Tong JN, Zhang KX, Yang H, Liao ZT, Song HJ, Chen J (2009b) Environmental and biotic turnover across the Permian–Triassic boundary on a shallow carbonate platform in western Zhejiang, South China. Aust J Earth Sci 56:775–797

Chen ZQ, Yang H, Luo M, Benton MJ, Kaiho K, Zhao LS, Huang YG, Zhang KX, Fang YH, Jiang HS, Qiu H, Li Y, Tu CY, Shi L, Zhang L, Feng XQ, Chen L (2015) Complete biotic and sedimentary records of the Permian–Triassic transition from Meishan section, South China: ecologically assessing mass extinction and its aftermath. Earth Sci Rev 149:63–103

Chen ZQ, Fang YH, Wignall PB, Guo Z, Su SQ, Liu ZL, Wang RQ, Huang YG, Feng XQ (2022) Microbial blooms triggered pyrite framboid enrichment and oxygen depletion in carbonate platforms immediately after the latest Permian extinction. Geophys Res Lett 49:e2021GL096998. https://doi.org/10.1029/2021GL096998

Chen J, Xu YG (2019) Establishing the link between Permian volcanism and biodiversity changes: insights from geochemical proxies. Gondwana Res 75:68–96

Clarkson MO, Kasemann SA, Wood RA, Lenton TM, Daines SJ, Richoz S, Ohnemueller F, Meixner A, Poulton SW, Tipper ET (2015) Ocean acidification and the Permo–Triassic mass extinction. Science 348:229–232

Cui Y, Li MS, van Soelen EE, Peterse F, Kürschner WM (2021) Massive and rapid predominantly volcanic CO_2 emission during the end-Permian mass extinction. Proc Nat Acad Sci 118:e2014701118. https://doi.org/10.1073/pnas.2014701118

Davydov VI (2021) Tunguska coals, Siberian sills and the Permian–Triassic extinction. Earth Sci Rev 212:103438. https://doi.org/10.1016/j.earscirev.2020.103438

Davydov VI, Karasev EV, Nurgalieva NG, Schmitz MD, Budnikov IV, Biakov AS, Kuzina DN, Silantiev VV, Urazaeva MN, Zharinova VV, Zorina SO, Gareev B, Vasilenko DV (2021) Climate and biotic evolution during the Permian–Triassic transition in the temperate Northern Hemisphere, Kuznetsk Basin, Siberia, Russia. Palaeogeogr Palaeoclimatol Palaeoecol 573:110432. https://doi.org/10.1016/j.palaeo.2021.110432

Dittert N, Henrich R (2000) Carbonate dissolution in the South Atlantic Ocean: evidence from ultrastructure breakdown in Globigerina bulloides. Deep-Sea Research I 47:603–620

Dupraz C, Visscher PT (2005) Microbial lithification in marine stromatolites and hypersaline mats. Trends Microbiol 13:429–438

Edmond JM, Huh Y (2003) Non-steady state carbonate recycling and implications for the evolution of atmospheric pCO_2. Earth Planet Sci Lett 216:125–139

Erwin DH (1993) The great Paleozoic crisis: life and death in the Permian. Columbia University Press, New York, p 327

Fan JX, Shen SZ, Erwin DH, Sadler PM, MacLeod N, Cheng QM, Hou XD, Yang J, Wang DX, Wang Y, Zhang H, Chen X, Xiang G, Zhang YC, Shi YK, Yuan DX, Chen Q, Zhang LN, Li C, Zhao YY (2020) A high-resolution summary of Cambrian to Early Triassic marine invertebrate biodiversity. Science 367:272–277

Fang Q, Hong HL, Chen ZQ, Yu JX, Wang CW, Yin K, Zhao LL, Liu Z, Cheng F, Gong NN, Furnes H (2017) Microbial proliferation coinciding with volcanism during the Permian–Triassic transition: new, direct evidence from volcanic ashes, South China. Palaeogeogr Palaeoclimatol Palaeoecol 474:164–186

Farley KA, Mukhopadhyay S (2001) An Extraterrestrial Impact at the Permian–Triassic boundary? Science 293:2343.https://doi.org/10.1126/science.293.5539.2343a

Feng QL, Algeo TJ (2014) Evolution of oceanic redox conditions during the Permo–Triassic: evidence from radiolarian deepwater facies. Earth-Sci Rev 137:34–51

Fielding CR, Frank TD, McLoughlin S, Vajda V, Mays C, Tevyaw AP, Winguth A, Winguth C, Nicoll RS, Bocking M, Crowley JL (2019) Age and pattern of the southern high-latitude continental end-Permian extinction constrained by multiproxy analysis. Nat Commun 10:385.https://doi.org/10.1038/s41467-018-07934-z

Foster WJ, Heindel K, Richoz S, Gliwa J, Lehrmann DJ, Baud A, Kolar-Jurkovšek T, Aljinović D, Jurkovšek B, Korn D, Martindale RC, Peckmann J (2020) Suppressed competitive exclusion enabled the proliferation of Permian/Triassic boundary microbialites. Deposition Rec 6:62–74

Foster WJ, Hirtz JA, Farrell C, Reistroffer M, Twitchett RJ, Martindale RC (2022) Bioindicators of severe ocean acidification are absent from the end-Permian mass extinction. Sci Rep 12:1202.https://doi.org/10.1038/s41598-022-04991-9

Frogner P, Gislason SR, Óskarsson N (2001) Fertilizing potential of volcanic ash in ocean surface water. Geology 29:487–490

Garbelli C, Angiolini L, Shen SZ (2017) Biomineralization and global change: a new perspective for understanding the end-Permian extinction. Geology 45:19–22

Gastaldo RA, Kamo SL, Neveling J, Geissman JW, Looy CV, Martini AM (2020) The base of the Lystrosaurus Assemblage Zone, Karoo Basin, predates the end-Permian marine extinction. Nat Commun 11:1428.https://doi.org/10.1038/s41467-020-15243-7

Georgiev SV, Stein HJ, Yang G, Hannah JL, Böttcher ME, Grice K, Holman AI, Turgeon S, Simonsen S, Cloquet C (2020) Late Permian–Early Triassic environmental changes recorded by multi-isotope (Re–Os–N–Hg) data and trace metal distribution from the Hovea-3 section, Western Australia. Gondwana Res 88:353–372

Ghosh N, Basu AR, Bhargava ON, Shukla UK, Ghatak A, Garzione CN, Ahluwalia AD (2016) Catastrophic environmental transition at the Permian–Triassic Neo-Tethyan margin of Gondwanaland: geochemical, isotopic and sedimentological evidence in the Spiti Valley, India. Gondwana Res 34:324–345

Grasby SE, Shen WJ, Yin RS, Gleason JD, Blum JD, Lepak RF, Hurley JP, Beauchamp B (2017) Isotopic signatures of mercury contamination in latest Permian oceans. Geology 45:55–58

Grasby SE, Bond DPG, Wignall PB, Yin R, Strachan LJ, Takahashi S (2021) Transient Permian–Triassic euxinia in the southern Panthalassa deep ocean. Geology 49:889–893

Grice K, Cao CQ, Love GD, Böttcher ME, Twitchett RJ, Grosjean E, Summons RE, Turgeon SC, Dunning W, Jin YG (2005) Photic zone euxinia during the Permian–Triassic superanoxic event. Science 307:706–709

Hallam A, Wignall PB (1999) Mass extinctions and sea-level changes. Earth Sci Rev 48:217–250

He WH, Shi GR, Feng QL, Campi MJ, Gu SZ, Bu JJ, Peng YQ, Meng YY (2007) Brachiopod miniaturization and its possible causes during the Permian–Triassic crisis in deep water environments, South China. Palaeogeogr Palaeoclimatol Palaeoecol 252:145–163

He WH, Zhang Y, Zheng YE, Zhang KX, Gui BW, Feng QL (2008) Late Changhsingian (latest Permian) radiolarian fauna from Chaohu, Anhui and a comparison with its contemporary faunas of South China. Alcheringa 32:199–222

He WH, Zhang Y, Zhang Q, Zhang KX, Yuan AH, Feng QL (2011) A latest Permian radiolarian fauna from Hushan, South China and its geological implications. Alcheringa 35:471–496

He WH, Shi GR, Zhang Y, Yang TL, Zhang KX, Wu SB, Niu ZJ, Zhang ZY (2014) Changhsingian (latest Permian) deep-water brachiopod fauna from South China. J Syst Paleontol 12:907–960

He WH, Shi GR, Twitchett RJ, Zhang Y, Zhang KX, Song HJ, Yue ML, Wu SB, Wu HT, Yang TL, Xiao YF (2015a) Late Permian marine ecosystem collapse began in deeper waters: evidence from brachiopod diversity and body size changes. Geobiology 13:123–138

He WH, Zhang KX, Wu SB, Feng QL, Yang TL, Yue ML, Xiao YF, Wu HT, Zhang Y, Wang GD, Chen B (2015b) End-Permian faunas from Yangtze basin and its marginal region: implications for palaeogeographical and tectonic environments. Earth Sci J China Univ Geosci 40:275–289 [In Chinese with English abstract]

He WH, Shi GR, Xiao YF, Zhang KX, Yang TL, Wu HT, Zhang Y, Chen B, Yue ML, Shen J, Wang YB, Yang H, Wu SB (2017) Body-size changes of latest Permian brachiopods in varied palaeogeographic settings in South China and implications for controls on animal miniaturization in a highly stressed marine ecosystem. Palaeogeogr Palaeoclimatol Palaeoecol 486:33–45

He WH, Shi GR, Zhang KX, Yang TL, Shen SZ, Zhang Y (2019) Brachiopods around the Permian–Triassic boundary of South China. Springer, Singapore, 261 pp

He WH, Weldon EA, Yang TL, Wang H, Xiao YF, Wu HT, Zhang KX, Wang YB, Wu SB (2020) The palaeoenvironmental and palaeobiogeographical significance of the Late Permian deep-water brachiopod fauna from Dongpan, South China, including descriptions of *Micromartinia* He & Weldon gen. nov. (Micromartiniidae He & Weldon fam. nov.) and Minutomarginifera nom. nov. J Syst Paleontol 18:885–909

Hinojosa JL, Brown ST, Chen J, DePaolo DJ, Paytan A, Shen SZ, Payne JL (2012) Evidence for end-Permian ocean acidifi cation from calcium isotopes in biogenic apatite. Geology 40:743–746

Hoffmann LJ, Breitbarth E, Ardelan MV, Duggen S, Olgun N, Hassellöv M, Wängberg SÄ (2012) Influence of trace metal release from volcanic ash on growth of Thalassiosira pseudonana and Emiliania huxleyi. Mar Chem 132–133:28–33

Hu ZY, Li WQ, Zhang H, Krainer K, Zheng QF, Xia ZG, Hu WX, Shen SZ (2021) Mg isotope evidence for restriction events within the Paleotethys ocean around the Permian–Triassic transition. Earth Planet Sci Lett 556:116704. https://doi.org/10.1016/j.epsl.2020.116704

Huang YF, He WH, Liao W, Wang YB, Yi ZX, Yang H (2022) Two pulses of increasing terrestrial input to marine environment during the Permian–Triassic transition. Palaeogeogr Palaeoclimatol Palaeoecol 586:110753. https://doi.org/10.1016/j.palaeo.2021.110753

Immenhauser A (2009) Estimating palaeo-water depth from the physical rock record. Earth Sci Rev 96:107–139

Isozaki Y, Shimizu N, Yao JX, Ji ZS, Matsuda T (2007) End-Permian extinction and volcanism-induced environmental stress: the Permian–Triassic boundary interval of lower-slope facies at Chaotian, South China. Palaeogeogr Palaeoclimatol Palaeoecol 252:218–238

Isozaki Y (1997) Permo–Triassic boundary superanoxia and stratified superocean: records from lost deep sea. Science 276:235–238

Isozaki Y (2009) Integrated "plume winter" scenario for the double-phased extinction during the Paleozoic–Mesozoic transition: The G–LB and P–TB events from a Panthalassan perspective. J Asian Earth Sci 36:459–480

Jensen MM, Petersen J, Dalsgaard T, Thamdrup B (2009) Pathways, rates, and regulation of N_2 production in the chemocline of an anoxic basin, Mariager Fjord, Denmark. Mar Chem 113:102–113

Jiang L, Cai CF, Xiang L, Fan JX, Li KK, Cai LL, Algeo TJ, Zhang FF (2022) Kerogen-specific isotope variations during the end-Permian mass extinction in South China. Earth Sci Rev 226:103912. https://doi.org/10.1016/j.earscirev.2021.103912

Jin YG, Wang Y, Wang W, Shang QH, Cao CQ, Erwin DH (2000) Pattern of marine mass extinction near the Permian–Triassic boundary in South China. Science 289:432–436

Joachimski MM, Lai XL, Shen SZ, Jiang HS, Luo GM, Chen B, Chen J, Sun YD (2012) Climate warming in the latest Permian and the Permian–Triassic mass extinction. Geology 40:195–198

Joachimski MM, Alekseev AS, Grigoryan A, Gatovsky YA (2020) Siberian Trap volcanism, global warming and the Permian–Triassic mass extinction: new insights from Armenian Permian–Triassic sections. Geol Soc Am Bull 132:427–443

Jurikova H, Gutjahr M, Wallmann K, Flögel S, Liebetrau V, Posenato R, Angiolini L, Garbelli C, Brand U, Wiedenbeck M, Eisenhauer A (2020) Permian–Triassic mass extinction pulses driven by major marine carbon cycle perturbations. Nat Geosci 13:745–750

Kaiho K, Kajiwara Y, Nakano T, Miura Y, Kuwahata H, Tazaki K, Ueshima M, Chen ZQ, Shi GR (2001) End-Permian catastrophe by a bolide impact: evidence of a gigantic release of sulfur from the mantle. Geology 29:815–818

Kaiho K, Aftabuzzaman M, Jones DS, Tian L (2021) Pulsed volcanic combustion events coincident with the end-Permian terrestrial disturbance and the following global crisis. Geology 49:289–293

Kozur H (1993) Upper Permian radiolarians from the Sosio Valley Area, Western Sicily (Italy) and from the uppermost Lamar limestone of West Texas. Jahrbuch Der Geologischen Bundesanstalt Wien 136:99–123

Kump L (2018) Climate change and marine mass extinction. Science 362:1113–1114
Kump LR, Pavlov A, Arthur MA (2005) Massive release of hydrogen sulfide to the surface ocean and atmosphere during intervals of oceanic anoxia. Geology 33:397–400
Levin LA (2003) Oxygen minimum zone benthos: adaptation and community response to hypoxia. Oceanogr Mar Biol Annu Rev 41:1–45
Li GS, Wang YB, Shi GR, Liao W, Yu LX (2016) Fluctuations of redox conditions across the Permian–Triassic boundary—new evidence from the GSSP section in Meishan of South China. Palaeogeogr Palaeoclimatol Palaeoecol 448:48–58
Li GS, Liao W, Li S, Wang YB, Lai ZP (2021) Different triggers for the two pulses of mass extinction across the Permian and Triassic boundary. Sci Rep 11:6686.https://doi.org/10.1038/s41598-021-86111-7
Liao ZT (1979) Brachiopod assemblage zone of Changhsing stage and brachiopods from Permo–Triassic boundary beds in China. Acta Stratigraphica Sinica 3:200–208 [in Chinese]
Liao ZT (1984) New genus and species of Late Permian and earliest Triassic brachiopods from Jiangsu, Zhejiang and Anhui Provinces. Acta of Palaeontologica Sinica 23:276–285 [in Chinese, with English abstract]
Liu SA, Wu HC, Shen SZ, Jiang GQ, Zhang SH, Lv YW, Zhang H, Li SG (2017) Zinc isotope evidence for intensive magmatism immediately before the end-Permian mass extinction. Geology 45:343–346
Liu ZY, Selby D, Zhang H, Shen SZ (2020) Evidence for volcanism and weathering during the Permian–Triassic mass extinction from Meishan (South China) osmium isotope record. Palaeogeogr Palaeoclimatol Palaeoecol 553:109790. https://doi.org/10.1016/j.palaeo.2020.109790
Mahowald NM, Hamilton DS, Mackey KRM, Moore JK, Baker AR, Scanza RA, Zhang Y (2018) Aerosol trace metal leaching and impacts on marine microorganisms. Nat Commun 9:2614.https://doi.org/10.1038/s41467-018-04970-7
Mays C, Vajda V, Frank TD, Fielding CR, Nicoll RS, Tevyaw AP, McLoughlin S (2020) Refined Permian–Triassic floristic timeline reveals early collapse and delayed recovery of south polar terrestrial ecosystems. Geol Soc Am Bull 132:1489–1513
Nabbefeld B, Grice K, Twitchett RJ, Summons RE, Hays L, Böttcher ME, Asif M (2010) An integrated biomarker, isotopic and palaeoenvironmental study through the Late Permian event at Lusitaniadalen, Spitsbergen. Earth Planet Sci Lett 291:84–96
Olsson J, Stipp SLS, Dalby KN, Gislason SR (2013) Rapid release of metal salts and nutrients from the 2011 Grímsvötn, Iceland volcanic ash. Geochim Cosmochim Acta 123:134–149
Onoue T, Takahata N, Miura M, Sato H, Ishikawa A, Soda K, Sano Y, Isozaki Y (2019) Enhanced flux of extraterrestrial ^{3}He across the Permian–Triassic boundary. Progr Earth Planet Sci 6:18.https://doi.org/10.1186/s40645-019-0267-0
Petrash DA, Gingras MK, Lalonde SV, Orange F, Pecoits E, Konhauser KO (2012) Dynamic controls on accretion and lithification of modern gypsum-dominated thrombolites, Los Roques, Venezuela. Sed Geol 245–246:29–47
Raup DM (1986) Biological extinction in Earth history. Science 231:1528–1533
Regelous M, Regelous A, Grasby SE, Bond DPG, Haase KM, Gleiβner S, Wignall PB (2020) Tellurium in Late Permian–Early Triassic sediments as a proxy for Siberian flood basalt volcanism. Geochem Geophys Geosyst 21:e2020GC009064. https://doi.org/10.1029/2020GC009064
Renne PR, Basu AR (1991) Rapid eruption of the Siberian Traps flood basalts at the Permo-Triassic boundary. Science 253:176–179
Retallack GJ, Jahren AH (2008) Methane release from igneous intrusion of coal during Late Permian extinction events. J Geol 116:1–20
Retallack GJ (2021) Multiple Permian–Triassic life crises on land and at sea. Glob Planet Change 198:103415. https://doi.org/10.1016/j.gloplacha.2020.103415
Sano H, Wada T, Naraoka H (2012) Late Permian to Early Triassic environmental changes in the Panthalassic Ocean: record from the seamount-associated deep-marine siliceous rocks, central Japan. Palaeogeogr Palaeoclimatol Palaeoecol 363–364:1–10

Sepkoski JJ (1981) A factor analytic description of the Phanerozoic marine fossil record. Paleobiology 7:36–53

Shen SZ, Crowley JL, Wang Y, Bowring SA, Erwin DH, Sadler PM, Cao CQ, Rothman DH, Henderson CM, Ramezani J, Zhang H, Shen YA, Wang XD, Wang W, Mu L, Li WZ, Tang YG, Liu XL, Liu LJ, Zeng Y, Jiang YF, Jin YG (2011a) Calibrating the end-Permian mass extinction. Science 334:1367–1372

Shen YA, Farquhar J, Zhang H, Masterson A, Zhang TG, Wing BA (2011) Multiple S-isotopic evidence for episodic shoaling of anoxic water during Late Permian mass extinction. Nat Commun 2:210.https://doi.org/10.1038/ncomms1217

Shen J, Algeo TJ, Zhou L, Feng QL, Yu JX, Ellwood B (2012) Volcanic perturbations of the marine environment in South China preceding the latest Permian mass extinction and their biotic effects. Geobiology 10:82–103

Shen J, Yu JX, Chen JB, Algeo TJ, Xu GZ, Feng QL, Shi X, Planavsky NJ, Shu WC, Xie SC (2019a) Mercury evidence of intense volcanic effects on land during the Permian–Triassic transition. Geology 47:1117–1121

Shen J, Chen JB, Algeo TJ, Yuan SL, Feng QL, Yu JX, Zhou L, O'Connell B, Planavsky NJ (2019b) Evidence for a prolonged Permian–Triassic extinction interval from global marine mercury records. Nat Commun 10:1563https://doi.org/10.1038/s41467-019-09620-0

Shen SZ, Ramezani J, Chen J, Cao CQ, Erwin DH, Zhang H, Xiang L, Schoepfer SD, Henderson CM, Zheng QF, Bowring SA, Wang Y, Li XH, Wang XD, Yuan DX, Zhang YC, Mu L, Wang J, Wu YS (2019c) A sudden end-Permian mass extinction in South China. Geol Soc Am Bull 131:205–223

Shen J, Chen JB, Algeo TJ, Feng QL, Yu JX, Xu YG, Xu GZ, Lei Y, Planavsky NJ, Xie SC (2021) Mercury fluxes record regional volcanism in the South China craton prior to the end-Permian mass extinction. Geology 49:452–456

Shen J, Chen JB, Yu JX, Algeo TJ, Smith RMH, Botha J, Frank TD, Fielding CR, Ward PD, Mather TA (2023) Mercury evidence from southern Pangea terrestrial sections for end-Permian global volcanic effects. Nat Commun 14:6. https://doi.org/10.1038/s41467-022-35272-8

Sial AN, Chen JB, Lacerda LD, Korte C, Spangenberg JE, Silva-Tamayo JC, Gaucher C, Ferreira VP, Barbosa JA, Pereira NS, Benigno AP (2020) Globally enhanced Hg deposition and Hg isotopes in sections straddling the Permian–Triassic boundary: link to volcanism. Palaeogeogr Palaeoclimatol Palaeoecol 540:109537.https://doi.org/10.1016/j.palaeo.2019.109537

Silva-Tamayo JC, Lau KV, Jost AB, Payne JL, Wignall PB, Newton RJ, Eisenhauer A, Depaolo DJ, Brown S, Maher K, Lehrmann DJ, Altiner D, Yu MY, Richoz S, Paytan A (2018) Global perturbation of the marine calcium cycle during the Permian–Triassic transition. Geol Soc Am Bull 130:1323–1338

Song HJ, Tong JN, Chen ZQ (2009) Two episodes of foraminiferal extinction near the Permian–Triassic at the Meishan section, South China. Aust J Earth Sci 56:765–773

Song HJ, Wignall PB, Tong JN, Yin HF (2013) Two pulses of extinction during the Permian–Triassic crisis. Nat Geosci 6:52–56

Song HJ, Wignall PB, Song HY, Dai X, Chu DL (2019) Seawater temperature and dissolved oxygen over the past 500 million years. J Earth Sci 30:236–243

Sun YD, Joachimski MM, Wignall PB, Yan CB, Chen YL, Jiang HS, Wang LN, Lai XL (2012) Lethally hot temperatures during the Early Triassic greenhouse. Science 338:366–370

Svensen H, Planke S, Polozov AG, Schmidbauer N, Corfu F, Podladchikov YY, Jamtveit B (2009) Siberian gas venting and the end-Permian environmental crisis. Earth Planet Sci Lett 277:490–500

Takahashi S, Yamakita S, Suzuki N, Kaiho K, Ehiro M (2009) High organic carbon content and a decrease in radiolarians at the end of the Permian in a newly discovered continuous pelagic section: a coincidence? Palaeogeogr Palaeoclimatol Palaeoecol 271:1–12

Takahashi S, Kaiho K, Hori RS, Gorjan P, Watanabe T, Yamakita S, Aita Y, Takemura A, Spörli KB, Kakegawa T, Oba M (2013) Sulfur isotope profiles in the pelagic Panthalassic deep sea during the Permian–Triassic transition. Glob Planet Change 105:68–78

Takahashi S, Nakada R, Watanabe Y, Takahashi Y (2019) Iron-depleted pelagic water at the end-Permian mass extinction inferred from chemical species of iron and molybdenum in deep-sea sedimentary rocks. Palaeogeogr Palaeoclimatol Palaeoecol 516:384–399

Takahashi S, Hori RS, Yamakita S, Aita Y, Takemura A, Ikehara M, Xiong YJ, Poulton SW, Wignall PB, Itai T, Campbell HJ, Spörli BK (2021) Progressive development of ocean anoxia in the end-Permian pelagic Panthalassa. Glob Planet Change 207:103650. https://doi.org/10.1016/j.gloplacha.2021.103650

Twitchett RJ, Looy CV, Morante R, Visscher H, Wignall PB (2001) Rapid and synchronous collapse of marine and terrestrial ecosystems during the end-Permian mass extinction event. Geology 29:351–354

Wang Y, Shen SZ, Zhang YC, Wang XD, Wang W, Sadler PM, Erwin DH, Crowley JL, Henderson CM (2014) Quantifying the process and abruptness of the end-Permian mass extinction. Paleobiology 41:113–129

Wang XD, Cawood PA, Zhao LS, Chen ZQ, Lyu ZY, Ma B (2019) Convergent continental margin volcanic source for ash beds at the Permian–Triassic boundary, South China: constraints from trace elements and Hfisotopes. Palaeogeogr Palaeoclimatol Palaeoecol 519:154–165

Wang H, He WH, Xiao YF, Yang TL, Zhang KX, Wu HT, Huang YF, Peng XF, Wu SB (2023) Stepwise collapse of biotic communities and its relations to oxygen depletion along the north margin of during the Permian–Triassic transition. Palaeogeogr Palaeoclimatol Palaeoecol. https://doi.org/10.1016/j.palaeo.2023.111569

Weber ME, Pisias NG (1999) Spatial and temporal distribution of biogenic carbonate and opal in deep-sea sediments from the eastern equatorial Pacific: implications for ocean history since 1.3Ma. Earth Planet Sci Lett 174:59–73

Weber ME, von Stackelberg U (2000) Variability of surface sediments in the Peru Basin: dependence on water depth, productivity, bottom water flow, and seafloor topography. Mar Geol 163:169–184

Wignall PB, Hallam A (1992) Anoxia as a cause of the Permian–Triassic mass extinction: facies evidence from northern Italy and the western United States. Palaeogeogr Palaeoclimatol Palaeoecol 93:21–46

Wignall PB, Twitchett RJ (1996) Oceanic anoxia and the end Permian mass extinction. Science 272:1155–1158

Wu YS (2017) A latest Permian non-reef calcisponge fauna from Laibin, Guangxi, southern China and its significance. J Palaeogeogr 6:60–68

Wu SB, Wei M, Zhang KX (1986) Facies changes and controlling factors of the Late Permian Changxing limestone in the Changxing area. Geol Rev 32:419–425 [in Chinese with English abstract]

Wu HT, He WH, Shi GR, Zhang KX, Yang TL, Zhang Y, Xiao YF, Chen B, Wu SB (2018a) A new Permian-Triassic boundary brachiopod fauna from the Xinmin section, southwestern Guizhou, south China and its extinction patterns. Alcheringa 42:339–372

Wu HT, He WH, Weldon EA (2018b) Prelude of benthic community collapse during the end-Permian mass extinction in siliciclastic offshore sub-basin: Brachiopod evidence from South China. Glob Planet Change 163:158–170

Wu BJ, Luo GM, Joachimski MM, Wignall PB, Lei LD, Huang JH, Lai XL (2021) Carbon and nitrogen isotope evidence for widespread presence of anoxic intermediate waters before and during the Permian–Triassic mass extinction. Geol Soc Am Bull 134:1397–1413

Xiang L, Schoepfer SD, Zhang H, Yuan DX, Cao CQ, Zheng QF, Henderson CM, Shen SZ (2016) Oceanic redox evolution across the end-Permian mass extinction at Shangsi, South China. Palaeogeogr Palaeoclimatol Palaeoecol 448:59–71

Xiang L, Schoepfer SD, Yuan DX, Zheng QF, Zhang H (2022) Oceanic redox evolution across the end-Permian mass extinction at Penglaitan section, South China. Palaeoworld 31:93–102

Xiao YF, Suzuki N, He WH (2017) Water depths of the latest Permian (Changhsingian) radiolarians estimated from correspondence analysis. Earth Sci Rev 173:141–158

Xie SC, Algeo TJ, Zhou WF, Ruan XY, Luo GM, Huang JH, Yan JX (2017) Contrasting microbial community changes during mass extinctions at the Middle/Late Permian and Permian/Triassic boundaries. Earth Planet Sci Lett 460:180–191

Xiong CH, Wang JS, Huang P, Cascales-Miñana B, Cleal CJ, Benton MJ, Xue JZ (2021) Plant resilience and extinctions through the Permian to Middle Triassic on the North China block: a multilevel diversity analysis of macrofossil records. Earth Sci Rev 223:103846. https://doi.org/10.1016/j.earscirev.2021.103846

Yang ZY, Yin HF, Wu SB, Yang FQ, Ding MH, Xu GR (1987) Permian–Triassic Boundary Stratigraphy and Fauna of South China. Geological Publishing House, Beijing, 378 pp [in Chinese with English abstract]

Yang ZY, Wu SB, Yin HF, Xu GR, Zhang KX (1991) Permo–Triassic events of South China. Geological Publishing House, Beijing, 183 pp [in Chinese with English abstract]

Yin HF, Feng QL, Lai XL, Baud A, Tong JN (2007) The protracted Permo–Triassic crisis and multi-episode extinction around the Permian–Triassic boundary. Glob Planet Change 55:1–20

Yin HF, Jiang HS, Xia WC, Feng QL, Zhang N, Shen J (2014) The end-Permian regression in South China and its implication on mass extinction. Earth Sci Rev 137:19–33

Zhang KX, Tong JN, Yin HF, Wu SB (1997) Sequence stratigraphy of the Permian–Triassic boundary section of Changxing, Zhejiang, southern China. Acta Geol Sin 71:90–103

Zhang H, Cao CQ, Liu XL, Mu L, Zheng QF, Liu F, Xiang L, Liu LJ, Shen SZ (2016) The terrestrial end-Permian mass extinction in South China. Palaeogeogr Palaeoclimatol Palaeoecol 448:108–124

Zhang Y, Shi GR, Wu HT, Yang TL, He WH, Yuan AH, Lei Y (2017) Community replacement, ecological shift and early warning signals prior to the end-Permian mass extinction: a case study from a nearshore clastic-shelf section in South China. Palaeogeogr Palaeoclimatol Palaeoecol 487:118–135

Zhang FF, Shen SZ, Cui Y, Lenton TM, Dahl TW, Zhang H, Zheng QF, Wang WQ, Krainer K, Anbar AD (2020) Two distinct episodes of marine anoxia during the Permian–Triassic crisis evidenced by uranium isotopes in marine dolostones. Geochim Cosmochim Acta 287:165–179

Zhang H, Zhang FF, Chen JB, Erwin DH, Syverson DD, Ni P, Rampino M, Chi Z, Cai YF, Xiang L, Li WQ, Liu SA, Wang RC, Wang XD, Feng Z, Li HM, Zhang T, Cai HM, Zheng W, Cui Y, Zhu XK, Hou ZQ, Wu FY, Xu YG, Planavsky N, Shen SZ (2021) Felsic volcanism as a factor driving the end-Permian mass extinction. Sci Adv 7.https://doi.org/10.1126/sciadv.abh1390

Zhao TY, Algeo TJ, Feng QL, Zi JW, Xu GZ (2019) Tracing the provenance of volcanic ash in Permian–Triassic boundary strata, South China: Constraints from inherited and syn-depositional magmatic Zircons. Palaeogeogr Palaeoclimatol Palaeoecol 516:190–202

Zhou L, Kyte FT (1988) The Permian–Triassic boundary event: a geochemical study of three Chinese sections. Earth Planet Sci Lett 90:411–421

Zopfi J, Ferdelman TG, Jørgensen BB, Teske A, Thamdrup B (2001) Influence of water column dynamics on sulfide oxidation and other major biogeochemical processes in the chemocline of Mariager Fjord (Denmark). Mar Chem 74:29–51

Appendix A: Information on the Studied Sections

Section name and abbreviation	GPS or locality	Palaeogeographic setting (P–T transition)	Formations (Fm) covered and ages	Thickness of the measured section	Key references
Meishan (MS)	31° 04′ 50″ N, 119° 42′ 22″ E; north of Zhejiang	Ramp of carbonate platform	Yinkeng Fm (P_3–T_1); Changhsing Fm (P_3)	Yinkeng Fm: 11.20 m; Changhsing Fm: 42.42 m	Ammonoids: Zhao et al. (1981) and Yang et al. (1987); conodonts: Mei et al. (1998), Jiang et al. (2007), Zhang et al. (2007) and Yuan et al. (2014) (updated here); foraminifers: Song et al. (2009)
Daoduishan (DDS)	31° 05′ 44″ N, 119° 43′ 50″ E; 2.5 km northeast of the Meishan D section	Ramp of carbonate platform	Yinkeng Fm (P_3–T_1); Changhsing Fm (P_3)	Yinkeng Fm: 0.68 m; Changhsing Fm: 7.10 m	Brachiopods: He et al. (2016, 2019) (updated here)
Huangzhishan (HZS)	30° 55′ 18″ N, 119° 59′ 21″ E; north of Zhejiang	Carbonate platform	Yinkeng Fm (P_3–T_1); Changhsing Fm (P_3)	Yinkeng Fm: 11.19 m; Changhsing Fm: 4.76 m	Brachiopods: He et al. (2015) (updated here); foraminifers, corals, crinoids, bryozoans, ostracods and bivalves in the upper part of the Changhsing Formation after Chen et al. (2009); conodonts referred to Chen et al. (2008, 2009)
Majiashan (MJS)	31° 37′ 33″ N, 117° 49′ 12″ E; central Anhui	Deep-water basin	Yinkeng Fm (P_3–T_1); Talung Fm (P_3)	Yinkeng Fm: 1.21 m; Talung Fm: 14.85 m	Brachiopods: He et al. (2015); radiolarians: He et al. (2008) and Gui et al. (2009); carbon isotope based on Fang, (2021)

(continued)

© The Editor(s) (if applicable) and The Author(s), under exclusive license to Springer Nature Singapore Pte Ltd. 2023

W. He et al. (eds.), *Stratigraphy Around the Permian–Triassic Boundary of South China*, New Records of the Great Dying in South China,
https://doi.org/10.1007/978-981-99-9350-5

(continued)

Section name and abbreviation	GPS or locality	Palaeogeographic setting (P–T transition)	Formations (Fm) covered and ages	Thickness of the measured section	Key references
Yangou (YG)	29° 11′ 45″ N, 117° 21′ 03″ E; northeast of Jiangxi	Carbonate platform	Taye Fm (P_3–T_1); Changhsing Fm (P_3)	Taye Fm: 8.09 m; Changhsing Fm: 11.10 m	Conodonts: Sun et al. (2012) (updated here); foraminifers: Tian et al. (2014a); gastropods: Sun et al. (2021); carbon isotopes are after Tian et al. (2014b)
Jingshan (JS)	30° 55′ 26″ N, 113° 06′ 00″ E; central Hubei	Deep-water basin	Taye Fm (T_1); Talung Fm (P_3)	Taye Fm: 8.95 m; Talung Fm: 13.25 m	Fossils and redox conditions after Li, (2016)
Chibi (CB)	29° 41′ 43″ N, 113° 52′ 02″ E; southeast of Hubei	Carbonate platform	Taye Fm (P_3–T_1); Changhsing Fm (P_3)	Taye Fm: 8.37 m; Changhsing Fm: 3.59 m	Fossils, redox conditions and carbon isotopes after Li, (2016)
Zhaixia (ZX)	29° 29′ 40″ N, 114° 08′ 00″ E; southeast of Hubei	Carbonate platform	Taye Fm (including microbialites) (P_3–T_1); Changhsing Fm (P_3)	Taye Fm: 15.88 m (microbialites: 8.62 m); Changhsing Fm: 4 m	Ostracods: Wan, (2021) and Liu, (2010); conodonts: Yang et al. (2006); carbon isotopes after Wang et al. (2019)
Xiejiaping (XJP)	30° 20′ 12″ N, 110° 35′ 53″ E; southwest of Hubei	Basin	Taye Fm (T_1); Changhsing Fm (P_3); Talung Fm (P_3); Wuchiaping Fm (P_3)	Taye Fm: 4.09 m; Changhsing Fm: 7.08 m; Talung Fm: 3.76 m; Wuchiaping Fm: 2.00 m	Brachiopods referred to He et al. (2014); other fossils based on Niu Zhijun (unpublished)
Shangsi (SS)	32° 19′ 10″ N, 105° 27′ 19″ E; northeast of Sichuan	Deep-water basin	Feixianguan Fm (P_3–T_1); Talung Fm (P_3); Wuchiaping Fm (P_3)	Feixianguan Fm: 14.39 m; Talung Fm: 42.13 m	Radiolarians: Yao and Kuwahara, (1999); other fossils in beds 11 to 21: Li et al. (1989) (updated here); lithologic feature in beds 11 to 21 referred to Yang et al. (1987)
Laolongdong (LLD)	29° 56′ 24″ N, 106° 33′ 36″ E; Chongqing	Carbonate platform	Feixianguan Fm (including microbialites) (P_3–T_1); Changhsing Fm (P_3)	Feixianguan Fm: 3.48 m (microbialites: 2.34 m); Changhsing Fm: 1.75 m	Fossils, redox conditions and carbon isotopes: Huang et al. (2022)
Liangfangya (LFY)	29° 30′ 14″ N, 106° 24′ 15″ E; west of Chongqing	Carbonate platform	Feixianguan Fm (P_3–T_1); Changhsing Fm (P_3); Lungtan Fm (P_3)	Feixianguan Fm: 0.45 m; Changhsing Fm: 91.10 m	Conodonts: Yuan and Shen, (2011); other fossils and lithologic feature: Shen and He, (1991) (fossils updated here)

(continued)

(continued)

Section name and abbreviation	GPS or locality	Palaeogeographic setting (P–T transition)	Formations (Fm) covered and ages	Thickness of the measured section	Key references
Rencunping (RCP)	29° 34′ 49″ N, 110° 06′ 02″ E; northwest of Hunan	Deep-water basin	Taye Fm (P_3–T_1); Talung Fm (P_3); Wuchiaping Fm (P_3)	Taye Fm: 9.16 m; Talung Fm: 37.98 m; Wuchiaping Fm: 5.69 m	Brachiopods: He et al. (2015) (updated here); cephalopods: Zhang et al. (2009)
Kangjiaping (KJP)	29° 24′ 41″ N, 110° 53′ 34″ E; northwest of Hunan	Carbonate platform	Taye Fm (including microbialites) (P_3–T_1); Changhsing Fm (P_3)	Taye Fm: 15.00 m (microbialites: 4.60 m); Changhsing Fm: 15.00 m	Foraminifers: Yang et al. (2013); conodonts and carbon isotopes: Wang et al. (2009); redox conditions: Liao, (2020)
Jinjibang (JJB)	28° 00′ 38″ N; 104° 39′ 24″ E, southeast of Sichuan	Terrestrial–marine transitional facies	Feixianguan Fm (T_1); Xuanwei Fm (P_3)	Feixianguan Fm: 4.00 m; Xuanwei Fm: 15.16 m	Fossils in the Xuanwei Formation referred to Li et al. (1982) (updated here)
Guanbachong (GBC)	27° 23′ 51″ N, 103° 34′ 02″ E; northeast of Yunnan	Terrestrial facies	Dongchuan Fm (P_3–T_1); Kayitou Fm (P_3); Xuanwei Fm (P_3)	Dongchuan Fm: 4.00 m; Kayitou Fm: 8.90 m; Xuanwei Fm: 60.85 m	Fossils: Zhang et al. (2016) (updated here); carbon isotopes: Zhang et al. (2016)
Kele (KL)	27° 14′ 35″ N, 104° 23′ 47″ E; west of Guizhou	Terrestrial facies	Kayitou Fm (P_3–T_1); Xuanwei Fm (P_3)	Kayitou Fm: 17.20 m; Xuanwei Fm: 106.56 m	
Xiaohebian (XHB)	26° 50′ 31″ N, 104° 1′ 29″ E; west of Guizhou	Terrestrial facies	Dongchuan Fm (T_1); Kayitou Fm (P_3–T_1); Xuanwei Fm (P_3)	Dongchuan Fm: 2 m; Kayitou Fm: 9.75 m; Xuanwei Fm: 12.4 m	Fossils: Chu et al. (2016) (updated here)
Chahe (CH)	26° 43′ 44″ N, 103° 49′ 11″ E; west of Guizhou	Terrestrial facies	Dongchuan Fm (P_3–T_1); Kayitou Fm (P_3); Xuanwei Fm (P_3)	Dongchuan Fm: 22.50 m; Kayitou Fm: 8.30 m; Xuanwei Fm: 86.85 m	Plant fossils: Chu et al. (2016) (updated here); spore and pollen: Yu et al. (2022)
Zhejue (ZJ)	26° 32′ 34″ N, 103° 55′ 13″ E; west of Guizhou	Terrestrial facies	Kayitou Fm (P_3–T_1); Xuanwei Fm (P_3)	Kayitou Fm: 1.60 m; Xuanwei Fm: 67.64 m	Spore and pollen: Yu et al. (2022)
Jiucaichong (JCC)	26° 38′ 27″ N, 103° 59′ 57″ E; west of Guizhou	Terrestrial facies	Dongchuan Fm (T_1); Kayitou Fm (P_3–T_1); Xuanwei Fm (P_3)	Dongchuan Fm: 2.00 m; Kayitou Fm: 7.15 m; Xuanwei Fm: 34.45 m	Fossils: Chu et al. (2016) (updated here)

(continued)

(continued)

Section name and abbreviation	GPS or locality	Palaeogeographic setting (P–T transition)	Formations (Fm) covered and ages	Thickness of the measured section	Key references
Jinzhong (JZ)	26° 43′ 56″ N, 104° 26′ 41″ E; west of Guizhou	Terrestrial–marine transitional facies	Dongchuan Fm (T_1); Kayitou Fm (P_3–T_1); Xuanwei Fm (P_3)	Dongchuan Fm: 17.50 m; Kayitou Fm: 58.20 m; Xuanwei Fm: 21.66 m	Fossils: Chu et al. (2016) (updated here)
Mide (MD)	26° 06′ 59″ N, 104° 25′ 05″ E; east of Yunnan	Terrestrial–marine transitional facies	Kayitou Fm (P_3–T_1); Xuanwei Fm (P_3)	Kayitou Fm: 11.08 m; Xuanwei Fm: 6.13 m	Spore and pollen: Yu et al. (2022)
Wadu (WD)	26° 01′ 31″ N, 104° 40′ 22″ E; west ofGuizhou	Terrestrial–marine transitional facies	Feixianguan Fm (P_3–T_1); Xuanwei Fm (P_3)	Feixianguan Fm: 46.00 m; Xuanwei Fm: 106.95 m	Fossils: Chu et al. (2016) (updated here)
Tucheng (TC)	26° 01′ 08″ N, 104° 31′ 16″ E; west of Guizhou	Terrestrial–marine transitional facies	Kayitou Fm (P_3–T_1); Xuanwei Fm (P_3)	Kayitou Fm: 6.40 m; Xuanwei Fm: 6.35 m	Fossils: Yu et al. (2022) (updated here)
Zhongzhai (ZZ)	26° 09′ 00″ N, 105° 17′ 24″ E; west of Guizhou	Shallow-water clastic shelf	Feixianguan Fm (P_3–T_1); Lungtan Fm (P_3)	Feixianguan Fm: 11.40 m; Lungtan Fm: 18.30 m	Conodonts: Zhang et al. (2014a); brachiopods: Zhang et al. (2013, 2014b, 2015) (updated here);
Xinmin (XM)	26° 22′ 35″ N, 105° 55′ ′ 05″ E; central Guizhou	Deep-water basin	Taye Fm (P_3–T_1); Talung Fm (P_3); Changhsing Fm (P_3)	Taye Fm: 18.07 m; Talung Fm: 10.33 m; Changhsing Fm: 21.95 m	Brachiopods: Wu et al. (2018) (updated here); conodonts: Zhang et al. (2014c)
Kejiao (KJ)	25° 49′ 37″ N, 106° 32′ 26″ E; south ofGuizhou	Deep-water basin	Luolou Fm (P_3–T_1); Talung Fm (P_3)	Luolou Fm: 26.14 m; Talung Fm: 48.34 m	Fossils referred to Wang et al. (2023)
Dajiang (DJ)	25° 33′ 59″ N, 106° 39′ 36″ E; south ofGuizhou	Carbonate platform	Taye Fm (including microbialites) (P_3–T_1); Wuchiaping Fm (P_3)	Taye Fm: 23.50 m (microbialites: 14.20 m); Wuchiaping Fm: 5.30 m	Foraminifers and redox conditions: Liao, (2020); conodonts: Jiang et al. (2014)
Zuodeng (ZD)	23° 27′ 00″ N, 106° 59′ 47″ E; west of Guangxi	Carbonate platform	Luolou Fm (including microbialites) (P_3–T_1); Heshan Fm (P_3)	Luolou Fm: 23.94 m (microbialites: 19.14 m); Heshan Fm: 4.57 m	Ostracods: Wan et al. (2019); carbon isotopes: Yang, (2006); redox conditions: Liao, (2020)
Penglaitan (PLT)	23° 41′ 41″ N, 109° 18′ 39″ E; central Guangxi	Late stage of rift basin	Shipao (P_3–T_1); Linhao (P_3); Heshan (P_3)	Shipao Fm: 10.48 m; Linhao Fm: 59.21 m; Heshan Fm: 4.01 m	Fossils: Shen et al. (2019) (updated here)

(continued)

(continued)

Section name and abbreviation	GPS or locality	Palaeogeographic setting (P–T transition)	Formations (Fm) covered and ages	Thickness of the measured section	Key references
Dongpan (DP)	22° 16′ 07″ N, 107° 41′ 47″ E; southwest of Guangxi	Deep-water basin	Luolou Fm (T_1); Talung Fm (P_3)	Luolou Fm: 5.26 m; Talung Fm: 9.89 m	Radiolarians: Feng et al. (2007); ostracods: Yuan et al. (2007); foraminifers: Gu et al. (2007); bivalves: He et al. (2007) and Yang et al. (2015); ammonoids: Bu et al. (2006); brachiopods: He et al. (2020); carbon isotopes: Zhang et al. (2006)

Note Ages for the formations in Column 4 are updated according to the age analysis in Chap. 4

References

Bu JJ, Wu SB, Zhang HL, Meng YY, Zhang F, Zhang LY (2006) Permian-Triassic cephalopods from Dongpan Section, Guangxi, and its geological significanc. Geol Sci Technol Inf 25:47–51

Chen J, Henderson CM, Shen SZ (2008) Conodont succession around the Permian–Triassic Boundary at the Huangzhishan section, Zhejiang and its stratigraphic correlation. Acta Palaeontologica Sinica 47:91–114 [in English with Chinese abstract]

Chen ZQ, Tong JN, Zhang KX, Yang H, Liao ZT, Song HJ, Chen J (2009) Environmental and biotic turnover across the Permian–Triassic boundary on a shallow carbonate platform in western Zhejiang, South China. Aust J Earth Sci 56:775–797

Chu DL, Yu JX, Tong JN, Benton MJ, Song HJ, Huang YF, Song T, Tian L (2016) Biostratigraphic correlation and mass extinction during the Permian–Triassic transition in terrestrial-marine siliciclastic settings of South China. Glob Planet Change 146:67–88

Fang YH (2021) Late Permian to Early Triassic redox variations of South China and unusual biosedimentary responses. Doctoral Dissertation of China University of Geosciences, pp 1–145 [in Chinese with English abstract]

Feng QL, Gu SZ, He WH, Jin YX (2007) Latest Permian Entactinaria (Radiolaria) from southern Guangxi, China. J Micropalaeontol 26:19–37

Gu SZ, Feng QL, He WH (2007) The last Permian deep-water fauna: latest Changhsingian small foraminifers from southwestern Guangxi, South China. Micropaleontology 53:311–330

Gui BW, Feng QL, Aihua Y (2009) Late Changhsingian (latest Permian) radiolarians from Chaohu, Anhui. J Earth Sci 20:797–810

He WH, Feng QL, Weldon EA, Gu SZ, Meng YY, Zhang F, Wu SB (2007) A Late Permian to Early Triassic bivalve fauna from the Dongpan section, southern Guangxi, South China. J Paleontol 81:1009–1019

He WH, Shi GR, Twitchett RJ, Zhang Y, Zhang KX, Song HJ, Yue ML, Wu SB, Wu HT, Yang TL, Xiao YF (2015) Late Permian marine ecosystem collapse began in deeper waters: evidence from brachiopod diversity and body size changes. Geobiology 13:123–138

He WH, Shi GR, Yang TL, Zhang KX, Yue ML, Xiao YF, Wu HT, Chen B, Wu SB (2016) Patterns of brachiopod faunal and body-size changes across the Permian–Triassic boundary: evidence from the Daoduishan section in Meishan area, South China. Palaeogeogr Palaeoclimatol Palaeoecol 448:72–84

He WH, Shi GR, Zhang KX, Yang TL, Shen SZ, Zhang Y (2019) Brachiopods around the Permian–Triassic boundary of South China. Springer, Singapore, 261pp

He WH, Shi GR, Zhang Y, Yang TL, Zhang KX, Wu SB, Niu ZJ, Zhang ZY (2014) Changhsingian (latest Permian) deep-water brachiopod fauna from South China. J Syst Paleontol 12:907–960

He WH, Weldon EA, Yang TL, Wang H, Xiao YF, Wu HT, Zhang KX, Wang YB, Wu SB (2020) The palaeoenvironmental and palaeobiogeographical significance of the Late Permian deep-water brachiopod fauna from Dongpan, South China, including descriptions of Micromartinia He & Weldon gen. nov. (Micromartiniidae He & Weldon fam. nov.) and Minutomarginifera nom. nov. J Syst Paleontol 18:885–909

He WH, Zhang Y, Zheng YE, Zhang KX, Gui BW, Feng QL (2008) A late Changhsingian (latest Permian) radiolarian fauna from Chaohu, Anhui and a comparison with its contemporary faunas of South China. Alcheringa 32:199–222

Huang YF, He WH, Liao W, Wang YB, Yi ZX, Yang H (2022) Two pulses of increasing terrestrial input to marine environment during the Permian–Triassic transition. Palaeogeogr Palaeoclimatol Palaeoecol 586:110753. https://doi.org/10.1016/j.palaeo.2021.110753

Jiang HS, Lai XL, Luo GM, Aldridge RA, Zhang KX, Wignall PB (2007) Restudy of conodont zonation and evolution across the P/T boundary at Meishan section, Changxing, Zhejiang, China, environmental and biotic changes during the Paleozoic–Mesozoic transition. Glob Planet Change 55:39–55

Jiang HS, Lai XL, Sun YD, Wignall PB, Liu JB, Yan CB (2014) Permian–Triassic Conodonts from Dajiang (Guizhou, South China) and their implication for the age of microbialite deposition in the aftermath of the end-Permian mass extinction. J Earth Sci 25:413–430

Li GS (2016) Fluctuations of redox conditions and ecological responses across the Permian-Triassic boundary in different paleogeographic settings of Middle & Lower Yangtze Region. Doctoral Dissertation of China University of Geosciences, pp 1–119 [in Chinese with English abstract]

Li ZJ, Zhu JN, Hu YF (1982) The new views on the correlation for the Late Permian coal-bearing strata in Junlian area, South Sichuan. J Stratigr 6:174–182 [in Chinese]

Li ZS, Zhan LP, Zhu XF, Zhang JH, Huang HQ, Xu DY, Yan Z, Li HM (1989) Study on the Permian-Triassic biostratigraphy and event stratigraphy of northern Sichuan and southern Shaanxi. Geological Publishing House, Beijing, 435 pp [in Chinese]

Liao W (2020) Palaeoenvironment change in Shallow-marine carbonate platform across the Permian-Triassic boundary in South China and its possible cause. Doctoral Dissertation of China University of Geosciences, pp 1–92 [in Chinese with English abstract]

Liu H (2010) Ostracod fauna near the Permian–Triassic boundary in Chongyang, Hubei Province and its palaeoenvironmental significance. Dissertation submitted to China University of Geosciences for the Master Degree of Science, pp 1–52 [in Chinese with English abstract]

Mei SL, Zhang KX, Wardlaw BR (1998) A refined succession of Changhsingian and Griesbachian neogondolellid conodonts from the Meishan section, candidate of the global stratotype section and point of the Permian–Triassic boundary. Palaeogeogr Palaeoclimatol Palaeoecol 143:213–226

Shen SZ, He XL (1991) Brachiopod assemblages in the Changxing Formation of Zhongliangshan in Chongqing. J Stratigr 15:189–196 [in Chinese]

Shen SZ, Ramezani J, Chen J, Cao CQ, Erwin DH, Zhang H, Xiang L, Schoepfer SD, Henderson CM, Zheng QF, Bowring SA, Wang Y, Li XH, Wang XD, Yuan DX, Zhang YC, Mu L, Wang J, Wu YS (2019) A sudden end-Permian mass extinction in South China. Geol Soc Am Bull 131:205–223

Song HJ, Tong JN, Chen ZQ (2009) Two episodes of foraminiferal extinction near the Permian–Triassic boundary at the Meishan section, South China. Aust J Earth Sci 56:765–773

Sun DY, Tong JN, Xiong YL, Tian L, Yin HF (2012) Conodont Biostratigraphy and evolution across Permian-Triassic Boundary at Yangou Section, Leping, Jiangxi Province, South China. J Earth Sci 23:311–325

Sun X, Qiu XC, Tihelka E, Yang H, Sun DY, Tong JN, Tian L (2021) A diverse gastropod fauna from the shallow marine carbonate platform of the Yangou section (South China) in the immediate aftermath of the Permian–Triassic mass extinction. Geol J 56:6199–6209

Tian L, Tong JN, Song HJ, Liang L, Yang LR, Song HY, Wang CG, Zhao XM, Sun DY (2014a) Foraminferal evolution and formation of oolitic limestone near Permian–Triassic boundary at Yangou section, Jiangxi Province. Earth Sci J China Univ Geosci 39:1573–1586 [in Chinese with English abstract]

Tian L, Tong JN, Sun DY, Xiong YL, Wang CG, Song HJ, Song HY, Huang YF (2014b) The microfacies and sedimentary responses to the mass extinction during the Permian–Triassic transition at Yangou Section, Jiangxi Province, South China. Sci China Earth Sci 57:2195–2207

Wan JY, Yuan AH, Crasquin S, Jiang HS, Yang H, Hu X (2019) High-resolution variation in ostracod assemblages from microbialites near the Permian–Triassic boundary at Zuodeng, Guangxi region, . Palaeogeogr Palaeoclimatol Palaeoecol 535.https://doi.org/10.1016/j.palaeo.2019.109349

Wan JY (2021) Response of ostracod assemblages from microbialites in South China to the end-Permian mass extinction. Doctoral Dissertation of China University of Geosciences, pp 1–152 [in Chinese with English abstract]

Wang QX, Tong JN, Song HJ, Yang H (2009) Ecological evolution across the Permian/Triassic boundary at the Kangjiaping Section in Cili County, Hunan Province, China. Sci China Ser D Earth Sci 2009(52):797–806 [in Chinese with English abstract]

Wang T, Burne RV, Yuan AH, Wang YB, Yi ZX (2019) The evolution of microbialite forms during the Early Triassic transgression: a case study in Chongyang of Hubei Province, South China. Palaeogeogr Palaeoclimatol Palaeoecol 519:209–220

Wang H, He WH, Xiao YF, Yang TL, Zhang KX, Wu HT, Huang YF, Peng XF, Wu SB (2023) Stepwise collapse of biotic communities and its relations to oxygen depletion along the north margin of Nanpanjiang Basin during the Permian–Triassic transition. Palaeogeogr Palaeoclimatol Palaeoecol. https://doi.org/10.1016/j.palaeo.2023.111569

Wu HT, He WH, Shi GR, Zhang KX, Yang TL, Zhang Y, Xiao YF, Chen B, Wu SB (2018) A new Permian-Triassic boundary brachiopod fauna from the Xinmin section, southwestern Guizhou, south China and its extinction patterns. Alcheringa 42:339–372

Yang H, Zhang SX, Jiang HS, Wang YB (2006) Age and general characteristics of calcimicrobialite near the Permian-Triassic Boundary in Chongyang, Hubei Province. Earth Sci J China Univ Geosci 31:165–170 [in Chinese with English abstract]

Yang H (2006) Calcimicrobialites after end-Permian mass extinction in South China and its implications of paleo-environment. These for Master Degree of China University of Geosciences, pp 1–61 [in Chinese with English abstract]

Yang LR, Song HJ, Tong JN, Chu DL, Tian L (2013) The extinction pattern of fusulinids during the Permian–Triassic crisis at the Kangjiaping section, Cili, Hunan Province. Micropalaeontologica Sinica 30:353–366 [in Chinese with English abstract]

Yang TL, He WH, Zhang KX, Wu SB, Zhang Y, Yue ML, Wu HT, Xiao YF (2015) Palaeoecological insights into the Changhsingian-Induan (latest Permian–earliest Triassic) bivalve fauna at Dongpan, southern Guangxi, South China. Alcheringa 40:98–117

Yang ZY, Yin HF, Wu SB, Yang FQ, Ding MH, Xu GR (1987) Permian–Triassic boundary stratigraphy and fauna of South China. Geological Publishing House, Beijing, 378 pp [in Chinese with English abstract]

Yao A, Kuwahara K (1999) Middle-Late Permian Radiolarians from the Guangyuan-Shangsi Area, Sichuan Province, China. J Geosci Osaka City Univ 42:69–83

Yu JX, Broutin J, Lu ZS (2022) Plants and palynomorphs around the Permian–Triassic boundary of South China. Springer, Singapore, p 254

Yuan AH, Crasquin S, Feng QL, Gu SZ (2007) Latest Permian deep-water ostracods from southwestern Guangxi, South China. J Micropalaeontol 26:169–191

Yuan DX, Shen SZ, Henderson CM, Chen J, Zhang H, Feng HZ (2014) Revised conodont-based integrated high-resolution timescale for the Changhsingian Stage and end-Permian extinction interval at the Meishan sections, South China. Lithos 204:220–245

Yuan DX, Shen SZ (2011) Conodont succession across the Permian–Triassic boundary of the Liangfengya section, Chongqing, South China. Acta Palaeontologica Sinica 50:420–438 [in Chinese with English abstract]

Zhang F, Feng QL, He WH, Meng YY, Gu SZ (2006) Multidisciplinary stratigraphy across the Permian–Triassic boundary in deep-water environment of the Dongpan section, south China. Norw J Geol 86:125–131

Zhang H, Cao CQ, Liu XL, Mu L, Zheng QF, Liu F, Xiang L, Liu LJ, Shen SZ (2016) The terrestrial end-Permian mass extinction in South China. Palaeogeogr Palaeoclimatol Palaeoecol 448:108–124

hang KX, Tong JN, Shi GR, Lai XL, Yu JX, He WH, Peng YQ, Jin YL (2007) Early Triassic conodont-palynological biostratigraphy of the Meishan D Section in Changxing, Zhejiang Province, South China. Palaeogeogr Palaeoclimatol Palaeoecol 252:4–23

Zhang N, Jiang HS, Zhong WL, Huang HH, Xia WC (2014a) Conodont biostratigraphy across the Permian-Triassic boundary at the Xinmin section, Guizhou, South China. J Earth Sci 25:779–786

Zhang Y, He WH, Shi GR, Zhang KX, Wu HT (2015) A new Changhsingian (Late Permian) brachiopod fauna from the Zhongzhai section (South China) Part 3: Productida. Alcheringa 39:295–314

Zhang Y, He WH, Shi GR, Zhang KX (2013) A new Changhsingian (Late Permian) Rugosochonetidae (Brachiopoda) fauna from the Zhongzhai section, southwestern Guizhou Province, South China. Alcheringa 37:223–247

Zhang Y, Shi GR, He WH, Zhang KX, Wu HT (2014b) A new Changhsingian (Late Permian) brachiopod fauna from the Zhongzhai section (South China), Part 2: Lingulida, Orthida, Orthotetida and Spiriferida. Alcheringa 38:480–503

Zhang Y, Zhang KX, Shi GR, He WH, Yuan DX, Yue ML, Yang TL (2014c) Restudy of conodont biostratigraphy of the Permian–Triassic boundary section in Zhongzhai, southwestern Guizhou Province, South China. J Asian Earth Sci 80:75–83

Zhang ZY, He WH, Zhang Y, Yang TL, Wu SB (2009) Late Permian–earliest Triassic ammonoid sequences from the Rencunping section, Sangzhi County, Hunan Province, South China and their regional correlation. Geol Sci Technol Inf 28:23–30 [in Chinese with English abstract]

Zhao JK, Sheng JZ, Yao ZQ, Liang XL, Chen CZ, Rui L, Liao ZT (1981) The Changhsingian and the Permian–Triassic Boundary in South China. Nanjing Inst Geol Palaeontol Bull 2:1–112 [in Chinese with English abstract]

Appendix B: Index and Source (Provider) for Sections

© The Editor(s) (if applicable) and The Author(s), under exclusive license to Springer Nature Singapore Pte Ltd. 2023

W. He et al. (eds.), *Stratigraphy Around the Permian–Triassic Boundary of South China*, New Records of the Great Dying in South China,
https://doi.org/10.1007/978-981-99-9350-5

The manufacturer's authorised representative in the EU is Springer Nature Customer Service Centre GmbH, Europaplatz 3, 69115 Heidelberg, Germany. If you have any concerns regarding our products, please contact ProductSafety@springernature.com

Printed and bound by CPI Group (UK) Ltd, Croydon, CR0 4YY
01/10/2026
02225957-0008